Progress in Inflammation Research

Series editors

Michael J. Parnham
Fraunhofer IME & Goethe University
Frankfurt, Germany

Achim Schmidtko
Goethe University
Frankfurt, Germany

More information about this series at http://www.springer.com/series/4983

Ruth Lyck • Gaby Enzmann
Editors

The Blood Brain Barrier and Inflammation

Springer

Editors
Ruth Lyck
Theodor Kocher Institute
University of Bern
Bern
Switzerland

Gaby Enzmann
Theodor Kocher Institute
University of Bern
Bern
Switzerland

Series editors
Michael J. Parnham
Fraunhofer IME & Goethe University
Frankfurt
Germany

Achim Schmidtko
Goethe University
Frankfurt
Germany

Progress in Inflammation Research
ISBN 978-3-319-45512-9 ISBN 978-3-319-45514-3 (eBook)
DOI 10.1007/978-3-319-45514-3

Library of Congress Control Number: 2017936995

Printed on acid-free paper

This Springer imprint is published by Springer Nature
The registered company is Springer International Publishing AG
The registered company address is: Gewerbestrasse 11, 6330 Cham, Switzerland

Contents

General Introduction to Barrier Mechanisms in the Central Nervous System

Norman R. Saunders, Katarzyna M. Dziegielewska, Kjeld Møllgård, and Mark D. Habgood

Abstract There are five exchange interfaces between the peripheral circulation (blood), the cerebrospinal fluid (CSF) and the brain: (i) meninges, (ii) blood vessels, (iii) choroid plexuses, (iv) circumventricular organs and (v) ependyma (neuroependyma in embryos). All five interfaces have distinctive morphological and physiological properties; the first three are characterised by intercellular tight junctions that provide important structural basis for limiting molecular exchange across their interfaces. Cells that form these interfaces are also sites of extensive exchange mechanisms (transporters) that control entry and exit of a wide variety of molecules into the brain. Secretion of CSF by the choroid plexuses which flows through the ventricular system, and the exchange of substances between the CSF and brain is an important mechanism for the control of the characteristic composition of the brain interstitial fluid. Understanding of the complexity of barrier mechanisms is essential for evaluation of the effects of inflammatory conditions affecting the brain, whether in the adult or during development.

1 Introduction

It is becoming increasingly apparent that inflammation may play a significant role in a wide range of neurological disorders, both acute and chronic. The underlying question is to what extent different blood–brain barrier mechanisms may be affected

N.R. Saunders (✉)
Department of Pharmacology and Therapeutics, University of Melbourne,
Parkville, VIC 3010, Australia
e-mail: n.saunders@unimelb.edu.au

K.M. Dziegielewska • K. Møllgård • M.D. Habgood
Department of Pharmacology and Therapeutics, University of Melbourne,
Parkville, VIC 3010, Australia

Institute of Cellular and Molecular Medicine, The Panum Institute, University of Copenhagen,
København, Denmark
e-mail: kdzie@unimelb.edu.au; kjm@sund.ku.dk; mhabgood@unimelb.edu.au

R. Lyck, G. Enzmann (eds.), *The Blood Brain Barrier and Inflammation*,
Progress in Inflammation Research, DOI 10.1007/978-3-319-45514-3_1

by inflammation and whether they are involved in some of these disorders, either as a part of the primary cause or as a secondary consequence.

It will thus be essential to have a comprehensive knowledge of these mechanisms in the normal brain as a basis for studying and understanding pathological conditions. However, current understanding of normal mechanisms is incomplete, many questions remain unanswered, and the field requires much further work.

Historically "blood–brain barrier" is an old term dating back nearly a hundred years. It was first used by Lena Stern ("barrière hémato-encéphalique", [105]) although most in the literature attribute it to Ehrlich [34], Lewandowsky [66] or Goldmann [43]; none of whom actually used the term (see [95]). "Blood–brain barrier" was generally used, and still is by some to this day, solely to describe a structural mechanism preventing entry of substances into the brain, originally illustrated by the use of dyes administered parenterally. Stern herself, perhaps because of her training as both a medical doctor and physiologist, appreciated that the "barrier" had much wider functional implications. This was not fully appreciated until the second half of the twentieth century (see, e.g. [27, 28]).

In this introduction, we provide a summary of what is known of the structural and functional properties of the brain barrier interfaces both in the adult and in the developing brain, particularly where they may be relevant to understanding changes that may occur in inflammatory conditions.

2 Barriers of the Brain

It is now recognised that in adult brain there are five barrier interfaces between the blood (the periphery) and central nervous system (CNS) (see Fig. 1). The outer meningeal surface of the brain has at least three barrier structural components (Fig. 1a). Nabeshima et al. [80] provided an early ultrastructural description of this surface of the brain in several species, identified tight junctions in the layer of cells at the border of the arachnoid with the dura and designated this the arachnoid barrier layer. They also identified the vessels within the subarachnoid space as having extensive tight junctions. Møllgård and colleagues [15] have carried out a detailed immunohistochemical and confocal microscopical study of the barriers over the surface of the brain in fetal and adult human and rat material. These authors distinguish three separate barrier interfaces: (i) blood–arachnoid–outer CSF interface, as described by Nabeshima et al. [80]; (ii) blood–pia microvessel–outer CSF interface (sometimes incorrectly used as a surrogate for the blood–brain barrier itself, as pointed out by [15]); and (iii) the brain end feet–outer CSF interface (Fig. 1). Traditionally the meningeal barrier has been viewed purely as a physical barrier; however, it is now clear that particularly during brain development the arachnoid layers have a significant functional role in secreting growth factors and mitogens [30]. Detailed studies of the molecular properties of this layer, as recently performed for the blood–brain and blood–CSF barriers (see below), have yet to be undertaken.

The best-known and most studied barrier interfaces are the blood–brain barrier itself across the cerebral blood vessels (Fig. 1b) and the blood–cerebrospinal fluid

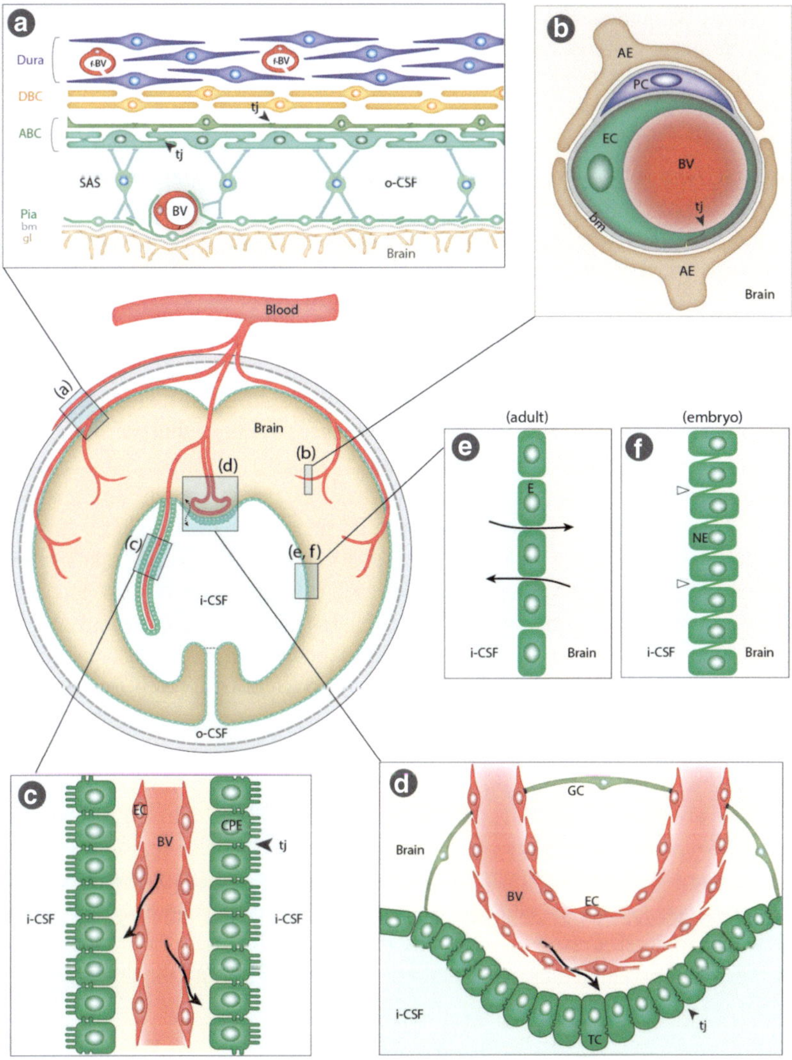

Fig. 1 Schematic diagram (*centre left*) of the five main barrier interfaces (**a**–**e**) in the brain and an additional one in the embryo (**f**). The barrier-forming cellular layers at each interface are coloured green. (**a**) *Meningeal barrier*: *o-CSF* outer cerebrospinal fluid, *SAS* subarachnoid space, *BV* blood vessels, *f-BV* fenestrated BV, *bm* basement membrane, *gl* glia limitans, *tj* tight junctions (*arrow heads*). (**b**) *Blood–brain barrier*: cerebral blood vessels (*BV*). Tight junctions between the endothelial cells (*EC*) restricting the paracellular cleft; *PC* pericytes, *AE* astroglial end feet. (**c**) *Blood–CSF barrier*: choroid plexuses. *CPE* epithelial cells, tight junctions (*arrowheads*). Blood vessels (*BV*) are fenestrated and do not form a barrier (*arrows*). (**d**) *Circumventricular organs*: tanycytes (*TC*), specialised ependymal cells of these brain areas connected by tight junctions (*arrowhead*); entry into the rest of the brain prevented by tight junctions between astroglial cells (*GC*). (**e**) *Ependyma in adult brain*. Apart from specialised tanycytes, ependymal cells are linked by gap junctions; there is no restricted exchange of even large molecules, such as proteins, between CSF and interstitial space of the brain (*solid arrows*). (**f**) *Embryonic CSF–brain barrier*. In early brain development, strap junctions (*open arrowheads*) are present between adjacent neuroepithelial cells (*NE*); these form a barrier restricting the movement of larger molecules, such as proteins, but not smaller molecules such as sucrose (From Saunders et al. [97])

barrier (Fig. 1c) across the choroid plexuses. Both these sites possess a critical structural feature, namely, tight junctions, between adjacent cells in the interfaces [14]. These junctions are responsible not only for a physical barrier function between the blood and the CNS, but they are also essential as an underpinning for the cellular transport properties of cells forming the interfaces. Without a diffusion restraint between cells of these interfaces, any cellular transport functions across the barriers would be ineffective. There is an important morphological difference in the site of the tight junctions in these two interfaces. At the blood–brain barrier, tight junctions are between endothelial cells of blood vessels, while in choroid plexuses they are between adjacent epithelial cells, and the blood vessels in the plexus stroma are fenestrated and do not hinder the movement of molecules from the blood into the extracellular space of the plexuses. Other cells that are thought to contribute to the properties of the blood–brain barrier are astrocytes, the end feet of which encircle the cerebral endothelial cells, and the pericytes situated within the basement membrane that surrounds the endothelial cells (Fig. 1b).

This complex of cells forming the blood–brain barrier is sometime referred to as the "neurovascular unit" [81]. An additional morphological feature, which is thought to contribute to the properties of the blood–brain barrier, is a lack of pinocytotic vesicles in the cerebral endothelial cells. Although little studied, this lack of pinocytotic vesicles does not appear to be a feature of epithelial cells in the choroid plexuses. These cells by contrast are able to transport macromolecules, such as proteins, from blood into CSF, which may involve endocytotic (at the blood side of the choroid plexus) and exocytotic (at the CSF side) mechanisms [33, 67, 70, 100].

Another barrier interface is that of the circumventricular organs (pineal gland, area postrema, median eminence, subfornical organ) between the outer and inner CSF and the brain tissue (Fig. 1d). The blood vessels in the circumventricular organs (Fig. 1d) do not possess diffusional barrier properties. In these regions, a physical barrier between the CSF and the brain is provided by tanycytes [64], which are specialised ependymal cells connected by tight junctions at their apices; they are only found associated with circumventricular organs. Entry into brain parenchyma beyond the circumventricular organs is prevented by tight junctions between astroglial cells (Fig. 1d).

In the adult brain, there is an interface between the ventricular CSF and the brain tissue formed by ependymal cells that are derived developmentally from the neuroepithelium that lines the wall of the cerebral ventricles in the developing brain. Apart from the regions where circumventricular organs abut on the ventricular CSF and tanycytes provide a barrier interface, throughout the rest of the ventricular system, the ependymal cell layer (Fig. 1f) does not present a diffusional restraint on movement between the CSF and brain interstitial fluid of even large molecules such as proteins [14]. These adult ependymal cells are joined by gap junctions. However, in the embryo and early fetus, the situation is different (Fig. 1f). Here the cells of the neuroepithelium are joined by membrane specialisations called strap junctions [79]. These strap junctions, unlike tight junctions, have a spiral distribution along the long axis of the neuroepithelial cells perpendicular to the CSF surface of the neuroepithelium. This is in contrast to the belt-like arrangement of tight junctions at the apex of epithelial cells such as those in the choroid plexuses. These junctions have been shown to be impermeable to all but very small molecules (286 Da) in fetal sheep [40] and mouse embryos [113]. As brain development progresses, these junctions have been shown in the mouse to become permeable to increasingly larger molecules so that by postnatal day 20, there is no restriction on the permeability of molecules of the size of many plasma proteins (70 kDa, [113]).

3 Cellular Components of the Neurovascular Unit

The components of this brain barrier interface interact in ways that are not yet fully understood. But it is possible to define individual contributions of some of the different cell types and the manner in which they interact with each other. The most studied and best understood are of course the *endothelial cells* themselves. These provide numerous influx and efflux transporter mechanisms that are the key components in defining the internal milieu of the brain, provision of metabolically important molecules and exclusion or removal of potentially toxic molecules. These will be described in separate sections below. The molecular composition and structure of tight junctions are both described in the chapter by Wolburg and in Bauer et al. [9].

Astrocytes in the adult brain have foot processes that encircle almost the entire circumference of the blood vessels. This encirclement is a process that begins in rodents just before birth and accompanies vascularisation of the brain, which is complete by 3 weeks of age [16]. There has been a long-held and persistent [83] belief in the blood–brain barrier field that astrocytes are essential for the formation of tight junctions during brain development in spite of much evidence to the contrary (reviewed in [95]). This belief is based mainly on in vitro experiments in which tight junctions between isolated cerebral endothelial cells form more effectively in the presence of cocultured astrocytes [1]. The other frequently cited paper is that of Janzer and Raff [54]. These authors injected purified rat astrocytes into the anterior chamber of the eye of adult syngeneic rats. The astrocytes formed aggregates that were claimed to become vascularised by 48 h. Therefore, it was proposed that the astrocytes had induced the formation of tight junctions in the nonneural blood vessels that had entered the aggregates of astrocytes. However, these authors did not have any evidence for the formation of tight junctions as this would have required electron microscopy. When the studies were repeated by Holash et al. [50], who did use electron microscopy, results showed that aggregates of astrocytes in the anterior eye chamber of the rat were poorly vascularised and did not have ultrastructural characteristic of cerebral capillaries. It seems surprising that anyone would suggest that astrocytes would be responsible for tight junction formation in the early stages of brain vascularisation: blood vessels already have tight junctions and are impermeable to intravascular tracers prior to differentiation of astrocytes [24, 35]. In the early postnatal period of intense vascularisation in the rodent brain, it is possible that astrocytes may play a role in tight junction formation and/or stabilisation; this is suggested by the deletion experiments of Ma et al. [74]. In the adult brain, astrocytes are thought to be involved in modulation and maintenance of barrier properties [1, 2, 26].

Aquaporin 4 Astrocytic end feet contain a high concentration of aquaporin 4 (AQP4) water channels and are considered important for water homeostasis in the normal brain. Part of the evidence for this comes from experiments in which the AQP4 gene has been deleted (AQP4−/−); in these animals acute water intoxication caused osmotic oedema without disruption of the blood–brain barrier as assessed by electron microscopy and Evans blue injection [77]. On the other hand, AQP4-deficient mice subjected to permanent middle cerebral artery occlusion (to produce vasogenic oedema) for 24 h had much better survival and neurological outcomes than the wild-type mice [77]. However, in normal brain, it is unclear how the high concentration of AQP4 channels in astrocytic end feet is integrated into brain water homeostasis. No specific water channels have yet been identified in cerebral endothelial cells. It is unclear how much and by what route(s) water crosses the cerebral vascular endothelium. On the face of the available observations, there would appear to be a functional mismatch between the lack of water channels in the endothelial cells and a high concentration of AQP4 in astrocytic end feet. As discussed below, it has been proposed that water transfer across cerebral endothelial cells may be via ion co-transporters and the glucose uniport GLUT1 [75]. Possibly water exchange also occurs via Virchow–Robin spaces around penetrating arteries and the extension of this pathway via the extracellular matrix which surrounds cerebral blood vessels [55] although as yet there is no functional evidence for this.

Pericytes are a less studied component of the neurovascular unit. As illustrated in Fig. 1b, they lie outside the endothelial cells embedded in the basement membrane/extracellular matrix. They appear to play a role in controlling capillary blood flow [46] although this has been disputed [48]. However, this difference in interpretation of optical imaging studies of capillary flow appears to depend on a redefinition of pericytes as smooth muscle cells by the latter authors [5]. Pericytes are clearly important for some aspects of blood–brain barrier function as under pathological conditions, they undergo changes such as migration, proliferation or differentiation [31, 32]. Although it is frequently asserted that pericytes play an important role in vascular and tissue homeostasis and maintenance of the blood–brain barrier by their interactions with other constituents of the neurovascular unit [5, 31], the mechanisms involved are unclear; some of the signalling mechanisms thought to be important for communication between pericytes and other constituents of the neurovascular unit have been reviewed by ElAli et al. [37] who also comment on a possible role for pericytes in immune responses affecting the blood–brain barrier, a topic that has been reviewed by Hurtado-Alvarado et al. [52]. A more clearly defined role for pericytes is their contribution to barrier properties of blood vessels in the developing brain. Thus, PDGF-β-deficient mice lack pericytes around their cerebral vessels and show increased permeability during embryogenesis ([72], confirmed by [24]). In addition to the lack of pericytes, these animals show increased expression of genes associated with vascular permeability such as leukocyte adhesion molecules. Using non-lethal PDGFR-β mouse mutants, Armulik et al. [4] showed regulation of specific gene expression pattern in the blood–brain barrier and polarisation of astrocytic end feet around cerebral blood vessels. The mutants showed increased permeability to water and a range of molecular tracers, which appeared to be due to increased endothelial transcytosis, a process that in the intact blood–brain barrier is characteristically low. The endothelial cell tight junctions in the mutants appeared to be functionally and ultrastructurally intact.

An important component of the neurovascular unit is the *extracellular matrix*, which contributes to the endothelial and the parenchymal basement membranes. The former is produced by the endothelial cells and contains pericytes, whereas the latter is laid down by astrocytes and demarcates the perivascular space (Fig. 1b). A detailed description of this structure is to be found in an accompanying chapter by Sorokin.

4 Influx Mechanisms of the Blood–Brain Barrier

The main influx mechanisms in cerebral endothelial cells are summarised in Fig. 2. Only some of these are described here because of limitations of space.

Glucose The first described and probably the most important glucose transporter is GLUT-1 (SLC2a1, solute-linked carrier (SLC) transporters), and its gene *Slc2a1* is expressed in both the cerebral endothelial cells [25] and choroid plexus epithelial cells [68, 69]. GLUT-1 is a facilitative transporter (i.e. transports glucose down a concentration gradient at a higher rate than would occur by diffusion). GLUT-1 is the main member of this family in cerebral endothelial cells [38], but expression of the genes for *Slc2a3*, *a8*, *a12* and *a13* has also been described in mouse cerebral endothelial cells [25]. *Slc2a1* has been detected in rodent choroid plexus [69, 78] expressed at a slightly higher level in the embryo [69]. Other glucose-transporting genes in this family that have been identified in rat choroid plexus are *Slc2a3*, *a4*, *a8*, *a12*, *a13* and *a15*, but only *Slc2a12* is expressed at a level that is likely to be functionally significant at fivefold higher in the adult [69].

SLC5 transporters are sodium–glucose co-transporters [114]. *Slc5a1*, *a5* and *a6* have been identified in mouse cerebral endothelial cells [25], and *Slc5a6* and *a10* have been identified in mouse choroid plexus [68]. There is good evidence for glucose transport into the brain in both adults [84, 85] and neonates [109]. Much of this is probably mediated by GLUT-1 (SLC2a1); the critical importance of this transporter is demonstrated by the severe clinical effects that occur in people with GLUT-1 deficiency [53]. The functional importance of all the other glucose transporters that have been identified at the blood–brain and blood–CSF barriers is unclear, as current methods for studying glucose transport do not distinguish between their possible roles. This reflects a general problem where many transporters for the same substrate have evolved. It is clear that at least in the case of GLUT-1

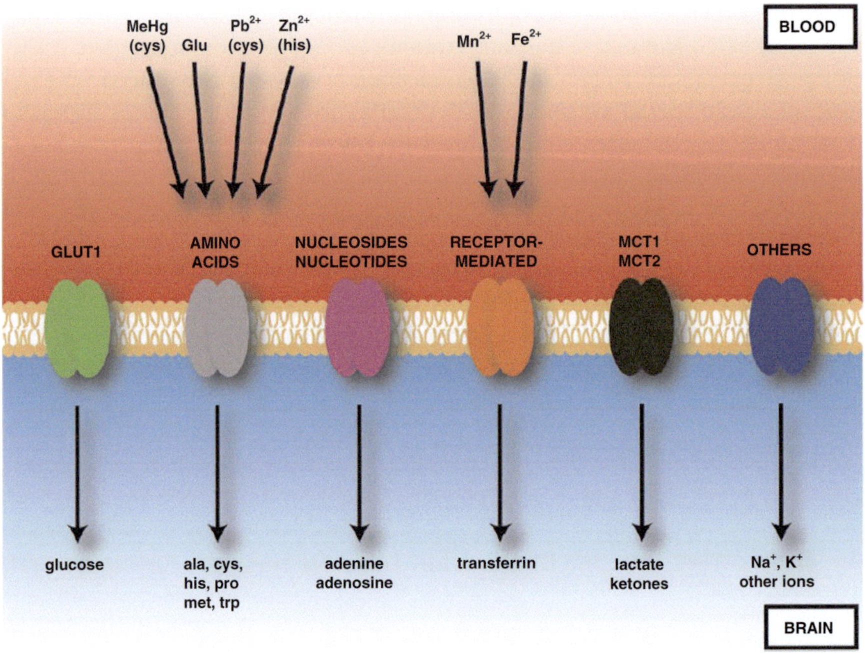

Fig. 2 Diagram of main inward transporters in cerebral endothelial cells. There is now detailed information on the molecular identity of the transporters involved in this wide range of transporter systems in the adult and developing brain [25, 38, 78]. Similar data are available for adult and developing choroid plexuses [68, 96]. Note that heavy metals bind to some amino acids and transferrin receptors. Because of vulnerability of developing brain to heavy metals, this transport may contribute to fetal or newborn neurotoxicity; *ala* alanine, *cys* cysteine, Fe^{2+} iron, *Glu* glutamate, *his* histidine, *MCT* monocarboxylate transporter, *MeHg* methyl mercury, *met* methionine, Mn^{2+} manganese, Pb^{2+} lead, *pro* proline, *trp* tryptophan, Zn^{2+} zinc (From Saunders et al. [93])

this apparent redundancy is insufficient to compensate for the loss of GLUT-1 function [53]. It is possible that studies of deletions of other glucose transporters may shed light on their contribution to glucose transport in the brain.

Amino Acids There is now substantial molecular information available on the expression of a large number of SLC transporters for amino acids both at the blood–brain and the blood–CSF (choroid plexus) barriers. Much less is known about amino acid transporters at the other brain interfaces. Early studies demonstrated the transport of a whole range of amino acids from blood into brain [84, 86]. Oldendorf [85] found that generally essential amino acids were transported to a much greater extent than non-essential amino acids. Amino acid carriers or transporters were originally classified according their electrical charge at physiological pH. Independent carriers for neutral, basic [90] and acidic amino acids [88] were defined. Subcategories were proposed when it was found that there was some degree of substrate specificity within one general class. There are of course many individual amino acids within each carrier group with competition within this group but not across carriers. With the advent of cloning and mRNA sequencing techniques, the molecular basis for the carrier systems became much clearer, with many more discovered than implied by the original classification. Christensen et al. [17] made a plea that the original designation should be retained because it is immediately apparent what the function of the amino acids transported is, e.g. EAAC1 for excitatory amino acid carrier, along with three designations: GLAST (for glutamate and aspartate transporter) and GLT1 and GLTP (glutamate transporters). These designations may not be particularly clear or helpful to the uninitiated, but they are still included in some databases. It is now recognised that amino acid transporters fall into the general category of SLCs. The main families are:

(i) SLC1A1 to A7, high-affinity glutamate and neutral amino acid transporter family [60].
(ii) SLC3A1 and A2; SLC7A1-A14. Mainly cationic amino acid transporters; some are transporters for large and small neutral amino acids [41].
(iii) SLC43A1-A3. Facilitator system L-amino acid transporters [11].

It is outside the scope of this introductory chapter to review these carrier systems in detail; current information can be found for both the blood–brain barrier and blood–CSF barriers [25, 68, 69, 78, 94, 96]. However, two general points are worth noting. Firstly, amino acid transporters are one of the many functional systems active within brain barriers that might be disrupted by inflammation. This therefore needs to be considered when evaluating results of experiments or clinical conditions in which there appears to be involvement of blood–brain barrier interfaces in inflammatory conditions. Secondly, it is clear that disorders of transport of some individual amino acids can result in significant neuropathology. For example, loss of function by mutations in the glutamate transporter SLC1A1 causes dicarboxylic aminoaciduria associated with mental retardation in humans [6]. The velocardiofacial syndrome (VCFS, Shprintzen syndrome) appears to be due to deletion of the gene for a cationic amino acid transporter, SLC7A4 [104]. It is notable that several SLCs may transport the same amino acid, yet a defective gene for one transporter can result in severe developmental defects. It seems reasonable to conclude that if inflammation interferes with one or more amino acid SLCs during development, there may well be serious subsequent developmental defects.

Monocarboxylate Transport A family of monocarboxylate transporters (MCTs), some of which are proton linked, is involved in transport of monocarboxylates (e.g. pyruvate, lactate and ketone bodies) across plasma membranes. These are the 14 members of the SLC16 family [44].

The equivalence of SLC16 family members nomenclature and their MCT designation is a bit of an illogical muddle, because MCTs were named in the order in which they were functionally characterised, whereas the SLC16 members were named as their cDNA sequences became available. Only for four is there good evidence of involvement in monocarboxylate transport in humans [45]; these are SLC16A1 (MCT1), SLC16A3 (MCT4), SLC16A7 (MCT2) and SLC16A8 (MCT3).

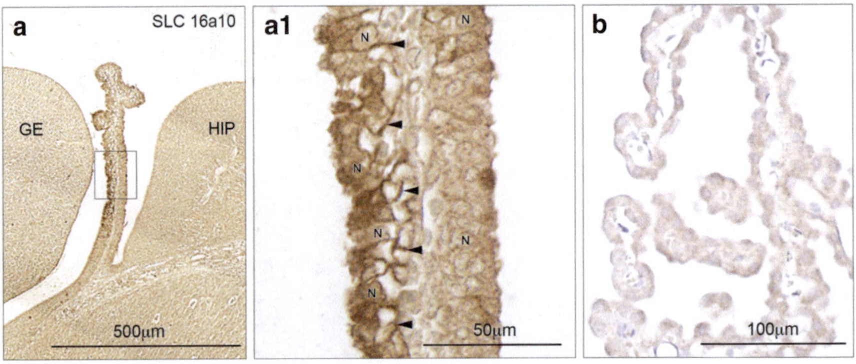

Fig. 3 Immunohistochemistry of influx transporters in sagittal and coronal sections from E15 and adult rat choroid plexus: SLC 16a10 (**a**, **a**1, **b**), The *rectangle* in (**a**), from E15 is shown at higher magnification in (**a1**). For comparison, adult choroid plexus stained for the same transporter is shown in (**b**). (**a**) The rostral-most leaflet of the lateral ventricular choroid plexus facing the ganglionic eminence (*GE*) demonstrates a very strong reactivity in contrast to the caudal-most leaflet facing the developing hippocampus (*HIP*). (**a1**) At higher magnification, immunostaining fills the entire cytoplasm apical to nuclei (*N*) and down along basolateral membrane (*arrowheads*) sparing empty glycogen spaces in epithelial cells of rostral leaflet. Caudal leaflet is also positively stained and shows a more uniform cytoplasmic reactivity. (**b**) Adult choroid plexus shows virtually no immunostaining for SLC 16a10. **a**, scale bar 500 μm. **a1**, scale bar 50 μm. **b**, scale bar 100 μm (From Saunders et al. [96])

SLC16A1 (MCT1) is involved in transport of monocarboxylates across the endothelial cells of the blood–brain barrier [44, 45] and epithelial cells of the choroid plexus [61] in the adult brain. Slc16a2 (MCT8), a6 (MCT7), a8 (MCT3), a9 (MCT9), a12 (MCT12) and a13 (MCT13) genes have been identified in adult mouse choroid plexus [78, 96]. Slc16a2 is a thyroid hormone transporter, which is expressed at similar levels in embryonic and adult choroid plexus; the others are all monocarboxylates transporters and expressed at a lower level in the rat embryonic plexus compared to the adult [96]. The only Slc16 transporter that is expressed at a higher level in mouse and rat embryonic choroid plexus is Slc16a10 [68, 96]; Slc16a10 transports tyrosine, the amino acid precursor of the thyroid hormones tri- and tetraiodothyronine. SLC16a10 has been shown to transport both triiodothyronine and tetraiodothyronine in transfected COS1 cells [42]. The finding of very high expression in rodent embryo choroid plexus [68, 96] together with demonstration of immunohistochemical presence of the protein (Fig. 3) suggests that it may play an important role in thyroid hormone transport in early brain development. The expression of monocarboxylate transporters appears to have been less studied in the embryonic brain barrier interfaces. Apart from a high expression of *Slc16a10* (MCT10) in embryonic mouse and rat choroid plexus [68, 96], nothing else seems to be known about monocarboxylate transporters in the embryonic choroid plexus. Triiodothyronine, T3, and thyroxine, T4, are essential for normal brain development. Inadequate delivery of T4 to the developing brain, usually due to iodine deficiency, results in cretinism [91, 100]. The choroid plexuses in the embryonic brain are prominent compared to vascularisation of the rest of the brain [57]; thus, it may be that MCT10 in the choroid plexuses along with TTR, a thyroid hormone carrier highly expressed throughout development, is the major mechanism by which thyroxine is delivered to the brain in early stages of its development. This complements the low developmental expression of *Slco1c1*, the main thyroid hormone transporter expressed at the blood–brain interfaces in the adult [63].

Developmental Regulation of Energy Substrate Transporters The finding of a high expression of glucose transporters in the embryonic brain, but not of those transporting lactate and other monocarboxylates [96], suggests that the embryonic brain may be similar to the adult brain in being exclusively dependent on glucose metabolism as a source of energy, as suggested previously by Vannucci and Vannucci [110] and unlike the postnatal brain in which monocarboxylate transport appears to be functionally significant, at least in rodents [65, 109].

Other Transport Systems Information on other important transported molecules (e.g. vitamins, metals, peptides and proteins) can be found in Davson and Segal [29] and Spector [102].

5 Efflux Mechanisms of the Brain Barrier Interfaces

ATP-binding cassette (ABC) efflux transporters are an important mechanism in the adult brain at both the blood–brain and blood–CSF barriers [47]. They actively exclude a large number of toxic but also potentially therapeutic compounds from the brain. At the blood–brain barrier interface (Fig. 4), the efflux transporters that have been best studied are ABCB1 (P-glycoprotein or MDR1) and ABCG2 (breast cancer resistance protein, BCRP). ABCC2 (multidrug resistance protein 2, MRP2) and ABCC4 (MRP4) have also been demonstrated at the blood–brain barrier interface [106]. In cerebral capillary endothelial cells (blood–brain barrier), PGP [92], BCRP [20], MRP2 [8], MRP4 [8] and MRP5 [82] are localised to the luminal membrane where they export compounds into the blood. At the blood–CSF interface, ABCC1 (multidrug resistance protein 1, MRP1) appears to be the predominant efflux transporter, but ABCC4 (MRP4) and ABCG2 (BCRP) are also present [36, 106]. Here MRP1 and MRP4 are localised to the basolateral membranes where they export compounds into the stroma of the plexus [36, 76]. The subcellular localisation of PGP in choroid plexus is not clear (see [97]). These transporters bind compounds that have been conjugated to transport motifs (glutathione, glucuronic acid or sulphated), which confers a wide range of substrate specificity and considerable overlap between transporters (see Fig. 4 and [73]). Some of the differences in mechanism by which the efflux transporters act are illustrated in Fig. 4. There are probably species differences in the level of expression and functional capacity of these various efflux transporters, and it is known that their expression changes with age during brain development at both interfaces [25, 36, 63, 98, 111]. For example,

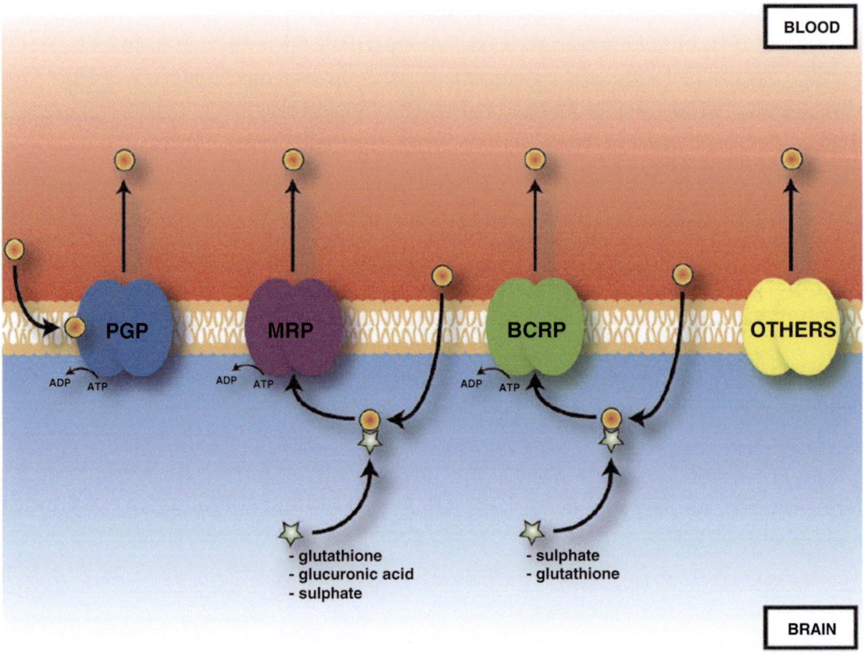

Fig. 4 Main outward transporters in cerebral endothelial cells. Some, e.g. PGP (P-glycoprotein), restrict entry into the cell. For others, e.g. MRP (multidrug resistance protein), ligand (drug or toxin) combines with glutathione, glucuronic acid or sulphate within cells before efflux (From Saunders et al. [93])

BCRP expression in rat choroid plexus is highest in embryos (20-fold compared to adult). In contrast there is practically no change in expression level for BCRP in the brain between the embryo and adult [36]. Possibly other members of this large group of transporters will be identified to be functional efflux transporters at one or more of the brain barrier interfaces. Knowledge of the presence and effectiveness of efflux transporters in the developing brain is essential for assessing what risk maternal exposure to drugs and toxins may pose to the embryonic and fetal brain as well as in premature infants. It is also essential to consider whether changes in these mechanisms might occur in pathological conditions such as inflammation.

6 Brain Barriers Regulate CSF Secretion

An essential component of the internal environment of both the adult and developing brain is the stability of the ionic composition of CSF, which is usually assumed to reflect that of brain interstitial fluid. Stability of the interstitial fluid and the CSF is important for the main function of the CNS, namely, generation and transmission of nerve impulses. One of the main functions of the epithelial cells of choroid plexuses in the lateral, third and fourth ventricles is secretion of CSF. The extent to which brain interstitial fluid may contribute to CSF within the ventricular system has been a matter of unresolved dispute for many years [29, 103].

The principal drivers of CSF secretion are intracellular carbonic anhydrase and Na/K-ATPase in the apical membrane of choroid plexus epithelial cells [29] with various ion channels in the basolateral and apical membranes also making an important contribution. Detailed electrophysiological and molecular studies by Brown and colleagues in Manchester and Damkier and colleagues in Aarhus are well and extensively reviewed in Damkier et al. [22]. A summary of the transporters and ion channels currently thought to be involved in CSF secretion in the adult and developing choroid plexus is shown in Fig. 5. The crucial information about the location of these channels and transporters is from the studies of Brown, Damkier and colleagues as indicated in Fig. 5 legend. Recent gene deletion studies of the Damkier group are beginning to unravel the functional importance of individual components of the complex array of transporters and channels involved in CSF secretion [18, 19, 23].

Developing Choroid Plexus Several solute carrier (*Slc*) gene products involved in CSF secretion are differentially expressed in development (Fig. 5). Thus, some of the transporters of sodium, the major extracellular cation, are enriched in the embryo (*Slc6a13*, *Slc4a4*) and others in the adult (e.g. *Slc5a5*, *Slc24a4*). Similarly, anion transporters for Cl^- and HCO_3^- are also fundamental for CSF secretion. These transporters belong to the *Slc4* family. Two that are upregulated in embryonic choroid plexus are *Slc4a1* (Cl -HCO_3 exchanger) and *Slc4a4* (Na^+-HCO_3^- co-transporter); see Saunders et al. [96]. In fetal rat and mouse choroid plexus, the levels of carbonic anhydrase and Na+/K+ATPase are much less than in the adult [57, 68]. Thus, in the adult *Atp1a2* (Na+/K+ATPase) was upregulated over greater than sevenfold, and the carbonic anhydrases (*CA5b*, *CA8*, *CA13*) were upregulated four to ninefold [68]. The low expression of Na+/K+ATPase and carbonic anhydrases in fetal choroid plexus probably accounts for the much lower CSF secretion rate in fetal [39] and neonatal brains [7, 56].

Functional studies on individual ion transporters in developing brain or choroid plexus have not been published, but it is possible to infer ion transport function from studies of CSF and plasma ion composition in the developing brain. It was shown many years ago that ion gradients between CSF and plasma are a characteristic of brain homeostasis. Some ion gradients are established very early in brain development, for example, Mg^{++} in early fetal life in monkeys [11] and Cl^- in fetal sheep [12] and neonatal rats [3]. This indicates that ion transport mechanisms are functional early in brain development. This is probably amongst the best evidence of the effective function of barrier mechanisms in the developing brain, but it has been largely ignored in favour of the implausible belief that the blood–brain barrier in the embryo or fetus is absent or otherwise defective see Saunders et al (2014).

CSF Turnover, Flow and Drainage The absolute volume and CSF secretion rate in the mammalian brain are directly related to the size of the brain (e.g. 0.325 μL.min^{-1} in mouse and 350 μLmin^{-1} in man), but the turnover of CSF is very similar as is the

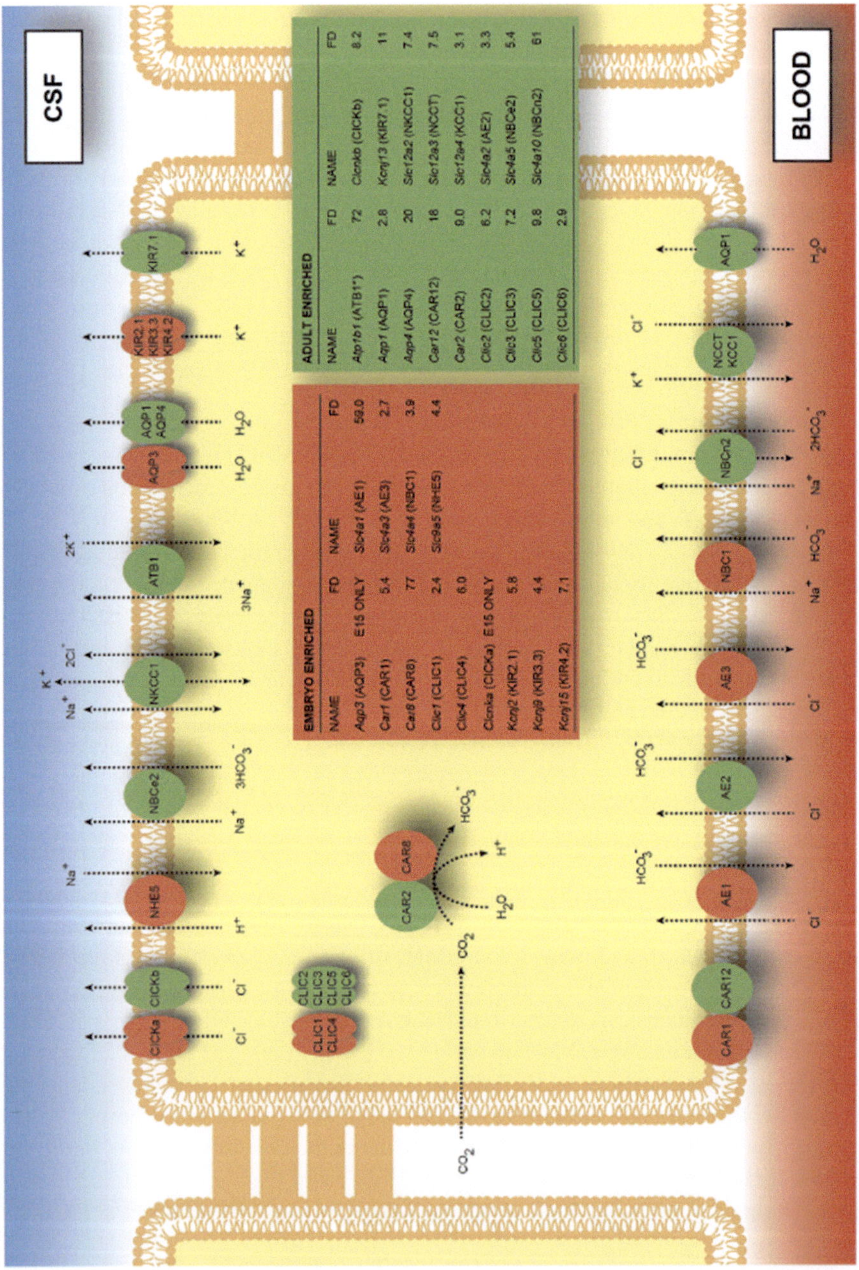
CSF
BLOOD
ClCKa
ClCKb
NHE5
NBCe2
NKCC1
ATB1
AQP3
AQP1 AQP4
KIR2.1 KIR3.3 KIR4.2
KIR7.1
CLIC1 CLIC4
CLIC2 CLIC3 CLIC5 CLIC6
CAR2
CAR8
CAR1
CAR12
AE1
AE2
AE3
NBC1
NBCn2
NCCT KCC1
AQP1
EMBRYO ENRICHED
NAME FD NAME FD
Aqp3 (AQP3) E15 ONLY Slc4a1 (AE1) 59.0
Car1 (CAR1) 5.4 Slc4a3 (AE3) 2.7
Car8 (CAR8) 77 Slc4a4 (NBC1) 3.9
Clic1 (CLIC1) 2.4 Slc9a5 (NHE5) 4.4
Clic4 (CLIC4) 6.0
Clcnka (ClCKa) E15 ONLY
Kcnj2 (KIR2.1) 5.8
Kcnj9 (KIR3.3) 4.4
Kcnj15 (KIR4.2) 7.1
ADULT ENRICHED
NAME FD NAME FD
Atp1b1 (ATB1*) 72 Clcnkb (ClCKb) 8.2
Aqp1 (AQP1) 2.8 Kcnj13 (KIR7.1) 11
Aqp4 (AQP4) 20 Slc12a2 (NKCC1) 7.4
Car12 (CAR12) 18 Slc12a3 (NCCT) 7.5
Car2 (CAR2) 9.0 Slc12a4 (KCC1) 3.1
Clic2 (CLIC2) 6.2 Slc4a2 (AE2) 3.3
Clic3 (CLIC3) 7.2 Slc4a5 (NBCe2) 5.4
Clic5 (CLIC5) 9.8 Slc4a10 (NBCn2) 61
Clic6 (CLIC6) 2.9

Fig. 5 Localisation of proteins for ion transporters, channels and associated enzymes and identification of their corresponding genes in adult and immature rat choroid plexus (data for localisation of proteins from Damkier et al. [21]). CSF secretion results from coordinated transport of ions and water from basolateral membrane to cytoplasm and then sequentially across apical membrane into ventricles (see [29]). Transporters, enzymes and their *genes* are AE2 (*Slc4a2*), NBC1 (*Slc4a4*), NCBn2 (*Slc4a10*), NHE5 (Slc9a5), ATB1 (*Atb1b1*, Na+/K+− ATPase, asterisk), ATB1 (*Atb1b1*), NKCC1 (*Slc12a2*), NCCT (*Slc12a3*) and KCC1 (*Slc12a4*). Aquaporin (AQP1/3/4) channels on CSF facing membrane mediate water flux into ventricles (Oshio K et al. [89]. Polarised distribution of carbonic anhydrase (CAR) and Na+/K+−ATPase, and aquaporins, enables net ion and water translocation to CSF ([57]; Brian OK, et al. [13]. Inset boxes show the fold differences for genes expressed at higher level in embryonic (red) or in adult (green) choroid plexus (From Liddelow et al. [71])

secretion rate related to the size of the choroid plexuses, in the range 0.38 %/min in man to 0.89 %/min in mouse [29]. This latter observation strongly suggests that most of the CSF originates from plexus secretion. Once secreted, there is a flow of CSF from the lateral ventricles through the third ventricle into the fourth ventricle via the aqueduct of Sylvius. The flow is generated in part by a pressure gradient from the lateral ventricles to the fourth ventricle but aided by cilia on the cells of the ependymal lining of the surface of the ventricles. As the CSF passes through the different ventricles, additional secreted fluid is added, and the composition of the CSF in different ventricles is slightly different [29]. CSF in the fourth ventricle leaves the inner ventricular system via the foramen of Magendie and the foramina of Luschka into the subarachnoid space, which surrounds the outer surface of the brain. The traditional view is that CSF flows into the dorsal subarachnoid space and from there is reabsorbed into the superior sagittal sinus on the venous side of the cerebral circulation via the arachnoid villi or arachnoid granulations in man [29]. However, this has been challenged by Johnston who has provided evidence that significant CSF drainage occurs along cranial and spinal nerves with absorption occurring into lymphatic vessels situated outside the CNS; the primary route appears to be via the cribriform plate into lymphatics in the submucosa of the respiratory and cribriform epithelia [58, 59].

Sink Effect of CSF This is an important functional concept introduced by Davson in the 1960s [28, 87]. Estimates of brain extracellular space (ECS) using standard markers such as inulin or $^{35}SO_4$ were only a few percent, i.e. much lower than in other tissues of the body. Oldendorf and Davson [87] proposed that this was because the CSF acted as a "drain" or "sink" removing the slowly penetrating markers as rapidly as they entered the CSF and brain (ECS) space. The experimental evidence for this was obtained by comparing the brain and CSF spaces when the marker was administered continuously intravenously or via ventriculocisternal perfusion or both; the combination gave values of around 9 % which was taken at the time as the best available estimate of brain ECS [87]. Subsequent estimates using a variety of techniques gave values of 15–20 % depending on the technique and brain region examined (see [107]). These values contrasted with the values estimated by electron microscopists as close to if not actually zero ([50, 115]; see [107]). It is as well to remember that all morphology is an artefact and the skill is in interpreting the artefacts in relation to functional data. Most tissue preservation techniques cause substantial shrinkage so it is perhaps not surprising that ECS volume in electron micrographs appeared to be negligible. This discrepancy between the EM and physiological estimates of ECS has recently been resolved by Knott and his colleagues who compared chemical fixation with a new ultrafast high-pressure freezing technique to preserve brain tissue. Using this new method, their estimate in the brain region they examined was 15.5 % [62]. This approach is important not only because it resolves the discrepancy between physiological and morphological estimates of brain ECS; it also gives a more realistic image of the relation between different cellular components in the brain. For example, chemical fixation shows complete encirclement of capillaries by astrocytic end feet, whereas rapid cryofixation revealed only a partial coverage of about two thirds. The functional significance of this is that solutes passing through the capillary wall from blood have more direct access to the neurons [62]. Brain ECS is reduced in some pathological conditions, e.g. ischaemia and/or hypoxia [112], but increased in others, e.g. experimental immune encephalomyelitis [99]. It is thus important to consider whether ECS volume might be altered in other inflammatory conditions. Detailed and scholarly reviews of the wide range of experimental evidence on the relation between CSF and brain ECS have been provided by Hladky and Barrand [49] and Spector et al. [103]. The review by Hladky and Barrand [49] is of particular value because it assesses the limitations of many studies in this field and suggests alternative explanations for some published data.

7 Conclusion

The brain barriers consist of multiple structural interfaces between the blood, CSF and brain, each with its own specific features. The cells of these interfaces contain a large variety of transporters that provide the distinctive aspects of the internal environment of the brain as well as supplying essential nutrients. Physical barrier functions are provided by the structural components of the brain interfaces, while efflux transporters provide additional specific protective mechanisms. Many of these mechanisms are present in embryonic brain and some may be more active than in the adult. In studying the effects of inflammation on the blood–brain barriers, it is important to consider that both the multiple interfaces and their various cellular functional mechanisms may be affected differentially.

References

1. Abbott NJ, Rönnbäck L, Hansson E (2006) Astrocyte-endothelial interactions at the blood–brain barrier. Nat Rev Neurosci 7(1):41–53
2. Alvarez JI, Katayama T, Prat A (2013) Glial influence on the blood brain barrier. Glia 61(12):1939–1958
3. Amtorp O, Sørensen SC (1974) The ontogenetic development of concentration differences for protein and ions between plasma and cerebrospinal fluid in rabbits and rats. J Physiol 243:387–400
4. Armulik A, Genové G, Mäe M et al (2010) Pericytes regulate the blood–brain barrier. Nature 468(7323):557–561
5. Attwell D, Mishra A, Hall CN et al (2016) What is a pericyte? J Cereb Blood Flow Metab 36(2):451–455
6. Bailey CG, Ryan RM, Thoeng AD et al (2011) Loss-of-function mutations in the glutamate transporter SLC1A1 cause human dicarboxylic aminoaciduria. J Clin Invest 121(1):446–453
7. Bass NH, Lundborg P (1973) Postnatal development of bulk flow in the cerebrospinal fluid system of the albino rat: clearance of carboxyl-(14C)inulin after intrathecal infusion. Brain Res 52:323–332

8. Bauer B, Hartz AMS, Lucking JR et al (2008) Coordinated nuclear receptor regulation of the efflux transporter, Mrp2, and the phase-II metabolizing enzyme, GSTpi, at the blood–brain barrier. J Cereb Blood Flow Metab 28:1222–1234
9. Bauer HC, Krizbai IA, Bauer H et al (2014) "You Shall Not Pass"-tight junctions of the blood brain barrier. Front Neurosci 8:392. doi:10.3389/fnins.2014.00392
10. Bito LZ, Myers RE (1970) The ontogenesis of haematoencephalic cation transport in the rhesus monkey. J Physiol (Lond) 208:153–170
11. Bodoy S, Fotiadis D, Stoeger C et al (2013) The small SLC43 family: facilitator system l amino acid transporters and the orphan EEG1. Mol Aspects Med 34(2–3):638–645
12. Bradbury MWB, Crowder J, Desai S et al (1972) Electrolytes and water in the brain and cerebrospinal fluid of the foetal sheep and guinea pig. J Physiol (Lond) 227:591–610
13. Brian OK, Tom P, Wang D (2010) Aquaporins: relevance to cerebrospinal fluid physiology and therapeutic potential in hydrocephalus. CSF Res 7:15.
14. Brightman MW, Reese TS (1969) Junctions between intimately apposed cell membranes in the vertebrate brain. J Cell Biol 40:48–77
15. Brøchner CB, Holst CB, Møllgård K (2015) Outer brain barriers in rat and human development. Front Neurosci 9:75. doi:10.3389/fnins.2015.00075
16. Caley DW, Maxwell DS (1970) Development of the blood vessels and extracellular spaces during postnatal maturation of rat cerebral cortex. J Comp Neurol 138:31–47
17. Christensen HN, Albritton LM, Kakuda DK et al (1994) Gene-product designations for amino acid transporters. J Exp Biol 196:51–57
18. Christensen HL, Nguyen AT, Pedersen FD et al (2013) Na(+) dependent acid–base transporters in the choroid plexus; insights from slc4 and slc9 gene deletion studies. Front Physiol 4:304. doi:10.3389/fphys.2013.00304
19. Christensen IB, Gyldenholm T, Damkier HH et al (2013) Polarization of membrane associated proteins in the choroid plexus epithelium from normal and slc4a10 knockout mice. Front Physiol 4:344. doi:10.3389/fphys.2013.00344
20. Cooray HC, Blackmore CG, Maskell L et al (2002) Localisation of breast cancer resistance protein in microvessel endothelium of human brain. Neuroreport 13:2059–2063
21. Damkier HH, Brown PD, Praetorius J (2010) Epithelial pathways in choroid plexus electrolyte transport. Physiology (Bethesda) 25(4):239–249
22. Damkier HH, Brown PD, Praetorius J (2013) Cerebrospinal fluid secretion by the choroid plexus. Physiol Rev 93(4):1847–1892
23. Damkier HH, Praetorius J (2012) Genetic ablation of Slc4a10 alters the expression pattern of transporters involved in solute movement in the mouse choroid plexus. Am J Physiol Cell Physiol 302(10):C1452–C1459
24. Daneman R, Zhou L, Kebede AA et al (2010) Pericytes are required for blood–brain barrier integrity during embryogenesis. Nature 468:562–566
25. Daneman R, Zhou L, Agalliu D et al (2010) The mouse blood–brain barrier transcriptome: a new resource for understanding the development and function of brain endothelial cells. PLoS One 5:e13741. doi:10.1371/journal.pone.0013741
26. Daneman R, Prat A (2015) The blood–brain barrier. Cold Spring Harb Perspect Biol 7(1):a020412. doi:10.1101/cshperspect.a020412
27. Davson H (1956) Physiology of the ocular and cerebrospinal fluids. Churchill, London
28. Davson H (1967) Physiology of the cerebrospinal fluid. Churchill, London
29. Davson H, Segal MB (1996) Physiology of the CSF and blood–brain barriers. CRC Press, Boca Raton
30. Decimo I, Fumagalli G, Berton V et al (2012) Meninges: from protective membrane to stem cell niche. Am J Stem Cells 1(2):92–105
31. Dore-Duffy P, Cleary K (2011) Morphology and properties of pericytes. Methods Mol Biol 686:49–68
32. Dore-Duffy P, Esen N, Serkin Z (2015) Chapter 5. The elusive multipotent microvascular pericytes. In: The blood–brain barrier in health and disease vol 1 morphology, biology and immune function. CRC Press, Boca Raton, pp 119–139

33. Dziegielewska KM, Habgood MD, Møllgård K et al (1991) Species-specific transfer of plasma albumin from blood into different cerebrospinal fluid compartments in the fetal sheep. J Physiol 439:215–237
34. Ehrlich P (1885) Das Sauerstoffbedürfnis des Organismus. Eine farbenanalytische Studie. Hirschwald, Berlin
35. Ek CJ, Habgood MD, Dziegielewska KM et al (2006) Functional effectiveness of the blood–brain barrier to small water-soluble molecules in developing and adult opossum (Monodelphis domestica). J Comp Neurol 496:13–26
36. Ek CJ, Wong A, Liddelow SA et al (2010) Efflux mechanisms at the developing brain barriers: ABC-transporters in the fetal and postnatal rat. Toxicol Lett 197:51–59
37. ElAli A, Thériault P, Rivest S (2014) The role of pericytes in neurovascular unit remodeling in brain disorders. Int J Mol Sci 15(4):6453–64574
38. Enerson BE, Drewes LR (2006) The rat blood–brain barrier transcriptome. J Cereb Blood Flow Metab 26(7):959–973
39. Evans CAN, Reynolds JM, Reynolds ML et al (1974) The development of a blood–brain barrier mechanism in foetal sheep. J Physiol 238(2):371–386
40. Fossan G, Cavanagh M, Evans CAN et al (1985) CSF-brain permeability in the immature sheep fetus: a CSF-brain barrier. Dev Brain Res 18:113–124
41. Fotiadis D, Kanai Y, Palacín M (2013) The SLC3 and SLC7 families of amino acid transporters. Mol Aspects Med 34(2–3):39–158
42. Friesema EC, Jansen J, Jachtenberg JW et al (2008) Effective cellular uptake and efflux of thyroid hormone by human monocarboxylate transporter 10. Mol Endocrinol 22(6):1357–1369
43. Goldmann EE (1909) Die äussere und innere Sekretion des gesunden und kranken Organismus im Lichte der 'vitalen Färbung'. Beiträg Klinische Chirurgie 64:192–265
44. Halestrap AP (2013) The SLC16 gene family – structure, role and regulation in health and disease. Mol Aspects Med 34(2–3):337–349. doi:10.1016/j.mam.2012.05.003
45. Halestrap AP (2013) Monocarboxylic acid transport. Compr Physiol 3(4):1611–1643. doi:10.1002/cphy.c130008
46. Hall CN, Reynell C, Gesslein B et al (2014) Capillary pericytes regulate cerebral blood flow in health and disease. Nature 508:55–60
47. Hartz AMS, Bauer B (2011) ABC transporters in the CNS – an inventory. Curr Pharm Biotechnol 12:656–673
48. Hill J, Rom S, Ramirez SH et al (2014) Emerging roles of pericytes in the regulation of the neurovascular unit in health and disease. J Neuroimmune Pharmacol 9(5):591–605
49. Hladky SB, Barrand MA (2014) Mechanisms of fluid movement into, through and out of the brain: evaluation of the evidence. Fluids Barriers CNS 11(1):26. doi:10.1186/2045-8118-11-26
50. Holash JA, Noden DM, Stewart PA (1993) Re-evaluating the role of astrocytes in blood brain barrier induction. Dev Dyn 197:14–25
51. Horstmann E, Meves H (1959) Die Feinstrucktur des moleculären Rindengraues und ihre physiologisches Bedeutung. Z Zellforschung 49:569–604
52. Hurtado-Alvarado G, Cabañas-Morales AM, Gómez-Gónzalez B (2014) Pericytes: brain-immune interface modulators. Front Integr Neurosci 7:80. doi:10.3389/fnint.2013.00080
53. Ito Y, Takahashi S, Kagitani-Shimono K et al (2015) Nationwide survey of glucose transporter-1 deficiency syndrome (GLUT-1DS) in Japan. Brain Dev 37(8):780–789
54. Janzer RC, Raff MC (1987) Astrocytes induce blood–brain barrier properties in endothelial cells. Nature 325:253–257
55. Jessen NA, Munk AS, Lundgaard I et al (2015) The glymphatic system: a beginner's guide. Neurochem Res 40(12):2583–2599
56. Johanson CE, Woodbury DM (1974) Changes in CSF flow and extracellular space in the developing rat. In: Vernadakis A, Weiner N (eds) Drugs and the developing brain. Plenum, New York, pp 281–287
57. Johansson PA, Dziegielewska KM, Liddelow SA et al (2008) The blood-CSF barrier explained: when development is not immaturity. Bioessays 30(3):237–248

58. Johnston M, Zakharov A, Papaiconomou C et al (2004) Evidence of connections between cerebrospinal fluid and nasal lymphatic vessels in humans, non-human primates and other mammalian species. Cerebrospinal Fluid Res 1(1):2. doi:10.1186/1743-8454-1-2
59. Johnston M, Zakharov A, Koh L et al (2005) Subarachnoid injection of Microfil reveals connections between cerebrospinal fluid and nasal lymphatics in the non-human primate. Neuropathol Appl Neurobiol 31(6):632–640
60. Kanai Y, Clémençon B, Simonin A et al (2013) The SLC1 high-affinity glutamate and neutral amino acid transporter family. Mol Aspects Med 34(2–3):108–120
61. Koehler-Stec EM, Simpson IA, Vannucci SJ et al (1998) Monocarboxylate transporter expression in mouse brain. Am J Physiol 275(3 Pt 1):E516–E524
62. Korogod N, Petersen CC, Knott GW (2015) Ultrastructural analysis of adult mouse neocortex comparing aldehyde perfusion with cryo fixation. Elife 4. doi:10.7554/eLife.05793
63. Kratzer I, Liddelow SA, Saunders NR et al (2013) Developmental changes in the transcriptome of the rat choroid plexus in relation to neuroprotection. Fluids Barriers CNS 10:25. doi:10.1186/2045-8118-10-25
64. Langlet F, Mullier A, Bouret SG et al (2013) Tanycyte-like cells form a blood-cerebrospinal fluid barrier in the circumventricular organs of the mouse brain. J Comp Neurol 521(15):3389–3405. doi:10.1002/cne.23355
65. Leino RL, Gerhart DZ, Drewes LR (1999) Monocarboxylate transporter (MCT1) abundance in brains of suckling and adult rats: a quantitative electron microscopic immunogold study. Brain Res Dev Brain Res 113(1–2):47–54
66. Lewandowsky M (1900) Zur Lehre von der Cerebrospinalflüssigkeit. Z Clin Med 40:480–494
67. Liddelow SA, Dziegielewska KM, Ek CJ et al (2009) Cellular transfer of macromolecules across the developing choroid plexus of Monodelphis domestica. Eur J Neurosci 29(2):253–266
68. Liddelow SA, Temple S, Møllgård K et al (2012) Molecular characterisation of transport mechanisms at the developing mouse blood-CSF interface: a transcriptome approach. PLoS One 7:e33554. doi:10.1371/journal.pone.0033554
69. Liddelow SA, Dziegielewska KM, Ek CJ et al (2013) Mechanisms that determine the internal environment of the developing brain: a transcriptomic, functional and ultrastructural approach. PLoS One 8:e65629. doi:10.1371/journal.pone.0065629.s005
70. Liddelow SA, Dzięgielewska KM, Møllgård K et al (2014) Cellular specificity of the blood-CSF barrier for albumin transfer across the choroid plexus epithelium. PLoS One 9(9):e106592. doi:10.1371/journal.pone.0106592
71. Liddelow SA, Dziegielewska KM, Ek CJ (2016) Correction: mechanisms that determine the internal environment of the developing brain: a transcriptomic, functional and ultrastructural approach. PLoS One 11(1):e0147680. doi:10.1371/journal.pone.0147680
72. Lindahl P, Johansson BR, Levéen P et al (1997) Pericyte loss and microaneurysm formation in PDGF-B-deficient mice. Science 277(5323):242–245
73. Löscher W, Potschka H (2005) Drug resistance in brain diseases and the role of drug efflux transporters. Nat Rev Neurosci 6:591–602
74. Ma S, Kwon HJ, Huang Z (2012) A functional requirement for astroglia in promoting blood vessel development in the early postnatal brain. PLoS One 7(10):e48001. doi:10.1371/journal.pone.0048001
75. MacAulay N, Zeuthen T (2010) Water transport between CNS compartments: contributions of aquaporins and cotransporters. Neuroscience 168(4):941–956
76. Maliepaard M, Scheffer GL, Faneyte IF et al (2001) Subcellular localization and distribution of the breast cancer resistance protein transporter in normal human tissues. Cancer Res 61:3458–3464
77. Manley GT, Fujimura M, Ma T et al (2000) Aquaporin-4 deletion in mice reduces brain edema after acute water intoxication and ischemic stroke. Nat Med 6(2):159–163
78. Marques F, Sousa JC, Coppola G et al (2011) Transcriptome signature of the adult mouse choroid plexus. Fluids Barriers CNS 8(1):10. doi:10.1186/2045-8118-8-10

79. Møllgård K, Balslev Y, Lauritzen B et al (1987) Cell junctions and membrane specializations in the ventricular zone (germinal matrix) of the developing sheep brain: a CSF-brain barrier. J Neurocytol 16:433–444
80. Nabeshima S, Reese TS, Landis DM et al (1975) Junctions in the meninges and marginal glia. J Comp Neurol 164:127–169
81. Neuwelt EA (2004) Mechanisms of disease: the blood–brain barrier. Neurosurgery 54:131–140; discussion 141–142
82. Nies AT, Jedlitschky G, König J et al (2004) Expression and immunolocalization of the multidrug resistance proteins, MRP1-MRP6 (ABCC1-ABCC6), in human brain. Neuroscience 129:349–360
83. O'Donnell ME (2015) Chapter 4. The neurovascular unit. In: The blood–brain barrier in health and disease vol 1 morphology, biology and immune function. CRC Press, Boca Raton, pp 86–118
84. Oldendorf WM (1971) Brain uptake of radiolabeled amino acids, amines, and hexoses after arterial injection. Am J Physiol 221:1629–1639
85. Oldendorf WH (1971) Uptake of radiolabeled essential amino acids by brain following arterial injection. Proc Soc Exp Biol Med 136:385–386
86. Oldendorf WH (1977) The blood–brain barrier. Exp Eye Res 25(Suppl):177–190
87. Oldendorf WH, Davson H (1967) Brain extracellular space and the sink action of cerebrospinal fluid. Measurement of rabbit brain extracellular space using sucrose labeled with carbon 14. Arch Neurol 17(2):196–205
88. Oldendorf WH, Szabo J (1976) Amino acid assignment to one of three blood–brain barrier amino acid carriers. Am J Physiol 230:94–98
89. Oshio K, Watanabe H, Song Y et al (2005) Reduced cerebrospinal fluid production and intracranial pressure in mice lacking choroid plexus water channel Aquaporin-1. FASEB J. 9(1):76–78.
90. Richter JJ, Wainer A (1971) Evidence for separate system for the transport of neutral and basic amino acids across the blood–brain barrier. J Neurochem 18:613–620
91. Rivas M, Naranjo JR (2007) Thyroid hormones, learning and memory. Genes Brain Behav 6(Suppl 1):40–44
92. Roberts LM, Black DS, Raman C et al (2008) Subcellular localization of transporters along the rat blood–brain barrier and blood-cerebral-spinal fluid barrier by in vivo biotinylation. Neuroscience 155:423–438
93. Saunders NR, Liddelow SA, Dziegielewska KM (2012) Barrier mechanisms in the developing brain. Front Neuropharmacol 3:46. doi:10.3389/fphar.2012.00046
94. Saunders NR, Daneman R, Dziegielewska KM et al (2013) Transporters of the blood–brain and blood-CSF interfaces in development and in the adult. Mol Aspects Med 34:742–752
95. Saunders NR, Dreifuss J-J, Dziegielewska KM et al (2014) The rights and wrongs of blood brain barrier permeability studies: a walk through 100 years of history. Front Neurosci 8:1–26. doi:10.3389/fnins.2014.00404/abstract
96. Saunders NR, Dziegielewska KM, Møllgård K et al (2015) Influx mechanisms in the embryonic and adult rat choroid plexus: a transcriptome study. Front Neurosci 9:123. doi:10.3389/fnins.2015.00123
97. Saunders NR, Habgood MD, Møllgård K et al (2016) The biological significance of brain barrier mechanisms: help or hindrance in drug delivery to the central nervous system? F1000Res 5. pii: F1000 Faculty Rev-313. doi:10.12688/f1000research.7378.1
98. Schumacher U, Møllgård K (1997) The multidrug-resistance P-glycoprotein (Pgp, MDR1) is an early marker of blood–brain barrier development in the microvessels of the developing human brain. Histochem Cell Biol 108:179–182
99. Simonová Z, Svoboda J, Orkand P et al (1996) Changes of extracellular space volume and tortuosity in the spinal cord of Lewis rats with experimental autoimmune encephalomyelitis. Physiol Res 45:11–22
100. Skeaff SA (2011) Iodine deficiency in pregnancy: the effect on neurodevelopment in the child. Nutrients 3(2):265–273

101. Smith DE, Streicher E, Milkovic K et al (1964) Observations on the transport of proteins by the isolated choroid plexus. Acta Neuropathol 3:372–386
102. Spector R (2009) Nutrient transport systems in brain: 40 years of progress. J Neurochem 111(2):315–320
103. Spector R, Robert Snodgrass S, Johanson CE (2015) A balanced view of the cerebrospinal fluid composition and functions: Focus on adult humans. Exp Neurol 273:57–68
104. Sperandeo MP, Borsani G, Incerti B et al (1998) The gene encoding a cationic amino acid transporter (SLC7A4) maps to the region deleted in the velocardiofacial syndrome. Genomics 49(2):230–236
105. Stern L, Gautier R (1918) Le passage dans le liquide céphalo-rachidien de substances introduites dans la circulation et leur action sur le système nerveux central chez les différentes espèces animales. R C R d Ia Soc de Phys et d'hist natur de Genève 35:91–94
106. Strazielle N, Ghersi-Egea J-F (2015) Efflux transporters in blood–brain interfaces of the developing brain. Front Neurosci 9:1–11. doi:10.3389/fnins.2015.00021
107. Syková E, Nicholson C (2008) Diffusion in brain extracellular space. Physiol Rev 88:1277–1340
108. Vannucci SJ, Seaman LB, Brucklacher RM et al (1994) Glucose transport in developing rat brain: glucose transporter proteins, rate constants and cerebral glucose utilization. Mol Cell Biochem 140(2):177–184
109. Vannucci SJ, Simpson IA (2003) Developmental switch in brain nutrient transporter expression in he rat. Am J Physiol Endocrinol Metab 285(5):E1127–E1134
110. Vannucci RC, Vannucci SJ (2000) Glucose metabolism in the developing brain. Semin Perinatol 24(2):107–115
111. Virgintino D, Errede M, Girolamo F et al (2008) Fetal blood–brain barrier P-glycoprotein contributes to brain protection during human development. J Neuropathol Exp Neurol 67:50–61
112. Vorísek I, Syková E (1997) Ischemia-induced changes in the extracellular space diffusion parameters, K+, and pH in the developing rat cortex and corpus callosum. J Cereb Blood Flow Metab 17(2):191–203
113. Whish S, Dziegielewska K, Møllgård K et al (2015) The inner CSF–brain barrier: developmentally controlled access to the brain via intercellular junctions. Front Neurosci 9:115. doi:10.3389/fnins.2015.00016
114. Wright EM (2013) Glucose transport families SLC5 and SLC50. Mol Aspects Med 34:183–196
115. Wyckoff RW, Young JZ (1956) The motorneuron surface. Proc R Soc Lond B Biol Sci 144(917):440–450

Topological Aspects of the Blood–Brain and Blood–Cerebrospinal Fluid Barriers and Their Relevance in Inflammation

Friederike Pfeiffer, Andreas F. Mack, and Hartwig Wolburg

Abstract In both multiple sclerosis and its animal model, experimental autoimmune encephalomyelitis, the blood–brain barrier (BBB) is known to be compromised at the level of postcapillary venules. In addition to this segment of blood circulation, recent research has focused on the role of the choroid plexus (CP), which is crossed by encephalitogenic T-cells to enter the cerebrospinal fluid (CSF) and to reach the subarachnoid space (SAS). Here, cytokines can activate local antigen-presenting cells to enhance the transmission of inflammatory cells from the subpial vasculature into the SAS to evoke meningitis- and encephalitis-related diseases. However, overcoming the endothelium in the postcapillary venules and the epithelium in the CP do not seem to be the only mechanisms in the induction process of inflammation. Rather, as there is continuity between the stroma of the CP and the SAS, this continuity could serve as a direct pathway for inflammatory cells. In this review, we describe the morphological properties of barrier-related cells in both the brain vasculature and the CP to draw attention to possible mechanisms of the processes of inflammation in the central nervous system. We focus on permeability-related structures, such as tight junctions in endothelial cells of the BBB and in endothelial and epithelial cells of the CP, which represent the main site of the blood–CSF barrier proper.

F. Pfeiffer
Werner Reichardt Centre for Integrative Neuroscience (CIN), University of Tübingen, Tübingen, Germany

A.F. Mack
Institute of Clinical Anatomy and Cell Analysis, University of Tübingen, Tübingen, Germany

H. Wolburg (✉)
Institute of Pathology and Neuropathology, University of Tübingen, Liebermeisterstraße 8, 72076 Tübingen, Germany
e-mail: h.wolburg@t-online.de

R. Lyck, G. Enzmann (eds.), *The Blood Brain Barrier and Inflammation*, Progress in Inflammation Research, DOI 10.1007/978-3-319-45514-3_2

1 Introduction

The original finding by Paul Ehrlich [16] that an intravenously administered dye did not stain the brain tissue, together with the complementary observation of his colleague Edwin Goldmann [28] that the very same dye, if applied into the cerebrospinal fluid (CSF), did stain the brain tissue, led to the concept of biological barriers between blood and brain (the blood–brain barrier [BBB]) and between blood and CSF (blood–CSF barrier [BCB]) [16, 28]. For a detailed description of the history of the BBB concept, see Saunders et al. [72]. Ehrlich and Goldmann used trypan blue as a tracer. The restricted access of the tracer to the brain after vascular administration led to the conclusion that there must be a barrier between the blood and the brain (the BBB), whereas the free access of trypan blue to the brain from the subarachnoid space (SAS) or the *cisterna magna* to the brain tissue made it clear that there is no CSF–brain barrier. Instead, because no dye spread into the circumventricular organs (CVOs) and the choroid plexus (CP), the existence of the BCB was compellingly demonstrated.

The cellular basis of these barriers was unclear for decades. Today, we know that in most vertebrates the barrier is located within the endothelium (endothelial BBB) and in the epithelial cells of the CP and the tanycytes of the CVOs (glial BCB). The endothelial cells of the BBB are not the only cells forming the barrier. It has been proposed that endothelial cells, astrocytes, pericytes, neurons, and the extracellular matrix (ECM) constitute what is now called the neurovascular unit (NVU) [2, 33]. The structures responsible for the restriction of the paracellular flux among endothelial cells (in the BBB) and among glial cells (in the BCB) were identified as tight junctions (TJs) [12, 31, 68, 86]. Today, molecules constituting the TJs have been identified as members of the occludin and claudin protein families [8, 32]. In particular, the TJs of the BBB have a unique molecular anatomy that is distinct from that of endothelial cell TJs in the body. In addition, there may be a complex crosstalk between tight and adherens junctions [79].

The BBB is under the control of brain-specific factors, which are widely undefined. The intracerebral control is indirectly indicated by the fact that BBB properties cannot be maintained when BBB endothelial cells are kept in culture, where they are not exposed to the influence of the brain environment (see below). In the brain, the perivascular space (PVS) is extremely narrow. At the level of capillaries, there is a single common basal lamina bordering the basal endothelial membrane and the astroglial endfoot membrane, forming the perivascular glial limiting membrane (for more details, see below). For clarity, we should state that in this chapter we use the term "basal lamina" for an ECM that is called by many authors "basement membrane." We want to avoid this term because the basement membrane is not a membrane, but a membrane-associated lamina. Pericytes are frequently embedded within, and completely surrounded by, the gliovascular basal lamina. At the capillary segment, no PVS is apparent. In the adjacent postcapillary venules and precapillary arterioles respectively the PVS gradually widens and is continuous with the Virchow–Robin space, which is best visible at the surface of the brain where blood vessels penetrate the brain parenchyma perpendicular to the brain surface. In this chapter, we focus on the topological properties of T-cell immigration into the central nervous system (CNS) during neuroinflammatory diseases such as multiple sclerosis (MS).

2 Brain Inflammation in Human Diseases and Animal Models

During inflammatory conditions in the CNS immune cells immigrate into the CNS and can be detected in the CNS parenchyma and in the CSF. This is a relatively new insight, because for many decades the CNS had been recognized as a site of immune privilege. The reasons for this view were the presence of the BBB, the absence of lymphatic vessels, and the inability to generate immune responses to immunogenic material [53, 65]. The endothelial BBB has often been regarded as the obvious place of entry into the CNS for the circulating immune cells. However, the CP has recently been considered to be an alternative entry site for circulating lymphocytes in the pathogenesis of inflammatory diseases [41, 67].

2.1 *Multiple Sclerosis*

The prevalence of MS varies around the world but there is a geographic preference for individuals living in northern latitudes, with Caucasians being especially vulnerable. Both environmental factors and genetic predisposition enhance the susceptibility to developing MS [59, 70]. Invading immune cells cause focal demyelinated lesions in the CNS, leaving axons partially demyelinated and also leading to a variable extent to axonal loss. In addition, glial scar formation is observed. The current view on the pathogenesis of MS assumes that inflammation is the driving force of demyelinating lesion formation. Therefore, extensive research efforts to ameliorate tissue injury have been directed toward modulating the immune response and thus preventing the formation of new plaques. Immune cells involved in the destruction of myelin may enter from PVSs, the meninges or the CSF [42, 65]. The inflammatory nature of MS lesions has been very well described by pathological studies. The inflammatory infiltrates are mainly composed of T lymphocytes, B lymphocytes, and of resident and immigrated antigen-presenting cells (APCs) such as macrophages and dendritic cells. Class I and II MHC antigens, adhesion molecules, chemokines and their receptors, and various immune effector molecules are expressed in the different forms of MS defined by different compositions of infiltrating cells. Clonal expansion of CD8 and CD4 T cells and also B cells indicates that the inflammatory reaction is mediated by a CNS antigen. Yet, the driving force of the inflammatory process is not yet fully discovered. In addition, there is still a debate as to whether neurodegeneration occurs independently of inflammation [42].

At what site do encephalitogenic T cells enter the brain to start the destructive cascade? To study leukocyte trafficking into the CNS, researchers depend on the one hand on human samples, and on the other hand on animal experiments. As access to human tissue from patients suffering from MS is limited, research greatly depends on animal models that display certain features of the human disease. The most commonly used animal model to investigate inflammation in MS is experimental autoimmune encephalomyelitis (EAE). Some of the current therapies used to treat MS have been successfully established and tested in EAE models.

2.2 An Animal Model for MS: Experimental Autoimmune Encephalomyelitis

Experimental autoimmune encephalomyelitis mimics MS in many regards; yet, there are some differences in the composition of inflammatory cells. Although it is generally accepted that MHC class I CD8 T cells play a dominant role in MS pathogenesis, EAE is mainly driven by MHC class II CD4 T-cells, because of the mode of immunization with myelin antigens in complete Freund's adjuvant. The role of CD8 T-cells in the development of EAE is still under investigation, but the differences in T-cell predominance may explain why some new experimental therapies were effective in EAE, but failed in MS [23]. To render investigations more complicated, CD8 T-cells also seem to play a protective role in both EAE and MS. In addition, the composition of immune cells mediating MS pathogenesis is variable among patients [23]. To overcome the limitations of the EAE model, new models have been designed that mimic the disease more closely. In a transgenic mouse model for relapse-remitting MS (RRMS), where mice carry a T cell receptor that is specific to myelin oligodendrocyte glycoprotein (MOG) peptide, the mice spontaneously develop relapse-remitting EAE [64]. In addition, attempts were made toward "humanized mice," where genes or cells from MS patients were transferred into mouse hosts [75]. So far, only single human genes have been transferred to mice, but it may be possible to transfer larger blocks or even entire chromosomes in the future [84].

2.3 Lymphocyte Entry into the CNS

Lymphocytes encounter different barriers to where they can enter the CNS: the BBB, the BCB, and the blood–leptomeningeal barrier [19, 60]. The molecular composition of TJs of these barriers differs [62], which results in different permeability properties under inflammatory conditions, but also different rates of lymphocyte extravasation. During EAE, inflammatory cuffs located at meningeal brain vessels or around the ventricles were accompanied by dye leakage into the brain tissue. Only some of the parenchymal cuffs showed dye leakage and thus BBB

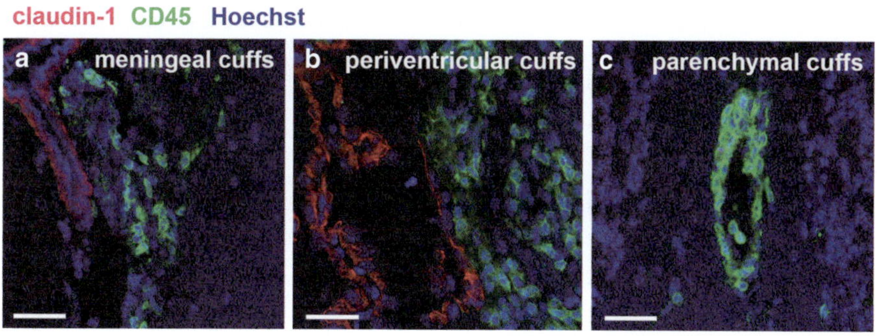

Fig. 1 Inflammatory cuffs in the chronic phase of experimental autoimmune encephalomyelitis (EAE; day 50, score 1 (according to Pfeiffer et al. [62]). Hoechst dye (stains nuclei) was injected intravenously to visualize leakage of the vessels. Examples of cuffs with dye leakage: (**a**) claudin-1 is diffusely expressed at the meningeal vessels; the cuff shows leaked dye. (**b**) Tight junctions of the epithelium of the choroid plexus express claudin-1, and the cuff shows dye leakage. (**c**) Parenchymal vessels of wild-type mice do not express claudin-1. This cuff is accompanied by dye leakage. *Red*: claudin-1, *green*: CD45-positive inflammatory cells, *blue*: Hoechst dye, 534D in size. *Bars*: 50μm

permeability (Fig. 1). Sealing the BBB by ectopic expression of claudin-1 in transgenic mice reduced BBB leakage for blood-borne tracers and endogenous plasma proteins, but had no effects on immune cell trafficking across the BBB during EAE. Sealing the BBB in these experiments led to a reduction of disease burden in the chronic phase of EAE in this model [62], highlighting the importance of an intact BBB to the severity of the disease course.

2.4 *BBB Permeability During CNS Inflammation*

As with any chronic brain inflammation, vascular permeability is affected in MS patients. Generally, the BBB is more severely disturbed in active compared with inactive lesions and normal-appearing white matter, but changes are also observed at these sites. During inflammation, leukocyte migration and also proinflammatory cytokines account for disturbance of the BBB. Once chronic inflammation has been established, vascular changes associated with BBB dysfunction persist, even if the inflammation is cleared [42].

In humans, changes at the BBB can be observed as an early event in the development of new lesions in MS, which was associated with inflammation [38, 60]. Amelioration of clinical symptoms was observed when CNS inflammation is resolved and the BBB function restored [44], suggesting a causal relationship between immune cell trafficking and BBB function, as described in EAE. TJ abnormalities are associated with BBB leakage and demyelination in the white matter of MS patients [39] and are persistent in the progressive phases of the disease [45]. Whether abnormal TJs can also be observed in gray matter of patients is still debated [45, 80]. BBB alterations and loss of TJ integrity are also observed in the cerebral cortex of EAE mice [21]. Molecular alterations of BBB TJs during inflammation are discussed below.

2.5 Cortical Inflammation in MS

Postcapillary venules are not the only site of the vascular tree affected by the accumulation of encephalitogenic leukocytes. Lymph follicle-like structures could be observed within the perivascular or meningeal compartments during progressive MS [74]. Along these lines, the degree of meningeal T-cell inflammation correlated with the extent of axonal loss in the spinal cord [6]. This entrapment of inflammation in the brain and spinal cord may explain the failure of current treatments in the progressive stages of MS, as they mainly target the peripheral immune system.

In addition to white matter damage, cortical demyelination can be observed in MS. Interestingly, the majority of cortical demyelination occurs in subpial band-like plaques. They are mainly associated with outer portions of the cortex and penetrate to a variable extent into the cortical tissue. Subpial cortical plaques that show active demyelination are always associated with meningeal inflammation [42]. Recently, it was shown that active cortical lesions are associated with T- and B-cell infiltration in the meninges [14, 34, 49], a fact that may reflect the differing architecture of the meningeal vessels compared with postcapillary venules. Interestingly, active cortical lesions associated with meningeal inflammation arise at sites where the circulation of CSF is impaired, suggesting that soluble inflammatory factors might diffuse into the cortex to induce demyelination either directly or indirectly through microglial activation [42]. These results indicate that the meninges play a crucial role in establishing inflammation in the CNS.

2.6 The Two-Wave Model for Lymphocyte Entry

Preceding these findings in human tissue specimens, myelin-specific T-cells have been shown to migrate to the meninges, where antigen-presenting phagocytes activate them in the SAS during the preclinical phase of EAE [7, 30, 40]. After the activated autoreactive effector T-cells have entered the CNS, further immune cells are recruited [9]. In addition, an increase in immune cell numbers and inflammatory mediators in the meninges correlated with clinical relapses in a mouse model rather than BBB permeability and immune cell accumulation in the CNS [71]. According to these data, a two-wave model has been proposed [7, 40, 67]. In this model, immune cells would first transit the meninges as part of the immune surveillance. Leukocyte access to the parenchyma would only occur when meningeal inflammation persists. In the first wave, T-cells traverse the BCB from the systemic circulation into the CSF [67]. As described above, the endothelium of the CP is different from that of the BBB, although the epithelium contains TJs [91]. From the CSF, the T-cells are able to circulate through the ventricular system into the SAS of the meninges where they meet resident APCs. In case these T-cells encounter myelin-loaded APCs, they are activated and mediate the recruitment of additional inflammatory cells, leading to BBB impairment and influx of the second wave of peripheral leukocytes into the CNS [71, 73]. Apparently, proinflammatory mediators can diffuse across the pia mater into the parenchyma of the CNS, enabling them to act directly on oligodendrocytes, neurons, and microglia [26]. Taken together, it seems highly likely that the CP and the meninges are a gateway to the brain during the onset of inflammation. Future therapies should therefore also be aimed at blocking the first-wave response in the meninges from initiating the second-wave response within the CNS. Therefore, researchers have already started to investigate alternative methods of drug delivery, for example, the direct route from the nasal mucosa to the brain should reach the SAS and bypass the BBB, as described in 1979 [36, 37]. Along these lines, Duchi et al. [15] showed that the intranasal administration of glatiramer acetate, which is an approved treatment for MS, was more effective at treating EAE than subcutaneous administration.

2.7 *Neuromyelitis Optica*

In a related inflammatory disease of the human CNS, neuromyelitis optica (NMO) autoantibodies that stain astrocytic processes in the perivascular and superficial glia-limiting membrane have been identified to react with aquaporin-4 (AQP4) [46, 47]. These antibodies are destructive toward astrocytes, but also toward myelin, oligodendrocytes, and axons. In the meantime, T-cell-mediated autoimmune reactions toward AQP4 have also been identified in patients with NMO [81]. Thus, NMO seems to be an autoimmune disease directed against an astrocytic antigen, leading to primary astrocytic damage followed by demyelination and neurodegeneration.

3 Topology of the Blood–Brain and Blood-Cerebrospinal Fluid Barriers

3.1 *The Endothelial Side of the Barrier*

The BBB is essential for the homeostasis of the microenvironment in the neural parenchyma and thus for the normal function of the brain. Strictly speaking, the BBB is located in the TJs of the brain capillary endothelial cells [12, 68], which are compactly integrated into the brain neuropil. This is the view of the BBB of cell biologists, whereas the neuroimmunologists' view of the BBB focuses on the post-capillary venules. The morphological and molecular properties of the endothelial cell TJs of the BBB are different from those of endothelial cells outside the brain, and from the fenestrated blood vessels in the CP and CVOs as well. Fenestration of endothelial cells is generally a feature of high permeability [22]; nevertheless, fenestrated endothelial cells are also interconnected by TJs.

The differences among different types of endothelial cells can best be visualized by means of the freeze-fracture technique, which allows characterization of the molecular anatomy of membranes (Fig. 2). Briefly, the most important point concerning the freeze-fracture technique is the cleavage of the biological membrane, which consists of two lipid layers between which the membrane proteins are inserted. The cleavage is possible because the middle of the membrane is hydrophobic and therefore a kind of predetermined breaking point. After the frozen membrane is cleaved, the fracture faces are shadowed with platinum and carbon resulting in a freeze-fracture replica that can be placed on an electron microscopic grid and observed under a transmission electron microscope. Membranes can be visualized twofold: if we look from the cell interior to the split membrane, only the external half, the so-called E-face, is exposed. Behind the E-face, we have to imagine the extracellular space or the intercellular cleft. If we look from outside the cell, only the internal half, the so-called protoplasmic fracture face, or P-face, is exposed. Behind the P-face, we have to imagine the cytoplasm of the cell. In freeze-fracture replicas, TJs appear as a network of strands with a certain complexity. The high complexity of the TJ network in brain endothelial TJs in comparison with all other nonbrain endothelial TJs was the first distinct feature that is special to the BBB junctions and is detected by freeze-fracturing [54].

Another paradigm describing the appearance of TJs in freeze-fracture analysis concerned the association of strands with one or the other leaflet of the membrane [88]. In vivo, the epithelial TJ network is associated nearly completely with the P-face (Fig. 2d), whereas the BBB network of TJs is associated with both the inner (P-face) and the outer leaflet (E-face) of the junctional membrane (Fig. 2a). In vitro, where the influence of brain-derived factors is lost, the TJs are almost completely associated with the E-face (Fig. 2b): the morphology switched to that of nonbrain endothelial TJs (Fig. 2c, taken from Mühleisen et al. [52]), indirectly assuming that the TJs of the BBB are under the control of the brain microenvironment [88]. However, it is not understood whether and how alterations of TJs (e.g., during development) might correlate with functional properties, such as increasing electrical resistance or decreasing permeability [72]. Nevertheless, transfection with

Fig. 2 Freeze-fracture replicas of different types of tight junctions (TJs). (**a**) Endothelial cells of the blood–brain barrier (BBB) showing the typical TJs within the brain. The TJ particles are associated equally with both the P-face (PF) and the E-face (EF). (**b**) TJs of cultured BBB endothelial cells. The P-face association is strongly reduced; most particles are associated with the E-face. (**c**) Complementary replica of TJs between arterial endothelial cells from the rabbit carotid artery (taken from Mühleisen et al. [52]). *On the left* (**c'**), the P-face is shown, where no TJ particles are associated with the TJ ridges. At the identical site, the complementary replica *on the right* (**c''**) shows all TJ particles associated with the TJ grooves at the E-face. This situation resembles that in cultured BBB endothelial cells. *GJ* gap junction, *V* membranous vesicles. (**d**) Freeze-fracture replica from a mouse intestinal epithelium with the dense P-face-associated network of TJ strands. (**e**) Cultured HEK cells transfected with the cDNA of claudin-1. This transfection resulted in a completely P-face-associated TJ network resembling that in epithelia in vivo (from cooperation with the group of Ingolf Blasig, Berlin-Buch, Germany)

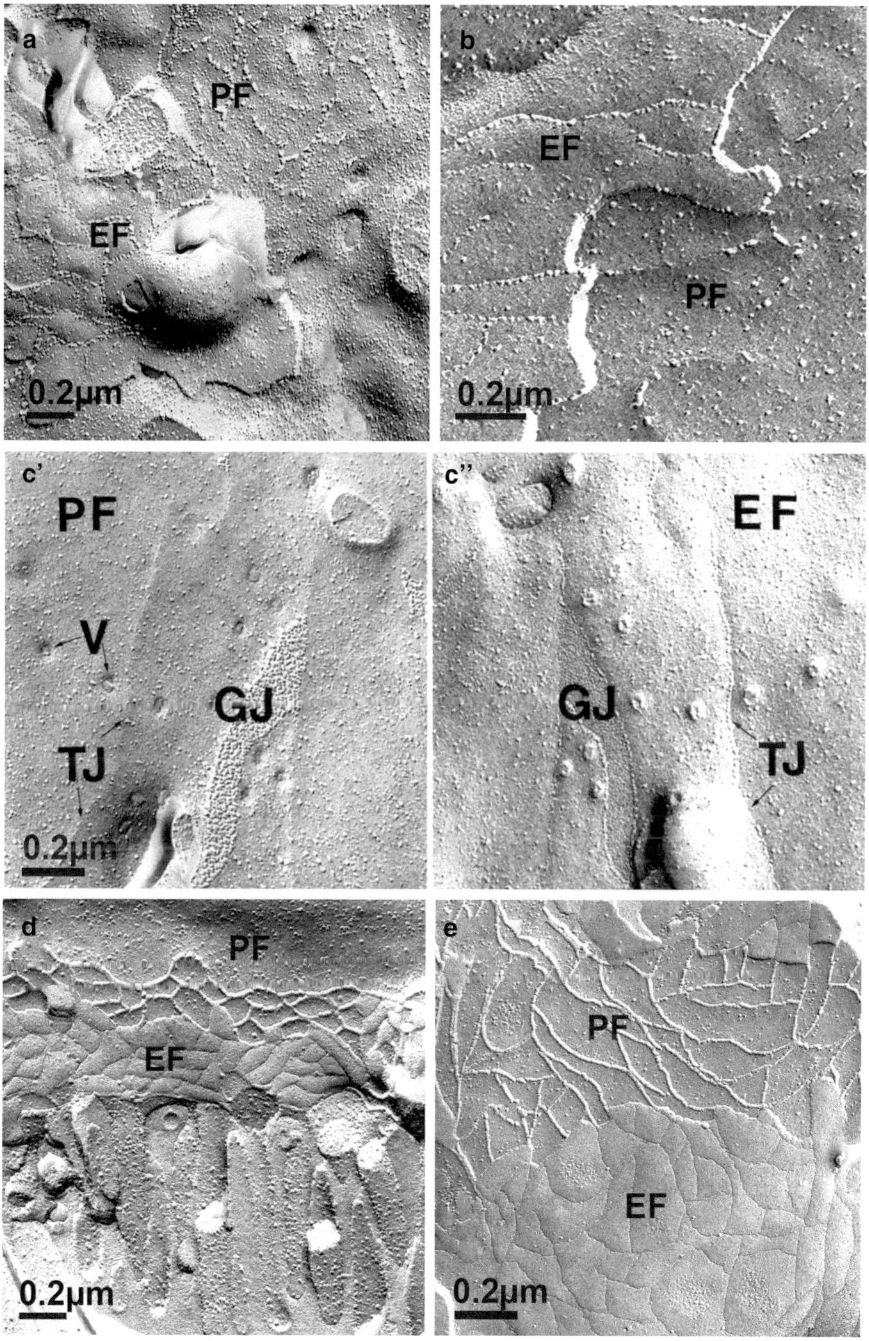
a
PF
EF
0.2µm
b
EF
PF
0.2µm
c'
PF
V
GJ
TJ
0.2µm
c''
EF
GJ
TJ
d
PF
EF
0.2µm
e
PF
EF
0.2µm

cDNA of claudin-5 resulted in the formation of E-face-associated TJs, and transfection of claudin-1 or claudin-3 in the formation of P-face-associated TJs (Fig. 2e) [24, 25, 63] suggesting that changing the stoichiometry of claudins might cause different expression of TJ networks, even under pathological conditions.

In the CVOs it is essential that the neurosecretory neurons obtain access to the vasculature to release their hormones, or vice versa, to “sense” signal molecules from the blood stream. To this end, the vessels have to be highly permeable, and indeed they have been found to be fenestrated.

3.2 The Glial Side of the Barriers

The quality of endothelial TJs depends on the brain microenvironment, which consists of astrocytes, pericytes, neurons, microglia, and the ECM. All these components are summarized in the term NVU, as introduced above [2, 33]. In the mature brain, astrocytes embrace vessels by sending end-feet toward the perivascular basal lamina [50], which is a constituent of the ECM.

The astroglial membranes contacting the basal lamina are characterized by the occurrence of orthogonal arrays of particles (OAPs; Figs. 3, 10). Like TJs, these particle arrays can be best investigated by freeze-fracturing. At spots where the astroglial membrane is not in direct contact with the basal lamina, the density of the OAPs is reduced (Fig. 3). Astrocyte polarity is defined by the ratio of OAPs in both astroglial membrane domains, the perivascular endfoot membrane and the non-end-

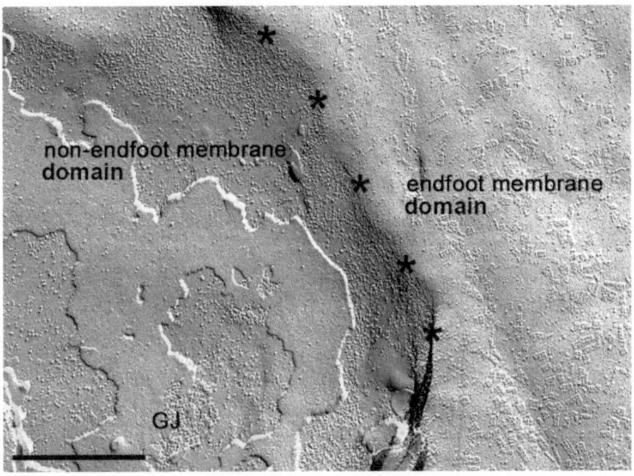

Fig. 3 Freeze-fracture replica of a superficial astrocytic endfoot membrane domain from the optic nerve of the rat. *On the right*, the density of the orthogonal arrays of particles (OAPs) is high; *on the left*, the membrane turns away from the contact with the basal lamina (*row of asterisks*), and here, at the non-endfoot membrane domain, there are almost no OAPs, indicating the importance of the basal lamina for the OAP-related polarization of the astrocyte. The gap junction (*GJ*) between the two adjacent astrocytic processes is evidence that this membrane domain cannot be covered by a basal lamina. *Bar*: 0.5μm

foot membrane. Interestingly, this glial OAP-related polarity is decreased during EAE [85] and other pathological conditions.

Today, it is well established that OAPs consist of the water channel protein AQP4 [93]. Thus, the OAP-related polarity correlates with an AQP4-related polarity. Aquaporins mediate water movements among the intracellular, interstitial, vascular, and ventricular compartments, which are under the strict control of osmotic and hydrostatic pressure gradients [5, 10, 89, 90]. For the relationship between AQP4 and the BBB, we have to state that the BBB characterized by a dense meshwork of endothelial P-face-associated TJ strands (see above) is somehow connected to highly polarized astrocytes. As we describe below, the BBB can then be characterized by a mutual relationship of two phenomena: complex TJs highly associated with the P-face, and the high AQP4/OAP-related polarity of glial cells. If this causative relationship is disturbed, as in manifold pathological processes, both TJ complexity and AQP4/OAP-related polarity are reduced. On the other hand, genetic deletion of AQP4 has been described not to disrupt the BBB [17]. Therefore, the role of AQP4 expression and AQP4 distribution in the maintenance of the BBB remains unclear.

The abluminal endothelial membrane directly faces the basal lamina, which also covers the astroglial end-feet, at least within the capillary bed. In the postcapillary venules, there are two different basal laminae, the endothelial and the glial laminae. Between these basal laminae there is a PVS [61] that is continuous with the Virchow–Robin space and plays an important role in CNS inflammation (see Engelhardt [18] and below).

The membrane domain of the astroglial end-feet is of particular importance for the brain physiology, because it is the place where potassium and water channels are co-localized. When K^+ increases in the extracellular space of the synaptic region, it is taken up by astroglial cells followed by a depolarization of the glial membrane potential. K^+ is then immediately redistributed in the astrocyte and extruded where the membrane potential is not depolarized ("spatial buffering"). The gate of extrusion of K^+ at the astroglial endfoot membrane is mainly the K^+ channel, Kir4.1. The uptake of K^+ is accompanied by osmotic water entry. Astrocytes swell during spatial buffering if water cannot leave the cell at those domains of the glial cell surface where the extracellular space is huge in comparison with the interstitial space of the neuropil. As a consequence, the astroglial endfoot is not only the site of K^+ extrusion, but the site of water outflow through AQP4 water channels as well [66].

The question arises as to what the mechanism is for developing and retaining the perivascular arrangement of AQP4 exactly in the endfoot membrane domain. The basal lamina is an obligatory constituent of each endothelial and epithelial cell, giving rise to a polarity characterized by basolateral and apical membrane domains [96]. The astroglial end-feet are covered by a glial basal lamina both at the superficial border and at the perivascular border (Fig. 10). The glial basal lamina contains many compounds of the ECM such as laminins, fibronectin, collagens, and diverse heparin sulfate proteoglycans such as perlecan or agrin. Agrin plays a role that has only been recognized in the last few years. Originally, this molecule was characterized to be responsible for the clustering of the acetylcholine receptors at the motor endplate [51]. Recent data suggest that agrin might play a role in the induction and/or maintenance of the polarity of astrocytes [56, 57, 83].

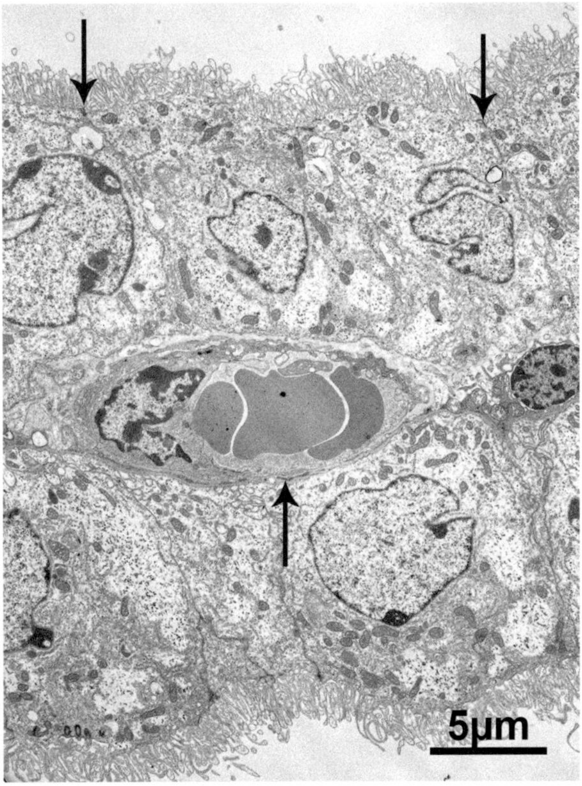

Fig. 4 Ultrathin section of the choroid plexus of the mouse brain showing TJ-connected epithelial cells (the *two upper arrows*) and blood vessels within the stroma of the choroid plexus, consisting of fenestrated endothelial cells interconnected by TJs as well (*lower arrow*)

The BBB in mammals is constitutively associated with highly polarized astrocytes, and this polarity might be essentially caused by agrin [76, 90]. However, agrin is not the only agent to polarize the astrocyte. There are other components located in the membrane of astrocytic end-feet such as the dystrophin–dystroglycan complex (DDC), which is in contact with both the ECM and the cytoskeleton. At the surface of the brain, in the superficial glial limiting membrane, lack of dystroglycan (i.e., in the GFAP-Cre/dystroglycan null mouse) [58] led to a reduction in OAPs, but not the AQP4 protein. This suggested that dystroglycan might play a role in AQP4 clustering, at least in superficial end-feet. Unexpectedly, in the perivascular glial limiting membrane and thus at the BBB, the lack of dystroglycan led to a veritable reduction in AQP4 expression and not to a reduction in the ability to cluster AQP4 molecules to OAPs [58]. However, the authors were not able to show an influence of dystroglycan on the TJ network at the BBB.

Another aspect of the glial side of the barrier is the epithelial border between blood and CSF, the BCB. It is formed by the CP epithelial cells, and the tanycytes and pituicytes of the CVOs, which also belong to the family of astroglial cells [95]. The difference between CP cells and CVO glial cells is that the main function of the CP is the production of the CSF, and to this end the CP vessels within the stroma have to be permeable to allow contact between blood and the epithelium of the CP (Fig. 4) [91].

These epithelial cells of the CP can be referred to as ependymoglial cells [95] as they are in continuity with the ependyma of the periventricular lining, but they are not identical. The main differences between the ependyma and CP are the presence of a basal lamina in the CP, the absence of a basal lamina in the ependyma, and the TJ barrier properties of the epithelium of the CP not found in the ependyma. Remarkably, the fenestrated blood vessels in the stroma of the CP are not free of TJs, despite their high permeability [48]. These TJs contain claudin-5, like all other endothelial cells, but in the freeze-fracture replica, the TJs are associated with the P-face and not with the E-face. This finding is puzzling and has not yet been explained, because all other claudin-5-positive TJs in peripheral and brain endothelial cells are associated with the E-face (Fig. 5).

Most neuroscientists do not seem to be aware of the fact that the basal lamina of the epithelial cells of the CP is continuous with the superficial glial limiting membrane. As a consequence, the pial tissue and the SAS are continuous with the stroma of the CP. This fact was already described and illustrated by Brightman and Reese [12] and emphasized by Reichenbach and Wolburg [69], and is of consequence in the context of leukocyte trafficking to the brain parenchyma under inflammatory conditions (see below) [12, 69]. The cerebral wall, delimited outside by the superficial glial limiting membrane and on the ventricular side by the ependyma, recedes abruptly to a single cell layer underlined by the basal lamina (Fig. 6) [87].

Thus, the epithelium of the CP can be topologically defined as a single cell layer adjacent to both the brain surface and the ventricle (for the situation of tanycytes and pituicytes, see Wolburg et al. [87, 95]). The ependyma can consistently be defined as a layer in contact with the ventricle, but not with the surface of the brain. The question arises as to whether or not this situation has an impact on our understanding of the inflammation in the brain (see below).

3.3 *Water Movements in the Brain*

Blood circulation and flow of the CSF are essential for the maintenance for ionic and water balance in the brain. Recently, some authors have suggested that perivascular water movement might play a major role in cerebral water balance. As mentioned above, these PVSs filled with fluid are continuous with the Virchow–Robin space and finally with the SAS. This space around larger blood vessels is located in between the astroglial end-feet and the vessel wall, thus behind the BBB. As this space communicates with the outer CSF, namely the SAS, some authors have suggested a re-circulation of the CSF. This water movement might have an important function for the removal of waste products from the brain parenchyma [35, 55]. As the brain parenchyma lacks lymphatic vessels and the corresponding drainage, this water movement involving astroglial end-feet has been termed the "glymphatic pathway."

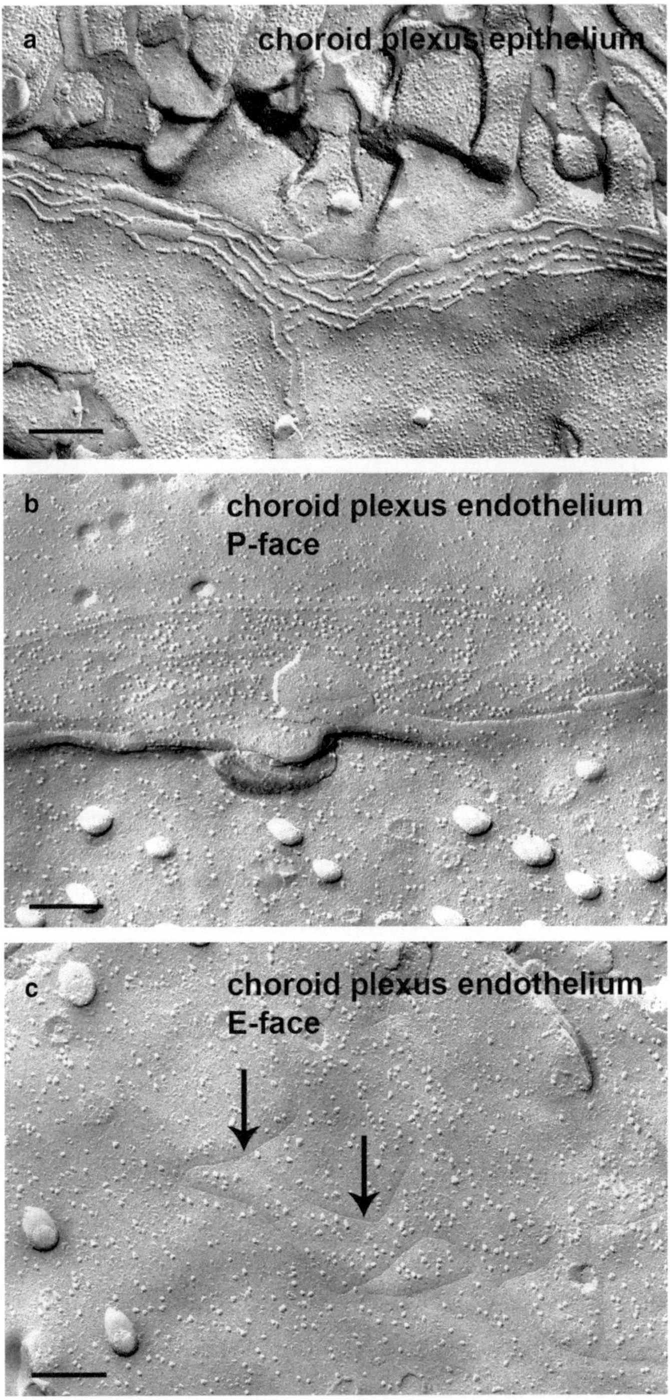

Fig. 5 Freeze-fracture replicas of (**a**) epithelial and (**b**, **c**) endothelial TJs of the choroid plexus of the mouse. The epithelial TJs are parallel-stranded; the endothelial TJs have particle rows on the P-face and particle-free grooves on the E-face. This observation is unexplained, because claudin-5 known to be expressed here is believed to form E-face-associated TJs. *Bars*: 200nm

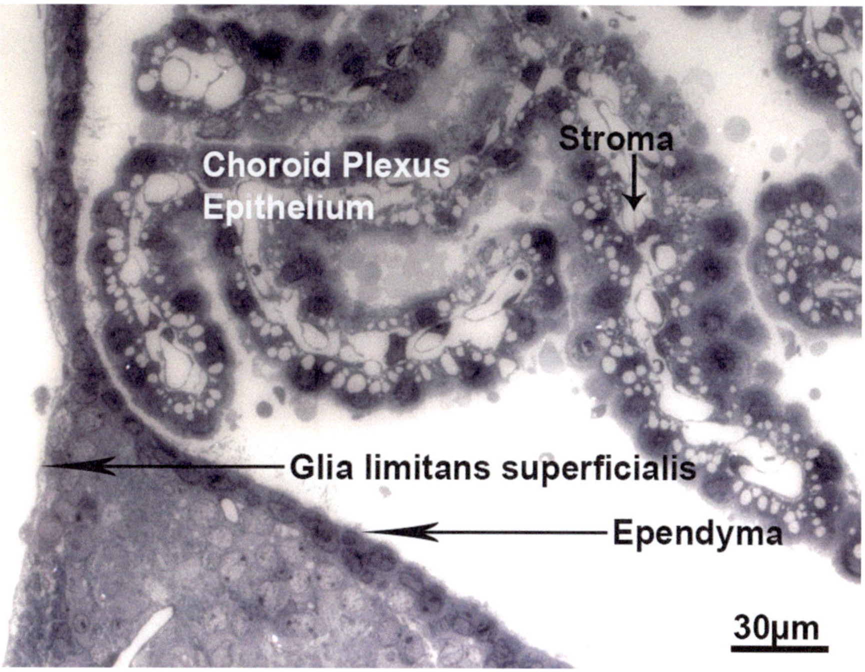

Fig. 6 Toluidine blue-stained semithin section of the choroid plexus of the mouse brain. Below, there is the cerebral wall, limited on the left by the superficial glial limited membrane. Where this limiting membrane approaches the ependyma, the cerebral wall is reduced to the monolayer of the plexus epithelium. Here, between the brain surface and the ventricle, there is only one single cell layer

4 Barrier-Related Cellular Characteristics of Epithelial Cells of the Choroid Plexus and Vascular Endothelial Cells During Inflammation

As discussed in the second paragraph of this chapter, immune cells breach the brain barriers and enter the CNS during inflammation. Yet, it remains unresolved whether the course and severity of an inflammatory disease are determined mainly by the immigration of encephalitogenic T-cells from vessels into brain, or via the CP.

Here, we want to look more closely at the anatomy of the brain and its interconnected spaces to reveal alternative and plausible routes for leukocyte trafficking into the CNS (see also Fig. 10).

The continuity of the SAS with the stroma of the CP, as explained above, has an important impact on our understanding of brain inflammation (Fig. 10). It is generally believed that during inflammation, T-cells, monocytes or bacteria overcome in a first step the fenestrated capillaries in the CP to enter the stroma, and in a second step the epithelial cells of the CP, with their TJ-based barrier, to enter the brain ventricle. From there, the cells can obtain access to the external CSF via the foramina of Luschka and Magendie to reach the SAS and the Virchow–Robin space. Reaching

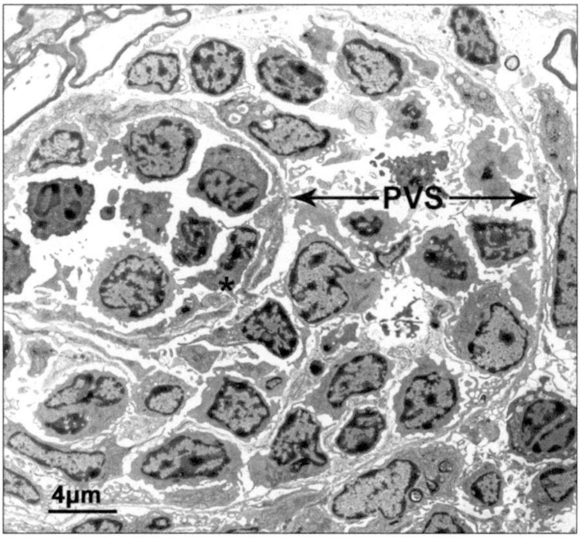

Fig. 7 Low magnification of an electron micrograph of a parenchymal cuff in the cerebellum of a mouse subjected to active EAE. The postcapillary venule is filled with various inflammatory cells. The cell marked by an *asterisk* is caught during transcellular transmigration into the perivascular space (*PVS*). The PVS reaches from the subendothelial basal lamina on the left to the perivascular glial limited membrane, as indicated by the *two-sided arrow*

the SAS and the Virchow–Robin space could of course also be achieved by overcoming pial vessels, but this is highly restricted by the BBB properties of these nonfenestrated vessels. Therefore, inflammatory cells could pass the "detour" of going through the fenestrated vessels in the CP to reach the stroma and thus the SAS. The use of this pathway, however, has not been shown experimentally, because the existence of inflammatory cells in the CSF was generally taken as evidence of their transepithelial transmigration through the epithelium of the CP. To our knowledge, nobody has y considered the alternative, more direct route from the stroma to the SAS so far. Instead, in vitro models of bacterial encephalitis used epithelial cells of the CP cultured on Transwell filters in two different modes: as a standard system, in which the CP cells grew in the upper compartment on the filter and were oriented apically to the upper ("CSF") compartment, or as an inverted system, in which the CP cells grew below the filter in the lower compartment, but were oriented basally to the upper ("blood") compartment [78]. A closer look on how polymorphonuclear neutrophils and monocytes transmigrate through CP cells in this model provided evidence for two processes: transcellular and paracellular diapedesis [77]. However, the alternative route from the stroma to the SAS was not considered in this experimental approach.

For diapedesis across the BBB, inflammatory immune cells can take the transcellular or the paracellular pathway [77]. The transcellular route requires well-coordinated membrane dynamics culminating in the formation of a pore across the endothelial cytoplasm. The paracellular route depends on the release or replacement of the complex architecture of junctional molecules. In recent years, a multitude of studies have provided evidence for the existence of both pathways [1, 13, 29, 65, 96]. Without doubt, electron microscopic analysis can most precisely reveal the site of diapedesis (Figs. 7 and 8) [20].

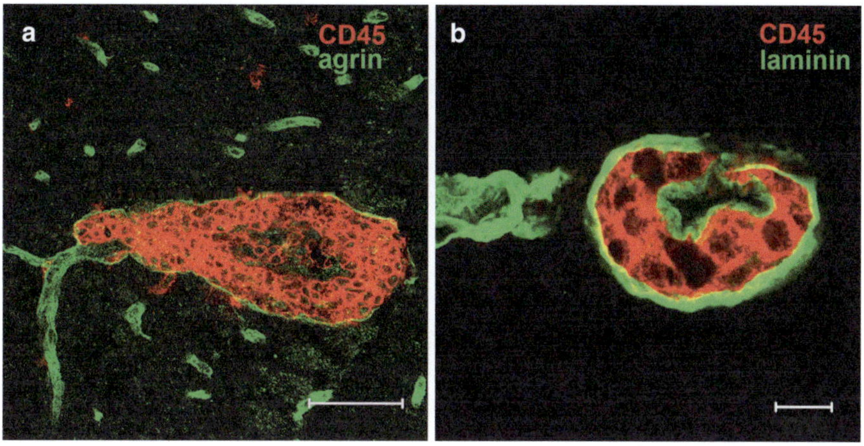

Fig. 8 Immunohistochemical micrographs showing a parenchymal cuff in the cerebellum of an EAE mouse 16 days after immunization. (**a**) Clinical score 3 (according to Pfeiffer et al. [62]), CD45-positive cells (*red*) in the PVS between the endothelial and glial basal lamina positive for agrin [29]. (**b**) 20 days after inflammation induction, clinical score 1 (according to Pfeiffer et al. [62]), CD45-positive cells (*red*) in the PVS between the endothelial and glial basal lamina positive for laminin [29]. *Bar* in **a**: 50 μm, in b: 10 μm

One detailed electron microscopic analysis elucidated the pathway of diapedesis of the T-cells across the inflamed BBB in situ in mice afflicted with acute EAE [92]. The main and most convincing observation was the consistent stability and maintenance of TJs. The high resolution of electron microscopy allowed identification of the site of transmigration that occurred at places directly adjacent to the intact and tightly closed endothelial junctions. In some cases, transcellular transmigration was also observed as emperipolesis. This classical term was formerly defined as a kind of uptake of a cell by another, without degradation of the ingested cell. Yet, the transmigrating cell was rather observed to be guided through the endothelium. There was a dramatic alteration of the endothelial cell morphology, which was again characterized first of all by the maintenance of TJs, but in addition and importantly by the formation of endothelial cavities resembling sluice-like chambers interpreted by the authors as "transmigration compartments." The luminal membrane was folded, forming luminally directed domains of the sluice chambers, which are immigrated by the mononuclear cells. Later on, the transmigration compartment closed in the luminal direction and opened in the abluminal direction, giving access to the subendothelial space. The former luminal membrane was now re-addressed as a novel abluminal membrane requiring a basal lamina. This complex morphological observation is far from being understood in molecular terms.

It is still unclear under what conditions the inflammatory cells take the para- or the transcellular route. In any case, it is striking that most if not all reports dealing with paracellular transmigration (together with or without transcellular transmigration) are in vitro studies. This is true for both transendothelial transmigration in the brain parenchyma and transepithelial transmigration in the CP. Interestingly,

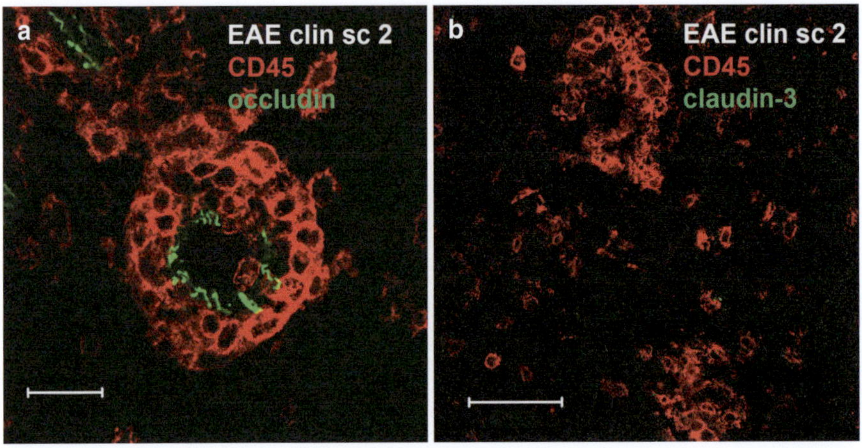

Fig. 9 Immunohistochemical staining of EAE inflammatory cuffs consisting of CD45-positive cells. The clinical score is 2 (according to Pfeiffer et al. [62]). (**a**): Occludin is present as a TJ molecule. (**b**) Claudin-3 seems to be down-regulated or degraded. *Bar* in **a**: 20 μm, in **b**: 50 μm

Abadier et al. [1] have described an influence of cell surface levels of endothelial ICAM-1 in determining the cellular route of T-cells [1]. In response to exposure to high levels of endothelial ICAM-1 the transcellular diapedesis was rapidly initiated, whereas exposure to intermediate levels evoked predominantly the paracellular transmigration. Whether this ICAM-1-related mechanism has some relevance in the in vivo situation is not known.

During inflammation of the brain, the gliovascular complex is doubly injured by impairment of the endothelial, barrier-related TJs and the reduction in the polarity of the astroglial cells by redistribution of AQP4.

Wolburg et al. [94] described the selective loss of claudin-3 during EAE and interpreted it as the cause of BBB leakage (Fig. 9b) [94]. Other proteins of TJs, such as claudin-5 or occludin, remained unchanged (Fig. 9a). However, Bennett et al. [11] observed a dramatic relocalization of the adapter protein ZO-1, located between the junction and the submembranous cytoskeleton, in coincidence with increased vascular permeability before the onset of clinical signs of the demyelinating disease [11]. Moreover, BBB disruption has been observed in EAE to precede the invasion of encephalitogenic T-cells [43]. The authors showed that the small molecule LY-317615 (enzastaurin) suppressed the transmigration of T-cells and induced the expression of the TJ proteins claudin-3 and claudin-5 and the TJ-associated protein ZO-1. Enzastaurin is an inhibitor of the protein kinase (PKC)-β, which in turn is involved in the expression of the angiogenic factor VEGF. Thus, PKCβ seems to stabilize the BBB under EAE conditions. Accordingly, overexpression of claudin-1 reduced BBB leakage in EAE and attenuated its clinical course, suggesting the key role of TJs in inflammation-related BBB dysfunction [4, 62]. Furthermore, administration of the plant-derived compound (RS)-glucoraphanin had a protective effect against the

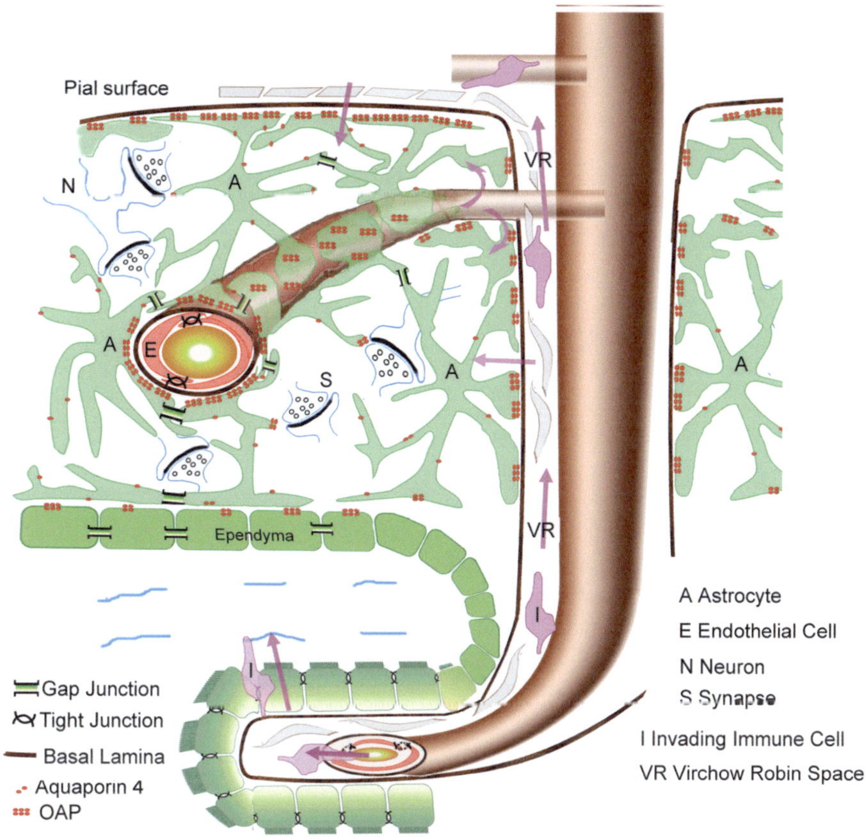

Fig. 10 Schematic diagram depicting the topological relationship between the BBB and the blood– cerebrospinal fluid barrier. Note that the basal lamina of the pial surface is continuous with the basal lamina and stroma of the epithelium of the choroid plexus. Possible entry sites of immune cells (*pink cells*) into the brain (*arrows*) have been indicated

disturbance of ZO-1, and claudin-1, -3, and -5 [27]. In a similar manner, idazoxan, an imidazoline-2 receptor ligand, has been described to protect the BBB against EAE-related damage by amelioration of the expression of ZO-1, occludin, claudin-5, and JAM [82]. Taken together, the rescue of the TJ appears promising in providing protection against an inflammation-dependent increase in BBB permeability.

The second paradigm of BBB damage during inflammation addressed here is the reduction of astrocyte polarity, as observed in EAE in vivo. As already introduced above, astrocytes are highly polarized cells in the sense that the molecular equipment of the endfoot membrane is completely different from that of the parenchymal membrane. For a brief reminder of the introduction, in the endfoot membrane there is a strong accumulation of OAPs, which represent the morphological correlate of the water channel protein AQP4. In the parenchymal membrane, the density of these OAPs is dramatically reduced (Fig. 3). However, AQP4 is not inserted into the membrane as a single protein or protein cluster, but is part of a molecular complex named the DDC consisting of AQP4, syntrophin, dystrophin, dystrobrevin, α- and β-dystroglycan. α-Dystroglycan is attached to the outer membrane surface and binds to laminin and agrin, and β-dystroglycan is a transmembrane protein. In addition, a further member of the DDC is the potassium channel Kir4.1 [89]. In EAE, Agrawal et al. [3] have demonstrated that β-dystroglycan is specifically involved in the penetration of inflammatory cells across the astroglial basal lamina [3]. β-Dystroglycan is selectively cleaved by matrix metalloproteinases (MMPs) 2 and 9 secreted by macrophages entering the CNS. Accordingly, MMP2 and MMP9 double knockout mice conferred resistance to leukocyte migration into the CNS across the glial basal lamina by preserving dystroglycan protein. We were able to support this finding by showing the specific loss of β-dystroglycan in the postcapillary venules during EAE [85], obviously by MMP2/9-dependent cleavage of β-dystroglycan. As at this identical location (postcapillary venule) the OAPs have reduced their polarity by increasing their density in the astroglial parenchymal membrane domain [85], the loss of β-dystroglycan seems to destabilize the DDC and allow the OAPs to populate "forbidden" membrane domains in the sense that normally this domain is not populated by OAPs. As reported earlier, the GFAP-Cre/dystroglycan null mouse, in which specifically the astrocytes are devoid of dystroglycan, revealed the involvement of dystroglycan in a twofold manner: by forming OAPs from AQP4 molecules at the superficial glial limited membrane and by inducing AQP4 expression at the level of the BBB. Although the astrocyte-specific dystroglycan deficiency cannot be compared with the MMP-dependent dystroglycan cleavage, the different experimental approaches prove the significance of dystroglycan for the stability of the DDC at the gliovascular border.

5 Conclusions

Inflammation of the CNS involves overcoming the endothelial BBB in the postcapillary venules, the glial BCB, or the meningeal barrier. The breakdown of the BBB is the classical pathway that includes deterioration of the ECM and the TJs. Indeed, both are highly dynamic structures that are under the close control of the brain microenvironment. It has been a seminal discovery to identify the TJs between the endothelial cells as the barrier proper between blood and brain, but up to now we have not succeeded in understanding the complex process of control and regulation of TJs of the BBB. The simple observation that epithelial, but not endothelial cells, are able to form a high resistance and low permeability barrier in vitro, sheds light on the importance of the brain microenvironment in the formation and maintenance of the barrier in vivo. This microenvironment consists of endothelial cells, pericytes, microglial cells, astrocytes, neurons and the ECM in between, which itself forms a microcosmos of its own. All of these different cellular and extracellular components operate together to establish the NVU.

An increase in the permeability of the different barriers in the CNS seems to be a stereotypic response of pathological processes such as stroke, inflammation, tumor, or neurodegenerative diseases. The disturbance of the barriers in both the postcapillary venules and the CP seems to be involved in the pathogenesis of inflammation. In particular, the migration path of inflammatory T-cells from fenestrated blood vessels into the stroma of the CP seems to be plausible and generally accepted. The pathway from there through the epithelium of the CP into the ventricle and from there through the foramina into the outer CSF in the SAS is assumed as well. The alternative pathway from the stroma directly to the SAS is possible because of the continuity of both of these spaces, which is a long-standing anatomical insight. However, this continuity seems to be forgotten by many authors in the field, and to be the basis for the pathway from the stroma directly to the adjacent SAS. This pathway could be of significant importance in the explanation of inflammation processes such as cortical MS, meningitis, or encephalitis.

Acknowledgements This review was broadly written on the basis of the former collaboration between the group of HW and that of Prof Dr Britta Engelhardt, Theodor Kocher-Institute of the University of Berne, Switzerland. In particular, Dr Karen Wolburg-Buchholz has contributed essential results cited in this book chapter. FP was formerly member of the group of Prof Engelhardt and was financially supported by the DFG (grant PF574/2-1) and the Integrating Project JUSTBRAIN (EU, FP7 No. HEALTH-2009-241861).

Literature

1. Abadier M, Haghayegh Jahromi N, Cardoso Alves L, Boscacci R, Vestweber D, Barnum S, Deutsch U, Engelhardt B, Lyck R (2015) Cell surface levels of endothelial ICAM-1 influence the transcellular or paracellular T-cell diapedesis across the blood–brain barrier. Eur J Immunol 45:1043–1058
2. Abbott NJ, Ronnback L, Hansson E (2006) Astrocyte-endothelial interactions at the blood–brain barrier. Nat Rev Neurosci 7:41–53
3. Agrawal S, Anderson P, Durbeej M, van Rooijen N, Ivars F, Opdenakker G, Sorokin LM (2006) Dystroglycan is selectively cleaved at the parenchymal basement membrane at sites of leukocyte extravasation in experimental autoimmune encephalomyelitis. J Exp Med 203:1007–1019
4. Alvarez JI, Saint-Laurent O, Godschalk A, Terouz S, Briels C, Larouche S, Bourbonniere L, Larochelle C, Prat A (2015) Focal disturbances in the blood–brain barrier are associated with formation of neuroinflammatory lesions. Neurobiol Dis 74:14–24
5. Amiry-Moghaddam M, Ottersen OP (2003) The molecular basis of water transport in the brain. Nat Rev Neurosci 4:991–1001
6. Androdias G, Reynolds R, Chanal M, Ritleng C, Confavreux C, Nataf S (2010) Meningeal T cells associate with diffuse axonal loss in multiple sclerosis spinal cords. Ann Neurol 68:465–476
7. Bartholomaus I, Kawakami N, Odoardi F, Schlager C, Miljkovic D, Ellwart JW, Klinkert WE, Flugel-Koch C, Issekutz TB, Wekerle H, Flugel A (2009) Effector T cell interactions with meningeal vascular structures in nascent autoimmune CNS lesions. Nature 462:94–98
8. Bauer HC, Krizbai IA, Bauer H, Traweger A (2014) "You Shall Not Pass"-tight junctions of the blood brain barrier. Front Neurosci 8:392
9. Becher B, Bechmann I, Greter M (2006) Antigen presentation in autoimmunity and CNS inflammation: how T lymphocytes recognize the brain. J Mol Med 84:532–543

10. Benfenati V, Ferroni S (2010) Water transport between CNS compartments: functional and molecular interactions between aquaporins and ion channels. Neuroscience 168:926–940
11. Bennett J, Basivireddy J, Kollar A, Biron KE, Reickmann P, Jefferies WA, McQuaid S (2010) Blood–brain barrier disruption and enhanced vascular permeability in the multiple sclerosis model EAE. J Neuroimmunol 229:180–191
12. Brightman MW, Reese TS (1969) Junctions between intimately apposed cell membranes in the vertebrate brain. J Cell Biol 40:648–677
13. Carman CV, Sage PT, Sciuto TE, de la Fuente MA, Geha RS, Ochs HD, Dvorak HF, Dvorak AM, Springer TA (2007) Transcellular diapedesis is initiated by invasive podosomes. Immunity 26:784–797
14. Choi SR, Howell OW, Carassiti D, Magliozzi R, Gveric D, Muraro PA, Nicholas R, Roncaroli F, Reynolds R (2012) Meningeal inflammation plays a role in the pathology of primary progressive multiple sclerosis. Brain 135:2925–2937
15. Duchi S, Ovadia H, Touitou E (2013) Nasal administration of drugs as a new non-invasive strategy for efficient treatment of multiple sclerosis. J Neuroimmunol 258:32–40
16. Ehrlich P (1885) Das Sauerstoff-Bedürfnis des Organismus. Eine farbenanalytische Studie. PhD thesis, Herschwal, Berlin, p 69–72
17. Eilert-Olsen M, Haj-Yasein NN, Vindedal GF, Enger R, Gundersen GA, Hoddevik EH, Petersen PH, Haug FM, Skare O, Adams ME, Froehner SC, Burkhardt JM, Thoren AE, Nagelhus EA (2012) Deletion of aquaporin-4 changes the perivascular glial protein scaffold without disrupting the brain endothelial barrier. Glia 60:432–440
18. Engelhardt B (2010) T cell migration into the central nervous system during health and disease: different molecular keys allow access to different central nervous system compartments. Clin exp Neuroimmunol 1:79–93
19. Engelhardt B, Ransohoff RM (2012) Capture, crawl, cross: the T cell code to breach the blood–brain barriers. Trends Immunol 33:579–589
20. Engelhardt B, Wolburg H (2004) Mini-review: transendothelial migration of leukocytes: through the front door or around the side of the house? Eur J Immunol 34:2955–2963
21. Errede M, Girolamo F, Ferrara G, Strippoli M, Morando S, Boldrin V, Rizzi M, Uccelli A, Perris R, Bendotti C, Salmona M, Roncali L, Virgintino D (2012) Blood–brain barrier alterations in the cerebral cortex in experimental autoimmune encephalomyelitis. J Neuropathol Exp Neurol 71:840–854
22. Esser S, Wolburg K, Wolburg H, Breier G, Kurzchalia T, Risau W (1998) Vascular endothelial growth factor induces endothelial fenestrations in vitro. J Cell Biol 140:947–959
23. Friese MA, Fugger L (2005) Autoreactive CD8+ T cells in multiple sclerosis: a new target for therapy? Brain 128:1747–1763
24. Furuse M, Fujita K, Hiiragi T, Fujimoto K, Tsukita S (1998) Claudin-1 and −2: novel integral membrane proteins localizing at tight junctions with no sequence similarity to occludin. J Cell Biol 141:1539–1550
25. Furuse M, Sasaki H, Tsukita S (1999) Manner of interaction of heterogeneous claudin species within and between tight junction strands. J Cell Biol 147:891–903
26. Gardner C, Magliozzi R, Durrenberger PF, Howell OW, Rundle J, Reynolds R (2013) Cortical grey matter demyelination can be induced by elevated pro-inflammatory cytokines in the subarachnoid space of MOG-immunized rats. Brain J Neurol 136:3596–3608
27. Giacoppo S, Galuppo M, Iori R, De Nicola GR, Bramanti P, Mazzon E (2014) The protective effects of bioactive (RS)-glucoraphanin on the permeability of the mice blood–brain barrier following experimental autoimmune encephalomyelitis. Eur Rev Med Pharmacol Sci 18:194–204
28. Goldmann EE (1913) Vitalfärbungen am Zentralnervensystem. Beitrag zur Physio-Pathologie des Plexus Choroideus und der Hirnhäute. Abh Preuss Akad Wiss Physik-Mathematik 1:1–60
29. Greenwood J, Heasman SJ, Alvarez JI, Prat A, Lyck R, Engelhardt B (2011) Review: leucocyte-endothelial cell crosstalk at the blood–brain barrier: a prerequisite for successful immune cell entry to the brain. Neuropathol Appl Neurobiol 37:24–39
30. Greter M, Heppner FL, Lemos MP, Odermatt BM, Goebels N, Laufer T, Noelle RJ, Becher B (2005) Dendritic cells permit immune invasion of the CNS in an animal model of multiple sclerosis. Nat Med 11:328–334

31. Gunzel D, Fromm M (2012) Claudins and other tight junction proteins. Compr Physiol 2:1819–1852
32. Haseloff RF, Dithmer S, Winkler L, Wolburg H, Blasig IE (2015) Transmembrane proteins of the tight junctions at the blood–brain barrier: structural and functional aspects. Semin Cell Dev Biol 38:16–25
33. Hawkins BT, Davis TP (2005) The blood–brain barrier/neurovascular unit in health and disease. Pharmacol Rev 57:173–185
34. Howell OW, Reeves CA, Nicholas R, Carassiti D, Radotra B, Gentleman SM, Serafini B, Aloisi F, Roncaroli F, Magliozzi R, Reynolds R (2011) Meningeal inflammation is widespread and linked to cortical pathology in multiple sclerosis. Brain 134:2755–2771
35. Iliff JJ, Wang M, Liao Y, Plogg BA, Peng W, Gundersen GA, Benveniste H, Vates GE, Deane R, Goldman SA, Nagelhus EA, Nedergaard M (2012) A paravascular pathway facilitates CSF flow through the brain parenchyma and the clearance of interstitial solutes, including amyloid beta. Sci Transl Med 4:147ra111
36. Illum L (2000) Transport of drugs from the nasal cavity to the central nervous system. Eur J Pharm Sci Off J Eur Fed Pharm Sci 11:1–18
37. Jackson RT, Tigges J, Arnold W (1979) Subarachnoid space of the CNS, nasal mucosa, and lymphatic system. Arch Otolaryngol 105:180–184
38. Kermode AG, Thompson AJ, Tofts P, MacManus DG, Kendall BE, Kingsley DP, Moseley IF, Rudge P, McDonald WI (1990) Breakdown of the blood–brain barrier precedes symptoms and other MRI signs of new lesions in multiple sclerosis. Pathogenetic and clinical implications. Brain 113(Pt 5):1477–1489
39. Kirk J, Plumb J, Mirakhur M, McQuaid S (2003) Tight junctional abnormality in multiple sclerosis white matter affects all calibres of vessel and is associated with blood–brain barrier leakage and active demyelination. J Pathol 201:319–327
40. Kivisakk P, Imitola J, Rasmussen S, Elyaman W, Zhu B, Ransohoff RM, Khoury SJ (2009) Localizing central nervous system immune surveillance: meningeal antigen-presenting cells activate T cells during experimental autoimmune encephalomyelitis. Ann Neurol 65:457–469
41. Kooij G, Kopplin K, Blasig R, Stuiver M, Koning N, Goverse G, van der Pol SM, van Het Hof B, Gollasch M, Drexhage JA, Reijerkerk A, Meij IC, Mebius R, Willnow TE, Muller D, Blasig IE, de Vries HE (2014) Disturbed function of the blood-cerebrospinal fluid barrier aggravates neuro-inflammation. Acta Neuropathol 128:267–277
42. Kutzelnigg A, Lassmann H (2014) Pathology of multiple sclerosis and related inflammatory demyelinating diseases. Handb Clin Neurol 122:15–58
43. Lanz TV, Becker S, Osswald M, Bittner S, Schuhmann MK, Opitz CA, Gaikwad S, Wiestler B, Litzenburger UM, Sahm F, Ott M, Iwantscheff S, Grabitz C, Mittelbronn M, von Deimling A, Winkler F, Meuth SG, Wick W, Platten M (2013) Protein kinase Cbeta as a therapeutic target stabilizing blood–brain barrier disruption in experimental autoimmune encephalomyelitis. Proc Natl Acad Sci U S A 110:14735–14740
44. Lassmann H, Wekerle H (2005) The pathology of multiple sclerosis. In: Compston A (ed) McAlpine's multiple sclerosis. Elsevier Churchill Livingstone, Edinburgh, pp 557–599
45. Leech S, Kirk J, Plumb J, McQuaid S (2007) Persistent endothelial abnormalities and blood–brain barrier leak in primary and secondary progressive multiple sclerosis. Neuropathol Appl Neurobiol 33:86–98
46. Lennon VA, Kryzer TJ, Pittock SJ, Verkman AS, Hinson SR (2005) IgG marker of optic-spinal multiple sclerosis binds to the aquaporin-4 water channel. J Exp Med 202:473–477
47. Lennon VA, Wingerchuk DM, Kryzer TJ, Pittock SJ, Lucchinetti CF, Fujihara K, Nakashima I, Weinshenker BG (2004) A serum autoantibody marker of neuromyelitis optica: distinction from multiple sclerosis. Lancet 364:2106–2112
48. Lippoldt A, Liebner S, Andbjer B, Kalbacher H, Wolburg H, Haller H, Fuxe K (2000) Organization of choroid plexus epithelial and endothelial cell tight junctions and regulation of claudin-1, −2 and −5 expression by protein kinase C. Neuroreport 11:1427–1431
49. Magliozzi R, Howell O, Vora A, Serafini B, Nicholas R, Puopolo M, Reynolds R, Aloisi F (2007) Meningeal B-cell follicles in secondary progressive multiple sclerosis associate with early onset of disease and severe cortical pathology. Brain 130:1089–1104

50. Mathiisen TM, Lehre KP, Danbolt NC, Ottersen OP (2010) The perivascular astroglial sheath provides a complete covering of the brain microvessels: an electron microscopic 3D reconstruction. Glia 58:1094–1103
51. McMahan UJ (1990) The agrin hypothesis. Cold Spring Harb Symp Quant Biol 55:407–418
52. Mühleisen H, Wolburg H, Betz E (1989) Freeze-fracture analysis of endothelial cell membranes in rabbit carotid arteries subjected to short-term atherogenic stimuli. Virchows Arch B Cell Pathol Incl Mol Pathol 56:413–417
53. Muldoon LL, Alvarez JI, Begley DJ, Boado RJ, Del Zoppo GJ, Doolittle ND, Engelhardt B, Hallenbeck JM, Lonser RR, Ohlfest JR, Prat A, Scarpa M, Smeyne RJ, Drewes LR, Neuwelt EA (2013) Immunologic privilege in the central nervous system and the blood–brain barrier. J Cereb Blood Flow Metab 33:13–21
54. Nagy Z, Peters H, Huttner I (1984) Fracture faces of cell junctions in cerebral endothelium during normal and hyperosmotic conditions. Lab Invest 50:313–322
55. Nedergaard M (2013) Neuroscience. Garbage truck of the brain. Sci 340:1529–1530
56. Noell S, Fallier-Becker P, Beyer C, Kroger S, Mack AF, Wolburg H (2007) Effects of agrin on the expression and distribution of the water channel protein aquaporin-4 and volume regulation in cultured astrocytes. Eur J Neurosci 26:2109–2118
57. Noell S, Fallier-Becker P, Deutsch U, Mack AF, Wolburg H (2009) Agrin defines polarized distribution of orthogonal arrays of particles in astrocytes. Cell Tissue Res 337:185–195
58. Noell S, Wolburg-Buchholz K, Mack AF, Beedle AM, Satz JS, Campbell KP, Wolburg H, Fallier-Becker P (2011) Evidence for a role of dystroglycan regulating the membrane architecture of astroglial endfeet. Eur J Neurosci 33:2179–2186
59. Noseworthy JH, Lucchinetti C, Rodriguez M, Weinshenker BG (2000) Multiple sclerosis. N Engl J Med 343:938–952
60. Obermeier B, Daneman R, Ransohoff RM (2013) Development, maintenance and disruption of the blood–brain barrier. Nat Med 19:1584–1596
61. Owens T, Bechmann I, Engelhardt B (2008) Perivascular spaces and the two steps to neuroinflammation. J Neuropathol Exp Neurol 67:1113–1121
62. Pfeiffer F, Schafer J, Lyck R, Makrides V, Brunner S, Schaeren-Wiemers N, Deutsch U, Engelhardt B (2011) Claudin-1 induced sealing of blood–brain barrier tight junctions ameliorates chronic experimental autoimmune encephalomyelitis. Acta Neuropathol 122:601–614
63. Piontek J, Fritzsche S, Cording J, Richter S, Hartwig J, Walter M, Yu D, Turner JR, Gehring C, Rahn HP, Wolburg H, Blasig IE (2011) Elucidating the principles of the molecular organization of heteropolymeric tight junction strands. Cell Mol Life Sci 68:3903–3918
64. Pollinger B, Krishnamoorthy G, Berer K, Lassmann H, Bosl MR, Dunn R, Domingues HS, Holz A, Kurschus FC, Wekerle H (2009) Spontaneous relapsing-remitting EAE in the SJL/J mouse: MOG-reactive transgenic T cells recruit endogenous MOG-specific B cells. J Exp Med 206:1303–1316
65. Ransohoff RM, Engelhardt B (2012) The anatomical and cellular basis of immune surveillance in the central nervous system. Nat Rev Immunol 12:623–635
66. Rash JE, Yasumura T, Hudson CS, Agre P, Nielsen S (1998) Direct immunogold labeling of aquaporin-4 in square arrays of astrocyte and ependymocyte plasma membranes in rat brain and spinal cord. Proc Natl Acad Sci U S A 95:11981–11986
67. Reboldi A, Coisne C, Baumjohann D, Benvenuto F, Bottinelli D, Lira S, Uccelli A, Lanzavecchia A, Engelhardt B, Sallusto F (2009) C-C chemokine receptor 6-regulated entry of TH-17 cells into the CNS through the choroid plexus is required for the initiation of EAE. Nat Immunol 10:514–523
68. Reese TS, Karnovsky MJ (1967) Fine structural localization of a blood–brain barrier to exogenous peroxidase. J Cell Biol 34:207–217
69. Reichenbach A, Wolburg H (2013) Astrocytes and ependymal cells. Kettenmann H, Ransom BR (eds) Neuroglia, 3rd edn. Oxford University Press, Oxford, p 35–49
70. Rolak LA (2003) Multiple sclerosis: it's not the disease you thought it was. Clin Med Res 1:57–60

71. Russi AE, Brown MA (2015) The meninges: new therapeutic targets for multiple sclerosis. Translational Res J Lab Clin Med 165:255–269
72. Saunders NR, Dreifuss JJ, Dziegielewska KM, Johansson PA, Habgood MD, Mollgard K, Bauer HC (2014) The rights and wrongs of blood–brain barrier permeability studies: a walk through 100 years of history. Front Neurosci 8:404
73. Sayed BA, Christy AL, Walker ME, Brown MA (2010) Meningeal mast cells affect early T cell central nervous system infiltration and blood–brain barrier integrity through TNF: a role for neutrophil recruitment? J Immunol 184:6891–6900
74. Serafini B, Rosicarelli B, Magliozzi R, Stigliano E, Aloisi F (2004) Detection of ectopic B-cell follicles with germinal centers in the meninges of patients with secondary progressive multiple sclerosis. Brain Pathol 14:164–174
75. Shultz LD, Ishikawa F, Greiner DL (2007) Humanized mice in translational biomedical research. Nat Rev Immunol 7:118–130
76. Steiner E, Enzmann GU, Lyck R, Lin S, Ruegg MA, Kroger S, Engelhardt B (2014) The heparan sulfate proteoglycan agrin contributes to barrier properties of mouse brain endothelial cells by stabilizing adherens junctions. Cell Tissue Res 358:465–479
77. Steinmann U, Borkowski J, Wolburg H, Schroppel B, Findeisen P, Weiss C, Ishikawa H, Schwerk C, Schroten H, Tenenbaum T (2013) Transmigration of polymorphnuclear neutrophils and monocytes through the human blood-cerebrospinal fluid barrier after bacterial infection in vitro. J Neuroinflammation 10:31
78. Tenenbaum T, Papandreou T, Gellrich D, Friedrichs U, Seibt A, Adam R, Wewer C, Galla HJ, Schwerk C, Schroten H (2009) Polar bacterial invasion and translocation of Streptococcus suis across the blood-cerebrospinal fluid barrier in vitro. Cell Microbiol 11:323–336
79. Tietz S, Engelhardt B (2015) Brain barriers: crosstalk between complex tight junctions and adherens junctions. J Cell Biol 209:493–506
80. van Horssen J, Brink BP, de Vries HE, van der Valk P, Bo L (2007) The blood–brain barrier in cortical multiple sclerosis lesions. J Neuropathol Exp Neurol 66:321–328
81. Varrin-Doyer M, Spencer CM, Schulze-Topphoff U, Nelson PA, Stroud RM, Cree BA, Zamvil SS (2012) Aquaporin 4-specific T cells in neuromyelitis optica exhibit a Th17 bias and recognize Clostridium ABC transporter. Ann Neurol 72:53–64
82. Wang XS, Fang HL, Chen Y, Liang SS, Zhu ZG, Zeng QY, Li J, Xu HQ, Shao B, He JC, Hou ST, Zheng RY (2014) Idazoxan reduces blood–brain barrier damage during experimental autoimmune encephalomyelitis in mouse. Eur J Pharmacol 736:70–76
83. Warth A, Kroger S, Wolburg H (2004) Redistribution of aquaporin-4 in human glioblastoma correlates with loss of agrin immunoreactivity from brain capillary basal laminae. Acta Neuropathol 107:311–318
84. Wekerle H, Flugel A, Fugger L, Schett G, Serreze D (2012) Autoimmunity's next top models. Nat Med 18:66–70
85. Wolburg-Buchholz K, Mack AF, Steiner E, Pfeiffer F, Engelhardt B, Wolburg H (2009) Loss of astrocyte polarity marks blood–brain barrier impairment during experimental autoimmune encephalomyelitis. Acta Neuropathol 118:219–233
86. Wolburg H, Lippoldt A (2002) Tight junctions of the blood–brain barrier: development, composition and regulation. Vascul Pharmacol 38:323–337
87. Wolburg H, Mack A (2014) Comment on the topology of mammalian blood-cerebrospinal fluid barriers. Neurol Psych Brain Res 20:70–72
88. Wolburg H, Neuhaus J, Kniesel U, Krauss B, Schmid EM, Ocalan M, Farrell C, Risau W (1994) Modulation of tight junction structure in blood–brain barrier endothelial cells. Effects of tissue culture, second messengers and cocultured astrocytes. J Cell Sci 107(Pt 5):1347–1357
89. Wolburg H, Noell S, Mack A, Wolburg-Buchholz K, Fallier-Becker P (2009) Brain endothelial cells and the glio-vascular complex. Cell Tissue Res 335:75–96
90. Wolburg H, Noell S, Wolburg-Buchholz K, Mack A, Fallier-Becker P (2009) Agrin, aquaporin-4, and astrocyte polarity as an important feature of the blood–brain barrier. Neuroscientist 15:180–193

91. Wolburg H, Paulus W (2010) Choroid plexus: biology and pathology. Acta Neuropathol 119:75–88
92. Wolburg H, Wolburg-Buchholz K, Engelhardt B (2005) Diapedesis of mononuclear cells across cerebral venules during experimental autoimmune encephalomyelitis leaves tight junctions intact. Acta Neuropathol 109:181–190
93. Wolburg H, Wolburg-Buchholz K, Fallier-Becker P, Noell S, Mack AF (2011) Structure and functions of aquaporin-4-based orthogonal arrays of particles. Int Rev Cell Mol Biol 287:1–41
94. Wolburg H, Wolburg-Buchholz K, Kraus J, Rascher-Eggstein G, Liebner S, Hamm S, Duffner F, Grote EH, Risau W, Engelhardt B (2003) Localization of claudin-3 in tight junctions of the blood–brain barrier is selectively lost during experimental autoimmune encephalomyelitis and human glioblastoma multiforme. Acta Neuropathol 105:586–592
95. Wolburg H, Wolburg-Buchholz K, Reichenbach A, Mack AF (2015) Ependymal cells. Reference module in biomedical sciences. http://dx.doi.org/10.1016/B978-0-12-801238-3.04586-4
96. Yousif LF, Di Russo J, Sorokin L (2013) Laminin isoforms in endothelial and perivascular basement membranes. Cell Adh Migr 7:101–110

The Contribution of the Extracellular Matrix to the BBB in Steady State and Inflammatory Conditions

Melanie-Jane Hannocks, Jula Huppert, Xueli Zhang, Eva Korpos, and Lydia Sorokin

Abstract The delicate neurons in the central nervous system are protected from the circulation by the blood-brain barrier (BBB), which prevents the transmigration of cells and harmful substances from entering the brain. Much attention has focused on the cellular components (endothelial cells and astrocyte endfeet) of this barrier, but it is becoming increasingly obvious that the noncellular components, specifically the basement membranes (BMs), play a crucial role in the integrity of the BBB. Not only do the BMs help to maintain barrier tightness through their actual presence, but also their distinct composition plays an important role in the regulation of their cellular counterparts. In this chapter we will describe the different BMs that are found in the BBB and present findings from an animal model of autoimmunity that have revealed considerable information on the function of the different BMs in maintaining BBB integrity.

1 The Blood-Brain Barrier

The neurons within the central nervous system (CNS) are extremely sensitive to any change in their microenvironment and, therefore, to keep them in a stabile milieu, the CNS is protected from the circulation by what is known as the blood-brain barrier (BBB). This barrier is established at the level of the microvasculature, where exchange of gases and the controlled transport of nutrients required for optimal neuronal transmission occurs and acts also to keep toxins and invading cells out of the CNS. The architecture of the BBB is unique and is constituted by both cellular and noncellular components. Highly specialized endothelial cells line the cerebral vessels and form the foundation of the BBB. Characteristic of these endothelial cells are complex tight junctions that interconnect adjacent cells and inhibit the free paracellular diffusion of solutes. They are also devoid of fenestrae and exhibit an extremely

M.-J. Hannocks • J. Huppert • X. Zhang • E. Korpos • L. Sorokin (✉)
Institute of Physiological Chemistry and Pathobiochemistry, University of Muenster, 48149 Muenster, Germany

Cells-In-Motion Cluster of Excellence, University of Muenster, 48149 Muenster, Germany
e-mail: sorokin@uni-muenster.de

R. Lyck, G. Enzmann (eds.), *The Blood Brain Barrier and Inflammation*, Progress in Inflammation Research, DOI 10.1007/978-3-319-45514-3_3

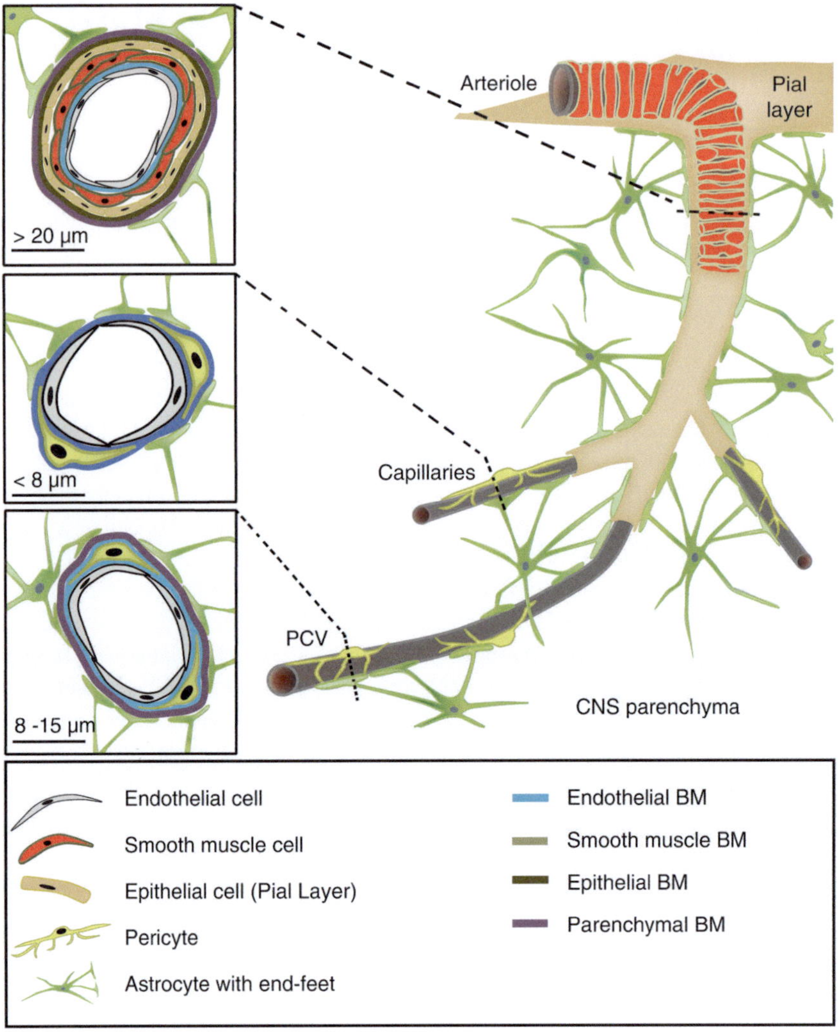

Fig. 1 Schematic representation depicting the cellular and BM components of the BBB. Inserts show the relationship of the different BMs to their cellular counterparts in different regions of the vascular bed: arterioles, capillaries and postcapillary venules (PCV)

low pinocytotic activity. At the abluminal side of the endothelial cells is the endothelial basement membrane (BM) (Fig. 1) in which pericytes are embedded and are considered critical for the development of the tight junctions [3, 8]. Unique to the vasculature of the CNS is the presence of a second BM subjacent to the ensheathing layer of astrocytic endfeet, known as the parenchymal BM [31], which together with the endothelial BM encloses the perivascular space. It has been proposed that signalling between the pericytes and astrocyte endfeet regulates the deposition of parenchymal BM components [3] as well as pericyte differentiation [47].

At sites where arterioles from the subarachnoid space penetrate into the CNS, the endothelial and parenchymal BMs are separated from each other by a layer of smooth muscle cells and a layer of pial cells and their associated BMs (Fig. 1). At the level of capillaries, the endothelial and parenchymal BMs appear fused in transmission electron microscopy, thereby, occluding a perivascular space. Our focus in this chapter is on the contribution of the extracellular matrix, and specifically of basement membranes, to the BBB.

2 The Extracellular Matrix of the CNS

As in all tissues, the extracellular matrix (ECM) of the CNS is of two types: basement membranes and the interstitial matrix. The interstitial matrix of the CNS, however, is unique as it lacks the fibrillar collagens typical of the stroma of most tissues. The reasons for this are the presence of a bony skull, which protects the brain from pressure or impact from outside forces, and the fact that the bulky collagen fibres would interfere in the precise intercommunication between neurons. Rather, the CNS interstitial matrix is a highly hydrated gel-like matrix composed of proteoglycans, hyaluronan, tenascins and link proteins that cushion the neuronal cell bodies and glial cells [19, 28]. By contrast, the basement membranes are sheet-like networks that are restricted to the vasculature and the pial surface of the brain and therefore interact directly with cellular components of the BBB. They are comprised of two independent networks, namely, the collagen IV network and the laminin network, that are linked together by the heparan sulphate proteoglycans, perlecan and agrin [5], and the nidogens [11]. Of all the BM components, the laminins are considered to be the biologically active components and show great diversity, with up to 18 different isoforms existing. Each isoform is characterized by a distinct α, β and γ chain composition that imparts biochemically and functionally distinct characteristics to the BM.

At the BBB, the endothelial BM underlies the endothelial monolayer and appears immediately subjacent to the parenchymal BM that underlies astrocyte endfeet at the CNS border. Endothelial BMs of most vessels in the CNS contain laminin 411 (composed of laminin $\alpha4$, $\beta1$, $\gamma1$ chains) and laminin 511 (composed of laminin $\alpha5$, $\beta1$, $\gamma1$ chains), while the parenchymal BM is characterized by the presence of laminin 211 [31]. At sites where the pial layer comigrates with arterioles into the CNS parenchyma [23], laminin $\alpha1$ is also detected at the outer border to the CNS parenchyma [31]. In addition to the laminins, the heparan sulphate proteoglycans also show differential localisation in the endothelial and parenchymal BMs, with perlecan occurring mainly in the endothelial BM and agrin occurring in the parenchymal BM [1]. Unpublished data from our laboratory also suggest differential localization of collagen IV isoforms in these two BMs. This molecular distinction between the endothelial and parenchyma BMs suggests that they contribute differently to BBB integrity and/or function.

Table 1 Laminin receptors expressed in the CNS [48]

Laminin isoforms	Laminin 511	Laminin 411	Laminin 211[a]
Receptors	Lu/B-CAM	MCAM	α-Dystroglycan
	Integrin α3β1		Integrin α3β1
	Integrin α6β1	Integrin α6β1	
	Integrin αvβ1		
	Integrin αvβ3		
	Integrin α5β1		

[a]Integrin α7β1 is also a major laminin 211 receptor but has not been reported to occur in the CNS

3 Laminin Receptors on Endothelium and Astrocyte Endfeet

In order to signal different functions to the cellular components of the BBB, the endothelial and parenchymal BM components must interact with cell surface receptors that belong either to the integrin or non-integrin class of receptors. The latter includes Lutheran blood group glycoprotein (Lu/B-CAM) [26, 38] and melanoma cell adhesion molecule (MCAM) [10], two members of the Ig (immunoglobulin) superfamily that occur on CNS endothelium, and α-dystroglycan, a component of the dystrophin glycoprotein complex expressed in the brain on astrocyte endfeet [1, 35] (Table 1).

By contrast, the integrins are a large family of heterodimeric receptors, each composed of an α and β subunit that contain extracellular, transmembrane and cytosolic domains [15]. Those reported to be expressed at the BBB and involved in BM binding include the β1 and β3 integrins, in particular, integrins α6β1 [31], αvβ1/αvβ3 [4], α1β1, and α5β1 [21] expressed on endothelial cells and integrin α3β1 [9], the precise localization of which (endothelial versus astroglial components) remains unclear (Fig. 2 and Table 1).

While it is not known precisely which receptor binds to which component in the endothelial and parenchymal BMs at the BBB, it is known from in vitro analyses that Lutheran blood group glycoprotein interacts only and with high affinity to laminin α5 [17, 26], and MCAM binds specifically to laminin α4 [10] (Table 1), suggesting that they probably contribute to the anchorage of the endothelial cells to the endothelial BM. Similarly, integrin α6β1, α3β1, α5β1 and αvβ3/αvb1 can all bind to laminin α5 [18, 30] and integrin α6β1 can also bind to laminin α4 [12] and, hence, may additionally contribute to cerebral endothelial cell anchorage to its BM. A recent study has shown that specific elimination of integrin β1 from endothelial cells reduces cell-cell adhesion due to changes in the cycling of the adherens junctional molecule VE-cadherin from the cell surface to intracellular endocytotic compartments in peripheral vessels [46]. This raises the possibility that β1 integrin-mediated binding to the endothelial BM may also affect endothelial junctional tightness. However, since cerebral endothelial cells have complex tight junctions, it is not clear whether adherens junction molecules, such as VE-cadherin, CD99 and CD99L2 [42], play as important a role in CNS vessels as they do in peripheral tissues. Whether changes in cycling or turnover of such junctional molecules could be affected by endothelial BM components has not yet been investigated.

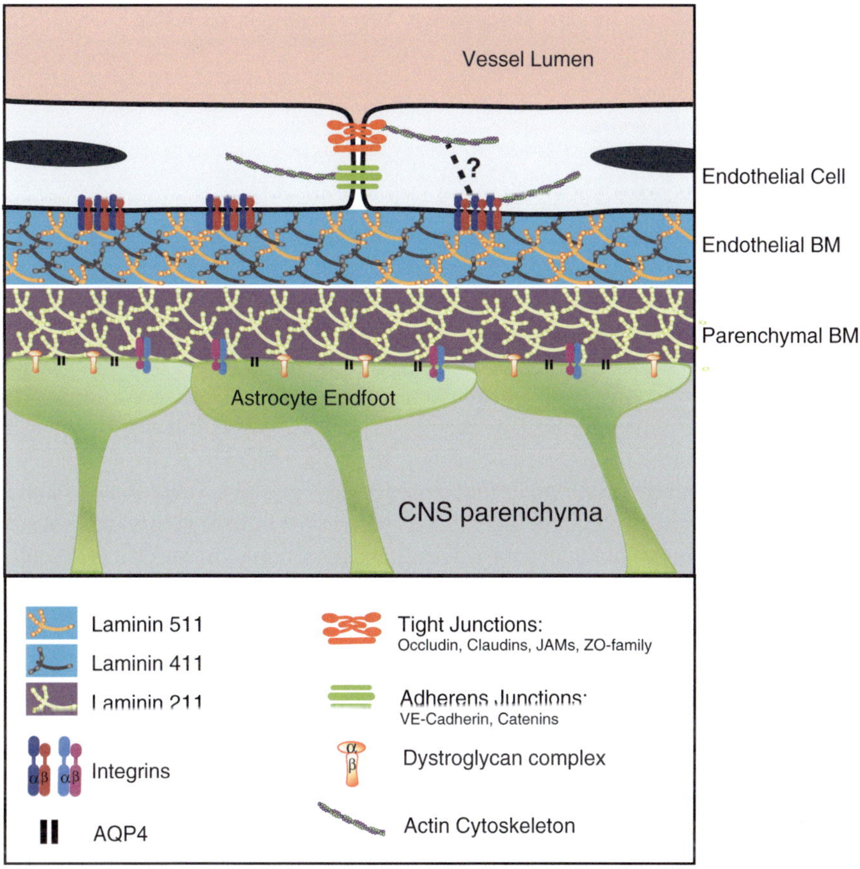

Fig. 2 Schematic representation depicting the possible associations between the BMs and their cellular counterparts at the level of PCVs. Immune cells transmigrate across the BBB at sites where there is little or no laminin α5 in the endothelial BM due to direct effects of laminin α5 on immune cell migration; whether laminins also affect endothelial cell tightness has not been investigated

However, one study has reported in vitro downregulation of the tight junction molecule claudin 5 in response to antibody inhibition of integrin β1 [25], supporting the concept that adhesion of the endothelial cells to their BM may contribute to BBB integrity by affecting junctional tightness.

In comparison to cerebral endothelium, astrocyte endfeet express relatively low levels of β1 integrins [31], and current data suggest that the main ECM receptor at this site is α-dystroglycan which binds with high affinity to laminin α2 and agrin [24, 36], thereby, anchoring the astrocyte endfeet to the parenchymal BM [1]. In addition, α-dystroglycan-mediated binding to the parenchymal BM acts to localize the water channel, aquaporin-4, specifically at the astrocyte endfeet and to prevent brain oedema [35] (Fig. 2).

4 Contributions of the Basement Membranes to BBB integrity

4.1 Genetic Defects

Although limited, there is some data on genetic diseases or gene elimination studies in mice that provides information on the significance of specific cell-matrix interactions at the BBB. The most significant are the reported mutations in the collagen α1 and α2 chains that constitute the collagen type IV heterotrimer and which result in microhaemorrhages within the brain in small vessel disease [13, 16, 29, 39]. Surprisingly, given the integral role of collagen type IV as a structural component of the BM, mice carrying such collagen type IV mutations do not have a lethal phenotype but can survive until birth or even beyond. Similarly, patients with collagen IV mutation do not have reduced life spans, but rather experience frequent microbleeds predominantly in the brain that cumulatively lead to neurological dysfunction, but which can also occur in other organs such as the kidney, skin and retina [14, 41]. This suggests that collagen IV $[\alpha 1]_2$ [α2], the most common isoform in vascular BMs, is not solely responsible for a functional BM but rather that it is the interplay of all BM components. This is further supported by analyses of mice lacking laminin α4 expression, one of the major laminin chains expressed by endothelium from the time of initial blood vessel formation [37], which also show perinatal bleeding in most organs including the CNS. This phenotype is rescued by the expression of laminin α5 around the time of birth, which normally does not occur until after week 3 [34]. Not only are endothelial BM components important but also those of the parenchymal BM: loss of or mutations in laminin α2 resulting in congenital muscular dystrophy can also be associated with abnormalities in the white matter as a result of increased vascular permeability [2]. Furthermore, astrocyte-specific deletion of laminin γ1 or dystroglycan in mice has also been described to lead to brain haemorrhaging [7] or gliosis [22], respectively. Animal studies involving the loss of integrin αv from glial cells have shown that haemorrhaging occurs during embryonic and neonatal development [20], stressing the need for cell-parenchymal BM interactions in BBB integrity. Interestingly, the deletion of integrin αv [20], α5 [40] or α6 [6] subunits from the endothelium has no reported effect on BBB function.

4.2 Autoimmune Diseases

While the BBB protects the delicate neurons from damage by the entry of noxious substances and invading cells, the brain is not exempt from autoimmune reactions or tissue damage due to injury or infection. Under such inflammatory conditions, the integrity of the BBB is challenged. Studies on experimental autoimmune encephalomyelitis (EAE), an animal model of autoimmunity to myelin that resembles several aspects of multiple sclerosis in humans, have revealed considerable information on the function of the BBB, in particular the BMs. Because the biochemical composition of the endothelial and parenchymal BMs is distinct, EAE provides an excellent model to study the relative contributions of these two BMs to the barrier function of the BBB to immune cells.

4.2.1 Leukocyte Transmigration Across the Endothelial BM

EAE is a T cell-mediated disease where T cells recognizing CNS antigens extravasate from the circulation at the level of postcapillary venules into the CNS to elicit an immune response and subsequent extravasation of monocytes and macrophages that contribute most significantly to demyelination events and disease symptoms (reviewed in [27]).

Before coming into contact with the endothelial BM, the leukocytes need to cross the endothelial monolayer involving multiple steps: attachment, rolling, adhesion, crawling and diapedesis [44]. The endothelial BM of postcapillary venules shows an ubiquitous distribution of laminin α4 and patchy distribution of laminin α5 [31], with leukocyte extravasation occurring exclusively at sites of little or no laminin α5 expression [31, 45]. Mice that lack laminin α4 have a compensatory uniform distribution of laminin α5 in all endothelial BMs [45] and display significantly reduced disease symptoms when employed in EAE studies due to decreased diapedesis of the T cells across the endothelial BM. In vitro studies have demonstrated that laminin α5 directly inhibits T cell migration [45], thereby, accounting for the reduced migration across CNS postcapillary venules in laminin α4 knockout mice. However, whether endothelial laminins can also affect expression or junctional localization of cell-cell adhesion molecules and, hence, endothelial barrier tightness is not clear. The tight junction-associated molecule claudin 3 [43] has been reported to be downregulated at sites of leukocyte extravasation in EAE, but whether a causal association exists between the absence of laminin α5 and the loss of claudin 3 at these sites has not been investigated. Using immunofluorescence staining, no association between the adherens junction molecules VE-cadherin, CD99, CD99L2 and ESAM and laminin α5 could be detected [45]. However, the relevance of adherens junctions in leukocyte transmigration across CNS postcapillary venules has not been demonstrated as it has been in peripheral tissues (reviewed in [42]).

4.2.2 Leukocyte Transmigration Across the Parenchymal BM

Once the leukocytes have crossed the endothelial BM, they accumulate within the perivascular space between the endothelial and parenchymal BMs (Fig. 3a). Interestingly, even though the leukocytes have already extravasated from the lumen of the postcapillary venules, there are no symptoms of EAE until the outer parenchymal BM is breeched and leukocytes enter into the CNS parenchyma. This highlights the importance of the parenchymal BM and associated astrocyte endfeet in the barrier properties of the BBB.

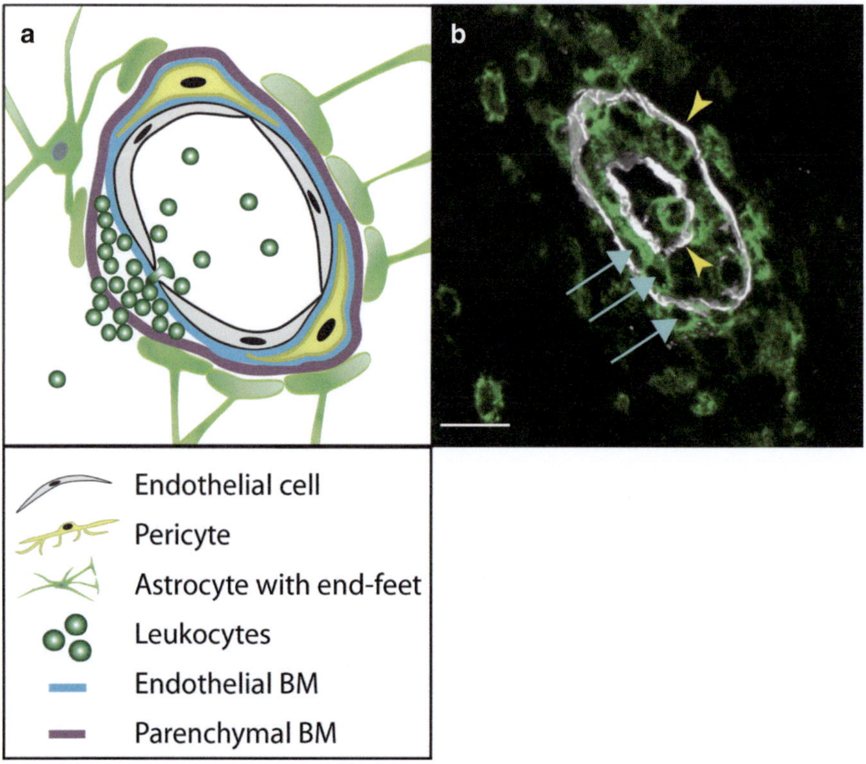

Fig. 3 Immune cell transmigration across the BBB. (**a**) Schematic representation of the PCV depicting immune cell diapedesis of the endothelial cell layer; accumulation of the immune cells between the endothelial and parenchymal BMs, forming a perivascular cuff; and finally immune cell infiltration into the CNS parenchyma where they illicit disease effects. (**b**) Immunofluorescent staining of a perivascular cuff. Pan-laminin antibody (*white*) and anti-CD45 antibody (*green*) visualize the BMs and leukocytes, respectively, on 5 μm mouse CNS sections. Arrow heads (*yellow*) mark the inner endothelial and outer parenchymal BMs, while arrows (*blue*) depict the leukocytes both within and outside the cuff. Scale bar = 15 μm

The transmigration of the leukocytes across the parenchymal BM is dependent on chemotactic signals [32] as well as the disruption of the dystroglycan-mediated anchorage of the astrocyte endfeet to the parenchymal BM [1]. Both steps are regulated by the focal and restricted activity of metalloproteinase (MMP)-2 and MMP-9 [1], which in turn are regulated by cytokines (TNF-α/IL-17 versus INF-γ) derived from the infiltrating leukocytes [32] (Fig. 3a). These events are hypothesized to facilitate the transmigration of the leukocytes across the parenchymal BM; however, precise mechanisms remain to be defined. It is clear, however, that MMP-2 and MMP-9 do not act to digest the BM components of the parenchymal BM (reviewed in [33]), although selective cleavage of defined bonds between ECM molecules or portions of ECM molecules not detectable with current tools cannot be excluded. The loss of astrocyte endfeet anchorage to the parenchymal

BM is likely to affect the intermolecular interactions between the BM components and thereby its barrier function, as well as potentially affecting astrocyte-astrocyte interactions.

5 Conclusion

The integrity of the BBB is vital to maintain the stable milieu for neuronal function. Much of the research to date has focused on the cellular components of the BBB, and only minor attention has been given to the noncellular components. However, it is now clear that both components are required and, in particular, it is the interaction of the two that maintains a tight barrier. The specific composition as well as the structural arrangement of the BMs plays a critical role. Genetic mutations in collagen IV or laminin α2 have shown that slight changes to the BM structure render the BBB less stable. While we have gained insight to the contribution of the endothelial and parenchymal BMs to the BBB from the EAE model, further research is needed to address whether the laminin α-chains affect endothelial adhesion and junctional molecules in addition to the migration modes of the extravasating immune cells. Whether these mechanisms are distinct to those as a result of tissue injury which, in contrast to autoimmune diseases, are antigen independent, is also not known. Understanding how BBB integrity is maintained and how the individual components function may eventually lead to developing anti-inflammatory therapies or even improving drug delivery.

References

1. Agrawal S, Anderson P, Durbeej M, van Rooijen N, Ivars F, Opdenakker G, Sorokin LM (2006) Dystroglycan is selectively cleaved at the parenchymal basement membrane at sites of leukocyte extravasation in experimental autoimmune encephalomyelitis. J Exp Med 203(4):1007–1019. doi:10.1084/jem.20051342
2. Alkan A, Sigirci A, Kutlu R, Aslan M, Doganay S, Yakinci C (2007) Merosin-negative congenital muscular dystrophy: diffusion-weighted imaging findings of brain. J Child Neurol 22(5):655–659. doi:10.1177/0883073807303219
3. Armulik A, Genove G, Mae M, Nisancioglu MH, Wallgard E, Niaudet C, He L, Norlin J, Lindblom P, Strittmatter K, Johansson BR, Betsholtz C (2010) Pericytes regulate the blood-brain barrier. Nature 468(7323):557–561. doi:10.1038/nature09522
4. Bader BL, Rayburn H, Crowley D, Hynes RO (1998) Extensive vasculogenesis, angiogenesis, and organogenesis precede lethality in mice lacking all alpha v integrins. Cell 95(4):507–519
5. Behrens DT, Villone D, Koch M, Brunner G, Sorokin L, Robenek H, Bruckner-Tuderman L, Bruckner P, Hansen U (2012) The epidermal basement membrane is a composite of separate laminin- or collagen IV-containing networks connected by aggregated perlecan, but not by nidogens. J Biol Chem 287(22):18700–18709. doi:10.1074/jbc.M111.336073
6. Bouvard C, De Arcangelis A, Dizier B, Galy-Fauroux I, Fischer AM, Georges-Labouesse E, Helley D (2012) Tie2-dependent knockout of alpha6 integrin subunit in mice reduces post-ischaemic angiogenesis. Cardiovasc Res 95(1):39–47. doi:10.1093/cvr/cvs153

7. Chen ZL, Yao Y, Norris EH, Kruyer A, Jno-Charles O, Akhmerov A, Strickland S (2013) Ablation of astrocytic laminin impairs vascular smooth muscle cell function and leads to hemorrhagic stroke. J Cell Biol 202(2):381–395. doi:10.1083/jcb.201212032
8. Daneman R, Zhou L, Kebede AA, Barres BA (2010) Pericytes are required for blood-brain barrier integrity during embryogenesis. Nature 468(7323):562–566. doi:10.1038/nature09513
9. De Arcangelis A, Mark M, Kreidberg J, Sorokin L, Georges-Labouesse E (1999) Synergistic activities of alpha3 and alpha6 integrins are required during apical ectodermal ridge formation and organogenesis in the mouse. Development 126(17):3957–3968
10. Flanagan K, Fitzgerald K, Baker J, Regnstrom K, Gardai S, Bard F, Mocci S, Seto P, You M, Larochelle C, Prat A, Chow S, Li L, Vandevert C, Zago W, Lorenzana C, Nishioka C, Hoffman J, Botelho R, Willits C, Tanaka K, Johnston J, Yednock T (2012) Laminin-411 is a vascular ligand for MCAM and facilitates TH17 cell entry into the CNS. PLoS One 7(7), e40443. doi:10.1371/journal.pone.0040443
11. Fox JW, Mayer U, Nischt R, Aumailley M, Reinhardt D, Wiedemann H, Mann K, Timpl R, Krieg T, Engel J et al (1991) Recombinant nidogen consists of three globular domains and mediates binding of laminin to collagen type IV. EMBO J 10(11):3137–3146
12. Fujiwara H, Kikkawa Y, Sanzen N, Sekiguchi K (2001) Purification and characterization of human laminin-8. Laminin-8 stimulates cell adhesion and migration through alpha3beta1 and alpha6beta1 integrins. J Biol Chem 276(20):17550–17558. doi:10.1074/jbc.M010155200
13. Gould DB, Phalan FC, Breedveld GJ, van Mil SE, Smith RS, Schimenti JC, Aguglia U, van der Knaap MS, Heutink P, John SW (2005) Mutations in Col4a1 cause perinatal cerebral hemorrhage and porencephaly. Science 308(5725):1167–1171. doi:10.1126/science.1109418
14. Gould DB, Phalan FC, van Mil SE, Sundberg JP, Vahedi K, Massin P, Bousser MG, Heutink P, Miner JH, Tournier-Lasserve E, John SW (2006) Role of COL4A1 in small-vessel disease and hemorrhagic stroke. N Engl J Med 354(14):1489–1496. doi:10.1056/NEJMoa053727
15. Hynes RO (1992) Integrins: versatility, modulation, and signaling in cell adhesion. Cell 69(1):11–25
16. Jeanne M, Jorgensen J, Gould DB (2015) Molecular and Genetic Analyses of Collagen Type IV Mutant Mouse Models of Spontaneous Intracerebral Hemorrhage Identify Mechanisms for Stroke Prevention. Circulation 131(18):1555–1565. doi:10.1161/CIRCULATIONAHA.114.013395
17. Kikkawa Y, Moulson CL, Virtanen I, Miner JH (2002) Identification of the binding site for the Lutheran blood group glycoprotein on laminin alpha 5 through expression of chimeric laminin chains in vivo. J Biol Chem 277(47):44864–44869. doi:10.1074/jbc.M208731200
18. Kikkawa Y, Sanzen N, Fujiwara H, Sonnenberg A, Sekiguchi K (2000) Integrin binding specificity of laminin-10/11: laminin-10/11 are recognized by alpha 3 beta 1, alpha 6 beta 1 and alpha 6 beta 4 integrins. J Cell Sci 113(Pt 5):869–876
19. Lau LW, Cua R, Keough MB, Haylock-Jacobs S, Yong VW (2013) Pathophysiology of the brain extracellular matrix: a new target for remyelination. Nat Rev Neurosci 14(10):722–729. doi:10.1038/nrn3550
20. McCarty JH, Lacy-Hulbert A, Charest A, Bronson RT, Crowley D, Housman D, Savill J, Roes J, Hynes RO (2005) Selective ablation of alphav integrins in the central nervous system leads to cerebral hemorrhage, seizures, axonal degeneration and premature death. Development 132(1):165–176. doi:10.1242/dev.01551
21. Milner R, Hung S, Erokwu B, Dore-Duffy P, LaManna JC, del Zoppo GJ (2008) Increased expression of fibronectin and the alpha 5 beta 1 integrin in angiogenic cerebral blood vessels of mice subject to hypobaric hypoxia. Mol Cell Neurosci 38(1):43–52. doi:10.1016/j.mcn.2008.01.013
22. Moore SA, Saito F, Chen J, Michele DE, Henry MD, Messing A, Cohn RD, Ross-Barta SE, Westra S, Williamson RA, Hoshi T, Campbell KP (2002) Deletion of brain dystroglycan recapitulates aspects of congenital muscular dystrophy. Nature 418(6896):422–425. doi:10.1038/nature00838
23. Nicholas DS, Weller RO (1988) The fine anatomy of the human spinal meninges. A light and scanning electron microscopy study. J Neurosurg 69(2):276–282. doi:10.3171/jns.1988.69.2.0276

24. Noell S, Fallier-Becker P, Deutsch U, Mack AF, Wolburg H (2009) Agrin defines polarized distribution of orthogonal arrays of particles in astrocytes. Cell Tissue Res 337(2):185–195. doi:10.1007/s00441-009-0812-z
25. Osada T, Gu YH, Kanazawa M, Tsubota Y, Hawkins BT, Spatz M, Milner R, del Zoppo GJ (2011) Interendothelial claudin-5 expression depends on cerebral endothelial cell-matrix adhesion by beta(1)-integrins. J Cereb Blood Flow Metab 31(10):1972–1985. doi:10.1038/jcbfm.2011.99
26. Parsons SF, Lee G, Spring FA, Willig TN, Peters LL, Gimm JA, Tanner MJ, Mohandas N, Anstee DJ, Chasis JA (2001) Lutheran blood group glycoprotein and its newly characterized mouse homologue specifically bind alpha5 chain-containing human laminin with high affinity. Blood 97(1):312–320
27. Ransohoff RM, Engelhardt B (2012) The anatomical and cellular basis of immune surveillance in the central nervous system. Nat Rev Immunol 12(9):623–635. doi:10.1038/nri3265
28. Rauch U (2007) Brain matrix: structure, turnover and necessity. Biochem Soc Trans 35(Pt 4):656–660. doi:10.1042/BST0350656
29. Renard D, Mine M, Pipiras E, Labauge P, Delahaye A, Benzacken B, Tournier-Lasserve E (2014) Cerebral small-vessel disease associated with COL4A1 and COL4A2 gene duplications. Neurology 83(11):1029–1031. doi:10.1212/WNL.0000000000000769
30. Sasaki T, Timpl R (2001) Domain IVa of laminin alpha5 chain is cell-adhesive and binds beta1 and alphaVbeta3 integrins through Arg-Gly-Asp. FEBS Lett 509(2):181–185
31. Sixt M, Engelhardt B, Pausch F, Hallmann R, Wendler O, Sorokin LM (2001) Endothelial cell laminin isoforms, laminins 8 and 10, play decisive roles in T cell recruitment across the blood-brain barrier in experimental autoimmune encephalomyelitis. J Cell Biol 153(5):933–946
32. Song J, Wu C, Korpos E, Zhang X, Agrawal SM, Wang Y, Faber C, Schafers M, Korner H, Opdenakker G, Hallmann R, Sorokin L (2015) Focal MMP-2 and MMP-9 activity at the blood-brain barrier promotes chemokine-induced leukocyte migration. Cell Rep 10(7):1040–1054. doi:10.1016/j.celrep.2015.01.037
33. Sorokin L (2010) The impact of the extracellular matrix on inflammation. Nat Rev Immunol 10(10):712–723. doi:10.1038/nri2852
34. Sorokin LM, Pausch F, Frieser M, Kroger S, Ohage E, Deutzmann R (1997) Developmental regulation of the laminin alpha5 chain suggests a role in epithelial and endothelial cell maturation. Dev Biol 189(2):285–300. doi:10.1006/dbio.1997.8668
35. Steiner E, Enzmann GU, Lin S, Ghavampour S, Hannocks MJ, Zuber B, Ruegg MA, Sorokin L, Engelhardt B (2012) Loss of astrocyte polarization upon transient focal brain ischemia as a possible mechanism to counteract early edema formation. Glia 60(11):1646–1659. doi:10.1002/glia.22383
36. Talts JF, Andac Z, Gohring W, Brancaccio A, Timpl R (1999) Binding of the G domains of laminin alpha1 and alpha2 chains and perlecan to heparin, sulfatides, alpha-dystroglycan and several extracellular matrix proteins. EMBO J 18(4):863–870. doi:10.1093/emboj/18.4.863
37. Thyboll J, Kortesmaa J, Cao R, Soininen R, Wang L, Iivanainen A, Sorokin L, Risling M, Cao Y, Tryggvason K (2002) Deletion of the laminin alpha4 chain leads to impaired microvessel maturation. Mol Cell Biol 22(4):1194–1202
38. Udani M, Zen Q, Cottman M, Leonard N, Jefferson S, Daymont C, Truskey G, Telen MJ (1998) Basal cell adhesion molecule/lutheran protein. The receptor critical for sickle cell adhesion to laminin. J Clin Invest 101(11):2550–2558. doi:10.1172/JCI1204
39. Vahedi K, Kubis N, Boukobza M, Arnoult M, Massin P, Tournier-Lasserve E, Bousser MG (2007) COL4A1 mutation in a patient with sporadic, recurrent intracerebral hemorrhage. Stroke 38(5):1461–1464. doi:10.1161/STROKEAHA.106.475194
40. van der Flier A, Badu-Nkansah K, Whittaker CA, Crowley D, Bronson RT, Lacy-Hulbert A, Hynes RO (2010) Endothelial alpha5 and alphav integrins cooperate in remodeling of the vasculature during development. Development 137(14):2439–2449. doi:10.1242/dev.049551
41. van der Knaap MS, Smit LM, Barkhof F, Pijnenburg YA, Zweegman S, Niessen HW, Imhof S, Heutink P (2006) Neonatal porencephaly and adult stroke related to mutations in collagen IV A1. Ann Neurol 59(3):504–511. doi:10.1002/ana.20715

42. Vestweber D (2015) How leukocytes cross the vascular endothelium. Nat Rev Immunol 15(11):692–704. doi:10.1038/nri3908
43. Wolburg H, Wolburg-Buchholz K, Kraus J, Rascher-Eggstein G, Liebner S, Hamm S, Duffner F, Grote EH, Risau W, Engelhardt B (2003) Localization of claudin-3 in tight junctions of the blood-brain barrier is selectively lost during experimental autoimmune encephalomyelitis and human glioblastoma multiforme. Acta Neuropathol 105(6):586–592. doi:10.1007/s00401-003-0688-z
44. Wolburg-Buchholz K, Mack AF, Steiner E, Pfeiffer F, Engelhardt B, Wolburg H (2009) Loss of astrocyte polarity marks blood-brain barrier impairment during experimental autoimmune encephalomyelitis. Acta Neuropathol 118(2):219–233. doi:10.1007/s00401-009-0558-4
45. Wu C, Ivars F, Anderson P, Hallmann R, Vestweber D, Nilsson P, Robenek H, Tryggvason K, Song J, Korpos E, Loser K, Beissert S, Georges-Labouesse E, Sorokin LM (2009) Endothelial basement membrane laminin alpha5 selectively inhibits T lymphocyte extravasation into the brain. Nat Med 15(5):519–527. doi:10.1038/nm.1957
46. Yamamoto H, Ehling M, Kato K, Kanai K, van Lessen M, Frye M, Zeuschner D, Nakayama M, Vestweber D, Adams RH (2015) Integrin beta1 controls VE-cadherin localization and blood vessel stability. Nat Commun 6:6429. doi:10.1038/ncomms7429
47. Yao Y, Chen ZL, Norris EH, Strickland S (2014) Astrocytic laminin regulates pericyte differentiation and maintains blood brain barrier integrity. Nat Commun 5:3413. doi:10.1038/ncomms4413
48. Yousif LF, Di Russo J, Sorokin L (2013) Laminin isoforms in endothelial and perivascular basement membranes. Cell Adh Migr 7(1):101–110. doi:10.4161/cam.22680

Pathophysiology of the Blood–Brain Barrier in Neuroinflammatory Diseases

Petra Majerova and Andrej Kovac

Abstract Alzheimer's disease, Parkinson's disease, Huntington's disease, multiple sclerosis, and amyotrophic lateral sclerosis are neurodegenerative disorders that result in progressive dysfunction and loss of neurons in the central nervous system (CNS). A strong link between neurodegeneration and chronic inflammation has recently been demonstrated. Neuropathological studies suggest that the neuroinflammatory responses might begin before significant neuronal loss, which supports the hypothesis that neuroinflammation might play an important role in the pathogenesis of most neurodegenerative disorders. Chronic neuroinflammation contributes to increased glial activation and proliferation, leading to the release of detrimental pro-inflammatory factors. The inflammatory processes promote changes in brain capillaries, such as loss of tight junction proteins, atrophy of pericytes, thickening of the basement membrane as a result of the accumulation of basement membrane proteins, and increased permeability to small molecules and plasma proteins. These changes accelerate transmigration of peripheral cells into the brain parenchyma. In this work, we discuss the role of neuroinflammation in neurodegenerative diseases. We review the impact of immune responses on the CNS, resulting in blood–brain barrier changes during neurodegeneration.

1 Introduction

Homeostasis of the central nervous system (CNS) is essential for its normal functioning and is maintained by the highly specialized brain endothelial structure, the blood–brain barrier (BBB). Astrocytes, neurons, pericytes, and microglia communicate with endothelial cells and are collectively referred to as the neurovascular unit. The BBB strictly controls the exchange of cells and molecules between blood and the CNS [30]. BBB disruption is associated with numerous pathological conditions that affect the CNS, such as ischemia, infections, epilepsy, tumors, and

P. Majerova • A. Kovac (✉)
Institute of Neuroimmunology, Slovak Academy of Sciences,
Dubravska cesta 9, 84510 Bratislava, Slovakia
e-mail: andrej.kovac@savba.sk

R. Lyck, G. Enzmann (eds.), *The Blood Brain Barrier and Inflammation*,
Progress in Inflammation Research, DOI 10.1007/978-3-319-45514-3_4

neuroinflammatory diseases including Alzheimer's disease (AD), Parkinson's disease (PD), Huntington's disease (HD), multiple sclerosis (MS), and amyotrophic lateral sclerosis (ALS).

Neuroinflammatory events may begin before significant loss of neural tissue during the process of neurodegeneration, which supports the hypothesis that neuroinflammation might be associated with the progress of neurodegenerative diseases and the modulation of pathogenesis. Whether inflammatory processes modulating BBB permeability precede the process of neurodegeneration or are the consequence of disease pathology remains to be demonstrated.

In neurodegenerative disorders associated with chronic neuroinflammation, immune response driven by glial cells triggers the disruption of the BBB. Inflammatory processes affect the BBB by increasing vascular permeability, enhancing migration of immune cells, altering transport systems, or influencing the role of the BBB as a signaling interface. These changes can range from mild and transient "BBB opening" to chronic breakdown, impairing neuronal activity and leading to neuronal damage and cognitive dysfunction [39]. Proinflammatory signaling molecules, such as cytokines, chemokines, and adhesion molecules produced by glial cells, neurons, and endothelial cells, respectively, cooperate to determine BBB properties and to control leukocyte–endothelial adhesion. These mediators play a prominent role in regulating blood-to-brain cell migration, perpetuating inflammation, and thus exacerbating the disease pathology [23, 104] (Fig. 15.1).

Although the role of neuroinflammation during neurodegeneration remains unclear, findings from experimental models and clinical studies have demonstrated a significant contribution of inflammation to pathological features and symptoms.

2 Multiple Sclerosis

Multiple sclerosis is a human chronic inflammatory disease of the CNS, leading to demyelination and neurodegeneration. MS, as an autoimmune disease, affects both the brain and the spinal cord. The most common form is relapsing-remitting MS, which affects more than 85 % of patients with MS. MS is more common in women than in men [20]. MS occurs in genetically predisposed young adults exposed to unknown environmental triggers [19]. Genome-wide association studies and meta-analysis identified 23 associated loci outside of the human leukocyte antigen genomic region [3, 53, 64].

Neuropathologically, MS is characterized by extensive focal and disseminated infiltration of mononuclear cells in the white and gray matter. Infiltration of autoreactive immune cells causes inflammatory response and neurodegenerative processes characterized by the development of multiple demyelinated plaques found in proximity to blood vessels, significant axonal damage and loss, and finally irreversible damage to the CNS [103]. Acute lesions display disruption of the BBB, as demonstrated by intravenous administration of gadolinium chelate diethylenetriamine pentaacetic acid, a contrast dye that can be visualized by magnetic resonance imaging

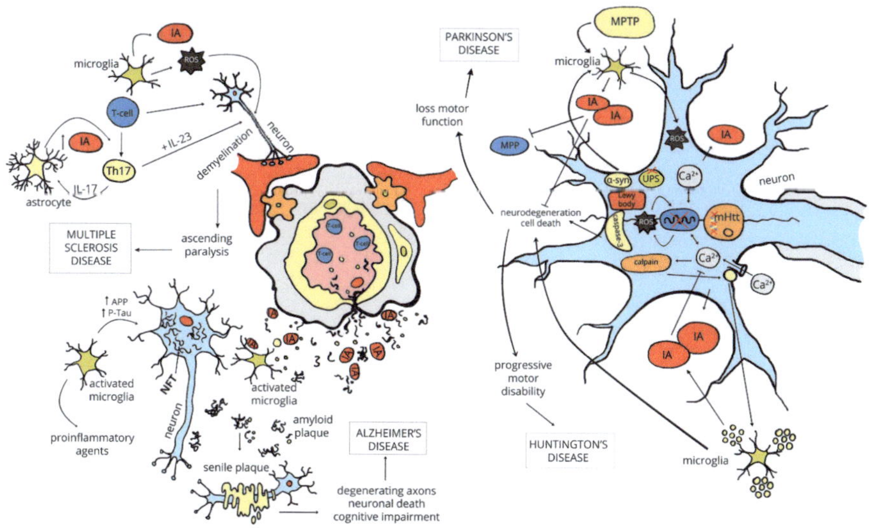

Fig. 15.1 Role of inflammatory processes in CNS diseases. Increased concentration of inflammatory agents (reactive oxygen species, cytokines, chemokines, etc.) is related to numerous neurodegenerative diseases such as Alzheimer´s disease, multiple sclerosis, Parkinson disease´s and Huntington´s disease. **In Multiple Sclerosis** T-cells proliferate and infiltrate the CNS through the upregulation of adhesions molecules on the brain endothelial cells. T-cells in the presence of cytokines differentiate into Th17 cells, which secrete IL 17, that can stimulate further production of inflammatory agents in astrocytes. T-cell contact induces expression of IL-6, reactive oxygen species and nitric oxide in astrocytes, which contribute to damaging myelin sheath on neurons and to fully development of MS. **In Alzheimer's disease**, amyloid-β peptides (Aβ) produced by cleavage of amyloid precursor protein (APP) and misfolded tau species, induce microgliosis, astrogliosis and trigger increased expression of inflammatory agents. Production of inflammatory molecules upregulate APP, further post-translational modifications of tau protein in neuronal cells and neurovascular unit changes. On the other hand, inflammatory agents such as cytokines could have a protective role, they could differentiate microglia into phagocytic cells capable of degrading Aβ and tau. **Huntington's disease** is associated with mutant form of Huntingtin (mHtt) protein. Toxic intracellular polyglutamine inclusions increase the intracellular Ca^{2+} due to NMDA receptor binding, lead to mitochondrial dysfunction with ROS production, and to axonal transport disruption due to mHtt/HAP1 complexes. Subsequently, increased amount of intracellular Ca^{2+} activated enzymes such as caspases and calpains, which finally cleaved mHtt into toxic N-terminal fragments and triggering apoptosis. Microglia cells expressing mHtt contribute to neuronal degeneration. Pathogenesis of **Parkinson's disease** is characterized by abnormal intracellular accumulation of insoluble alpha-synuclein aggregations in the form of Lewy bodies in dopaminergic (DA) neurons due to a mutation. Compare to HD, neuronal death is a result of mitochondrial dysfunction with ROS production, an intracellular increase of Ca^{2+}, oxidative stress and alterations in the ubiquitinproteasomal system. Created alpha-synuclein aggregates trigger microglia cells to produce ROS

(MRI), and postmortem evidence of focal microvascular leakage [75, 82]. Whether BBB dysfunction precedes immune cell infiltration or is the consequence of perivascular leukocyte accumulation remains to be established.

Recruitment of $CD4^+$ T cells into the cerebral interstitium is the most significant consequence of BBB inflammation in MS. In physiological conditions, only a few peripheral immune cells are present in CNS. Nevertheless, the luminal side of BBB is in constant contact with patrolling cells and this immune surveillance is critical for the organism to respond to any pathological process in the CNS [74]. Studies using a rat model for experimental autoimmune encephalomyelitis showed T-cell binding and diapedesis through leptomeningeal vessels and through the BBB [5]. Acute inflammatory lesions are infiltrated mainly by $CD4^+$ and $CD8^+$ T-and B-cells. Active (demyelinating) lesions at a later stage show an abundance of macrophages and reactive proliferating astrocytes.

The migration of immune cells from blood into the brain parenchyma occurs through a process involving tethering, rolling, adhesion, and finally extravasation across the BBB. The capture and rolling are mediated by the selectin family of adhesion molecules and their sulfated, sialylated, and fucosylated glycoprotein ligands [89]. The most efficient tethering molecules are P-selectin and L-selectin. Their most important ligand is P-selectin glycoprotein ligand-1 (PSGL-1), which is glycosylated sialomucin expressed on leukocytes. In vivo studies using mice deficient in PSGL-1 showed that PSGL-1 is the predominant P-selectin ligand expressed during inflammation. The anchoring of rolling leukocytes is achieved by interactions between antigen-4 and vascular cell adhesion molecule (VCAM-1) [99]. Leukocytes extravasate the BBB through tiny spaces into the brain parenchyma [32]. The process is regulated by proinflammatory cytokines [117] and leads to pathological lesions of MS (sclerotic plaques). The plaques growing by radial expansion result in abnormalities in normal-appearing white matter [38, 72].

The interaction of T-cell receptors on migrated $CD4^+$ T lymphocytes with myelin antigens, presented by major histocompatibility complex (MHC) class II expressed on brain-resident microglia and astrocytes leads to the activation of glia, subsequent to an immune attack on the myelin–oligodendrocyte complex and a destructive inflammatory response. The increase in the local concentration of proinflammatory mediators, such as cytokines and chemokines, reactive oxygen species (ROS), and enzymes, induces alterations of the endothelium of the BBB, leads to leukocyte–endothelium interactions, enhanced leukocyte transmigration across the BBB, and perpetuating inflammation, thus exacerbating the MS pathology [82].

The further leukocyte migration may be stimulated by reduced junctional integrity and may contribute to structural modifications of endothelial junctions and thus increased BBB permeability during inflammatory processes.

In active lesions, immune active T-cells, microglia, and astrocytes release Th1 cytokines, including interferon-gamma (IFN-γ), tumor necrosis factor-alpha (TNF-α), interleukin-1-beta (IL-1β), and interleukin-6 (IL-6), that initiate and sustain inflammatory responses. The cytokines induce increased expression of endothelial selectins and immunoglobulin superfamily molecules: intercellular adhesion molecule-1 (ICAM-1) and VCAM-1. IFN-γ can alter the organization of the tight (occludin) and adherens junctions (vascular endothelial cadherin [VE-cadherin]) of endothelial cells, and TNF-α and IL-1β induce expression of nitric oxide synthase, together promoting injury of the BBB [75, 97].

The cytokines act as the main stimuli for chemokine production. Elevated levels of CCL (2, 3, 4, 5, 7, 8) and CXCL10 have been described in MS patients [113].

Chemokines then change low affinity, selectin-mediated interaction of leukocytes with endothelial cells into the higher affinity, integrin-mediated interaction that leads to transendothelial migration of blood-borne cells. Taken together, in MS, cytokines, chemokines, and adhesion molecules cooperate to control leukocyte–endothelial adhesion and transmigration of blood-borne cells through the BBB, thus escalating the disease process.

Alterations of BBB integrity not only involve the alterations of the tight junctions, but also include changes in expression of the ATP-binding cassette transporters. The P-glycoprotein (P-gp) is upregulated on astrocytes and downregulated on endothelial cells within the active and inactive MS lesions, whereas ABCG2 is unaltered on endothelial cells in active lesions and increased in chronic lesions [69].

In summary, observations derived from in vitro experiments, animal models, and patient studies support the hypothesis that BBB disruption represents an early event in MS pathogenesis, preceding the infiltration of blood-borne cells that leads to myelin degradation and destruction of the CNS.

2.1 *Alzheimer's Disease*

Alzheimer's disease, the most common form of dementia, is characterized by cerebrovascular and neuronal dysfunctions leading to a progressive decrease in cognitive functions [7]. On the histopathological level, AD is defined by the presence of extracellular amyloid plaques composed of amyloid-beta (Aβ) peptide aggregates and neurofibrillary tangles formed of hyperphosphorylated, truncated, and aggregated tau protein [51, 54, 87]. In addition to the classic neuropathological features, accumulation of activated immune cells has been documented in the AD brain, indicating a contribution of neuroinflammation to the pathogenesis of this disease [122].

Microglia are the key players in the brain immune system. The loss of cellular branching, transition from ramified to round shape morphology, and modified expression of numerous cell surface receptors are characteristic of activated microglia that are present in areas affected by AD pathology. Clusters of reactive microglia with upregulated expression of a variety of inflammatory cytokines (IL-1β, IL-6, and TNF-α) are often associated with amyloid plaques [96], and in and around neurofibrillary tangles [24, 98]. Activated microglial cells showed increased expression of class II histocompatibility antigen near amyloid deposits in the senile plaque [81]. Microglia in AD also express high levels of MHC class I receptors [112], C3 and C1q [66], IL-1 or ferritin [92]. In AD changes in astrocytes occur. Glial fibrillary acidic protein (GFAP)-positive, hypertrophically activated astrocytes have been located in the proximity of senile plaques. The number of S100 calcium-binding protein B-positive astrocytes correlates with the number of neurofibrillary tangles [42]. However, no significant correlation between GFAP upregulation or excitatory amino-acid transporter 2 (EAAT2) downregulation and amyloid or tau pathology was observed [102].

Transgenic (Tg) animal models recapitulate many neuroinflammatory changes seen in humans. Dense clusters of activated microglia with hundreds of upregulated genes are associated with extracellular deposits of amyloid beta protein in APP23 amyloid Tg mouse. Mutations in one of them, TREM-2, have been linked to the development of dementia [40]. In P301S tau transgenic mice, microglia activation preceded tangle formation, immunosuppression with FK506-attenuated tau pathology, and increased lifespan of the animals [123]. We have shown that expression of truncated tau-induced inflammatory response manifested as upregulation of immune molecules, such as CD11a, CD11b, CD18, CD4, CD45, and CD68. The number of immune reactive microglia and astrocytes progressively increased with neurofibrillary tangle load, suggesting that activated glial cells might be involved in the immune response targeting tau pathology [126]. Reactive astrocytes have been found in the brain parenchyma of transgenic mice overexpressing the London mutant of the amyloid precursor protein, APP [V717I]. These reactive astrocytes produced an increased amount of proinflammatory molecules and upregulated expression of nitric oxide synthase [56].

Besides activation of immune cells, numerous cerebrovascular abnormalities, including endothelial and pericyte damage, reduced glucose transport, increased expression of proinflammatory molecules by activated cells and microvascular degeneration, were observed in AD [12, 127].

The idea that cerebrovascular changes might be the initial events of AD pathogenesis was proposed more than 30 years ago [48]. According to the two-hit vascular hypothesis, vascular changes lead to BBB dysfunction and cerebral hypoperfusion, initiating a cascade of events resulting in dementia [128].

In AD, the brain endothelium is often degenerated and this leads to the accumulation of Aβ on the outer side of the basement membrane of capillaries, promoting a local neuroinflammatory vascular response. A high number of AD patients exhibit vascular pathology and develop cerebral amyloid angiopathy (CAA) and cerebral infarcts. In patients with predominantly capillary CAA, loss of tight junction proteins of the BBB is accompanied by a massive inflammatory response [15].

Inflammatory changes in cerebrovascular endothelium are an integral part of AD pathology. There is an increased immunoreactivity for ICAM-1 and microvessel-associated monocyte chemoattractant protein (MCP-1) on the cerebrovascular endothelium of AD patients [41, 49]. In comparison with non-AD microvessels, the AD microvessels release significantly higher levels of a number of inflammatory factors including TNF-α, transforming growth factor-β (TGF-β), nitric oxide (NO), thrombin, cytokines such as IL-1β, IL-6, IL-8, and matrix metalloproteinases (MMPs) [49]. TGF-β1 is a multifunctional cytokine that has an intense effect on vasculogenesis, angiogenesis, and maintenance of vessel wall integrity. In AD, TGF-β1 has been detected to form part of senile plaques and neurofibrillary tangles [114]. Significantly higher levels of TGF-β1 were also found in serum and CSF of AD patients compared with nondemented elderly controls [17]. The chronic overexpression of TGF-β1 triggered an accumulation of basement membrane proteins and resulted in AD-like cerebrovascular amyloidosis and microvascular degeneration in a Tg mouse model, confirming its critical role in BBB changes seen in AD [121].

Many independent studies showed that various Aβ species are toxic to endothelial cells from the brain [90, 116] or other organs [9, 107, 110]. Treatment of endothelial cells with Aβ has been shown to induce activation of mitogen-activated protein kinases and increased production of proinflammatory cytokines and ROS [80].

Plasma-derived Aβ is transported through the BBB by the receptor for advanced glycation end products (RAGE) [22]. RAGE is upregulated in brain vasculature from AD, suggesting that it might play a role in the accumulation of Aβ within the brain. Interestingly, after interaction of Aβ with RAGE, endothelial cells upregulate expression of C-C chemokine receptor type 5 and MMP-2, which promotes T cells crossing the BBB [27, 78]. Elimination of Aβ from the brain is effected via the bulk flow of CSF and through the transcytosis process mediated by low-density lipoprotein receptor-related protein (LRP-1) and P-gp [33]. Systemic inflammation induced by injection of lipopolysaccharide (LPS) into mice downregulated the expression of transporters LRP-1 and P-gp, which correlated with impaired Aβ efflux [34, 63].

In contrast to Aβ, very little is known about the interaction between BBB and tau. In one of our studies, we showed that exposure of brain endothelial cells to tau does not evoke any significant responses. However, when glial cells were present, inflammatory mediators produced by these cells, such as NO, cytokines ,and chemokines, significantly modified endothelial properties, such as transendothelial electrical resistance and permeability for small molecules [71].

Whether the blood-borne immune cells infiltrate the brain in AD has been highly controversial. Several authors reported that the chronic neuroinflammation seen in neurodegeneration is provided exclusively by resident CNS cells without influx of leukocytes from the blood [28, 105]. Others described hematopoietic cells entering the brain in AD and possibly contributing to inflammatory processes [13, 37, 91, 111]. Recently, accumulating evidence has supported the notion that infiltrating peripheral cells play a significant and critical role in regulating amyloid depositions in the brain [45].

In summary, the inflammatory changes in the cerebrovascular endothelium are common in AD and despite intensive research, the exact mechanisms by which they contribute to the pathogenesis of AD are not completely understood. Moreover, it is clear that BBB and inflammation both play an important role in AD and it is therefore worth putting more effort into understanding their interplay in the course of this devastating neurological disease.

3 Amyotrophic Lateral Sclerosis

Amyotrophic lateral sclerosis is a neurodegenerative disease affecting upper and lower motor neurons in the brain and spinal cord resulting in progressive muscle atrophy, fatal paralysis, and death [94]. Most cases of ALS are sporadic. About 5–10 % are cases of genetically linked familiar ALS that can be caused by mutation in the Cu/Zn superoxide dismutase (SOD1) gene [68, 93].

The neurodegenerative process in ALS is accompanied by sustained inflammation in the brain and spinal cord [11]. In humans and animal models of ALS, gliosis with accumulation of a large number of microglia and astrocytes is observed in brain and spinal cord tissue [76, 124]. Astrocytes in ALS are defective in clearing glutamate because of a loss of EAAT2/GLT1 transporter. Approximately 60–70 % of ALS patients have up to 95 % loss of the EAAT2 protein in the motor cortex and spinal cord. Loss of EAAT2/GLT1 transporter was also described in SOD1 mice and correlates with neuronal loss [79].

The generation and wide use of transgenic rodent models expressing mutant SOD1 has significantly contributed to the understanding of ALS pathogenesis. A functional impairment of the BBB and the blood–spinal cord barrier (BSCB) that might contribute to disease pathogenesis and precede motor neuron death was described in the G93A SOD1 transgenic mouse strain, which carries a human mutant Cu/Zn superoxide dismutase transgene. SOD1 mutant mice display protein aggregates in the mitochondrial intermembrane space [120]. The mitochondria from SOD1 transgenic mice have altered calcium-buffering properties, which have an effect on calcium-mediated excitotoxicity, leading to neuronal death [21]. Other authors have also shown that overexpression of SOD1 in a transgenic mouse model attenuated BBB disruption by superoxide anion during ischemia [67]. Garbuzova-Davis et al. [43] demonstrated capillary alterations and increased albumin permeability in the brainstem and spinal cord at initial (presymptomatic) and late stages (symptomatic) of the disease in SOD1 mice. Electron microscopy showed highly vacuolated and degenerated endothelial cells, perivascular edema, downregulation of tight junction proteins, microhemorrhages, and swelling of astrocyte end-feet adjacent to capillaries. Compared with SOD1 transgenic mice, the SOD1 rat model of ALS demonstrated alterations of the capillaries, such as perivascular swollen astrocyte end-feet, reduced ZO-1 mRNA synthesis and IgG leakage only at a late (symptomatic) stage [86].

These observations were confirmed by Zhong et al. [125], who also showed microvascular barrier damage in the spinal cord preceded by neuroinflammation. Their analyses showed decreased expression of tight junction proteins such as ZO-1, occludin, claudin-5 before disease onset. On the other hand, markers of endothelial activation, such as ICAM-1, and inflammation, such as MCP-1 and cycloxygenase-2 (COX-2), remained unchanged.

Damage to BSCB and BBB was demonstrated in studies on *post-mortem* tissues from sporadic and familiar ALS patients. In the brains of ALS patients, inflammation and activation of immune cells are associated with neuronal death. Studies in the 1980s reported deposits of IgG and C3/C4 complement in the spinal cord and cortex in the brain from ALS patients, suggesting BSCB and BBB damage [25]. Engelhardt and Appel [31] observed perivascular inflammation and breakdown of the BBB, leading to leakage in the brain. They detected the presence of IgG in motor neurons and the presence of activated macrophages, mainly in the territory of degenerating pyramidal tracts and ventral horns. Henkel et al. [57] demonstrated decreased synthesis of the mRNA of occludin and ZO-1 in lumbar spinal cord tissue from ALS patients. Similarly, Garbuzova-Davis et al. [44] showed a significant decrease in the expression of ZO-1, occludin, and claudin-5 proteins in the white and gray matter of the medulla and the cervical spinal cord in patients with the sporadic form of ALS. Angiogenesis, compensating for vascular insufficiency, was also detected. Additionally, the increased expression of P-gp and breast cancer resistance protein (BCRP) was determined in the spinal cords of ALS patients and SOD1 animal models. This suggests that rather than dose adjustments, the combination of P-gp/BCRP inhibitors and anti-ALS therapies might be necessary [62].

The human ALS tissues showed abnormal perivascular accumulation of basement membrane protein collagen IV, possibly resulting from an imbalance between MMPs and the tissue inhibitors of metalloproteinases. This may, over a long period of time, alter the BBB/BSCB transport mechanisms [44]. However, studies showing opposing results are also available [83].

In summary, the inflammatory changes, together with BBB and BCSB damage, are widely observed in humans and animal models of ALS and should be considered a primary target for successful drug development.

3.1 Parkinson's Disease

Parkinson's disease is a complex progressive neurodegenerative disorder characterized by motor symptoms, including bradykinesia with resting tremor, rigidity, and gait disturbance [46]. The major neuropathological hallmarks of PD are progressive degeneration of dopaminergic neurons in the substantia nigra pars compacta (SNpc), the presence of α-synuclein (α-syn) inclusions called Lewy bodies, and chronic inflammation. The cause of PD is unknown, but chronic inflammation can act as an environmental factor and may increase the susceptibility to PD and finally promote the degeneration of dopaminergic neurons. PD can be triggered by diseases that induce systemic infections, such as pneumonia and respiratory and gastrointestinal infections [1].

Inflammatory responses manifested by glial reactions, T-cell infiltration, and increased expression of detrimental proinflammatory cytokines are recognized as prominent features of PD. Activated microglia can be seen in early stages of the disease and parallels the degeneration of dopaminergic neurons [47]. They are distributed not only in the SNpc and putamen, but also in other brain regions of PD patients and are associated with α-syn-positive Lewy neurites [61]. Accumulation of intrinsically disordered protein α-syn actively secreted or released by dying neurons to the extracellular space of the brain leads to microglial activation, CD4+ and CD8+ T-cell infiltration, and increased production of proinflammatory cytokines, such as IL-1α, IFN-γ, IL-1β, TNF-α, and IL-10 [55]. The higher levels of cytokines, mainly IL-1β, TNF-α, and IL-2, were also found in the CSF and serum of PD patients, indicating peripheral inflammation [6, 84, 85]. A massive astrogliosis is present in SNpc in some PD patients [58]; the majority of cases showed only a mild increase in the number of astrocytes and in immunoreactivity for GFAP [109].

The microgliosis and astrogliosis alter BBB permeability. Increased levels of cytotoxic peripheral CD4+ and CD8+ T-lymphocytes infiltrate the SNpc of PD patients and animal models [14]. Clinical studies also demonstrated progressive impairment of barrier integrity and IgG depositions surrounding degenerating neurons during PD progression [50, 70, 77, 88]. Additionally, deficiencies in cerebral blood flow have been demonstrated with PET imaging [2]. These findings support early work by Faucheux et al. [36], who found that PD patients have alterations in the histological appearance of endothelial cells within the SNpc.

Evidence from animal studies indicate a direct link among inflammatory processes, α-syn, and BBB permeability during PD pathogenesis. Peripheral inflammation induced by LPS injection does not have any effect on BBB permeability in α-syn knock-out mice; however, it significantly alters the barrier in wild-type animals [65]. A recent study showed that α-syn can be transported bi-directionally through the BBB, and LRP-1, but not P-gp, may be involved in its efflux from the brain. Interestingly, LPS-induced inflammation increased the uptake of α-syn in the blood-to-brain direction, indicating the possible role of blood-borne α-syn in brain pathology [106]. Increased expression of LRP-1 was observed in PD patients, suggesting that alteration in α-syn transport might contribute to PD pathogenesis [119].

The risk of PD may be influenced by environmental exposure and nongenetic factors. The role of environmental factors in PD development was first described in the 1980s. Various toxin-induced animal PD models, including the 6-hydroxydopamine rats and 1-methyl-1,2,3,6-tetrahydropyridine (MPTP) mice, also show BBB disruption, as demonstrated by increased permeability for FITC-albumin and horse radish peroxidase and decreased expression of the tight junction proteins ZO-1 and occludin [16, 18]. In PD models, neuroinflammation as a consequence of the action of environmental factors is integrally associated within the areas affected by pathology and may be a major contributor to the BBB changes, finally promoting neurodegeneration.

Recent publications showed that there is a decreased expression of P-gp in BBB disruption areas [4, 118]. As P-gp is one of the major efflux transporters at the BBB, the accumulation of xenobiotics, such as MPTP or 1,1-Bis(4-chlorophenyl)-2,2,2-trichloroethane (DDT), in the brain could be partially associated with P-gp reduction or dysfunction [115].

In summary, neuroinflammation and BBB changes are integral parts of PD and should be considered an important therapeutic target in future drug development programs.

3.2 *Huntington's Disease*

Huntington's disease is an autosomal dominant neurodegenerative disease linked to mutations in the *huntingtin* (htt) gene leading to degeneration of neurons, predominantly in the caudate putamen and cortex [52]. The mutant htt causes movement disturbances, psychiatric symptoms and cognitive decline. Although the mechanism by which mutant htt causes neurodegeneration remains unclear, evidence supports inflammation as being a key player in HD pathogenesis. It is possible that increased inflammation in HD brains is a consequence of neuronal death that is a direct result of mutant htt neurotoxicity. On the other hand, accumulation of mutant htt in glia may increase the vulnerability of neurons to excitotoxic stimuli and directly cause inflammation in the CNS [100].

The role of inflammation in HD pathogenesis was supported by microarray profiling, which revealed expression of inflammation-related genes in brain regions from HD patients [59]. Postmortem studies of HD brains revealed accumulation of activated microglial cells in regions affected by HD, especially in the basal ganglia and the frontal cortex [95]. The presence of immunoreactive microglia was seen in the presymptomatic stage of HD and increased as it progressed [108]. Increased microglial activation was also shown in a R6/2 mouse model of HD [101]. Astrocyte reactivity is an early feature of HD. GFAP immunoreactivity is detected in the striatum of presymptomatic carriers and it increases with disease progression [35]. Furthermore, the astrocytes from HD produce more VEGF through an IkB kinase-nuclear factor kB-dependent pathway [60]. Interestingly, no clear evidence for the activation of astrocytes in most models of HD exists.

Cytokines are increased in HD. Bjorkqvist et al. [8] determined increased amounts of proinflammatory cytokines, such as IL-6 and IL-8, in plasma samples and striatum. Both cytokines, IL-6 and IL-1b, are also increased in the R6/2 mouse. Studies with neutralizing antibody confirmed the hypothesis that IL-6 produced by peripheral immune cells might contribute to pathology in the R6/2 model [10]. IL-1b, another member of the proinflammatory cytokine family, is increased in HD sera and in brain lysates of the R6/2 model [29]. Increased production of proinflammatory cytokines together with impaired migration properties of microglia and peripheral monocytes [73] may lead to chronic exacerbated inflammation, and thus contribute to HD pathology.

Two recent studies investigated the impairment of BBB in HD. Drouin-Ouellet et al. [26] found that mutant huntingtin protein aggregates were present in components of the neurovascular unit of R6/2 mice and HD patients. This was accompanied by an increase in blood vessel density, in addition to BBB leakage in the striatum of R6/2 mice, which correlated with the decreased expression of occludin and claudin-5. The study revealed a significant increase in cerebral blood flow in the cortical gray matter of HD patients. The results published by Hsiao et al. [60] further broaden the field, by measuring the blood vessel density and vascular reactivity using MRI. The results in several different knock-in models indicate that vascular density and reactivity are noticeably changed when mutant htt is expressed in both neurons and astrocytes.

In summary, all the above-mentioned studies clearly demonstrated that BBB is compromised in both HD patients and animal models of the disease. However, further studies are needed to investigate at what stage of the disease this process begins.

4 Conclusion

It is becoming increasingly evident that neuroinflammation plays a crucial role in the development and progression of many neurodegenerative disorders. Chronic neuroinflammation associated with neuronal damage includes extended activation of microglia and astrocytes followed by increased secretion of detrimental proinflammatory cytokines and chemokines. The prolonged inflammation affects the BBB, which in turn supports the infiltration of blood-borne cells into the brain parenchyma that further intensifies the inflammatory process. In future research, suppression of the inflammatory events at the site of the BBB should be explored as a therapeutic strategy against neuroinflammatory diseases.

References

1. Arai H, Furuya T, Mizuno Y, Mochizuki H (2006) Inflammation and infection in Parkinson's disease. Histol Histopathol 21(6):673–678
2. Ballanger B, Lozano AM, Moro E, van Eimeren T, Hamani C, Chen R, Cilia R, Houle S, Poon YY, Lang AE, Strafella AP (2009) Cerebral blood flow changes induced by pedunculopontine nucleus stimulation in patients with advanced Parkinson's disease: a [(15)O] H2O PET study. Hum Brain Mapp 30(12):3901–3909. doi:10.1002/hbm.20815
3. Baranzini SE, Galwey NW, Wang J, Khankhanian P, Lindberg R, Pelletier D et al (2009) Pathway and network-based analysis of genome-wide association studies in multiple sclerosis. Hum Mol Genet 18(11):2078–2090. Epub 2009/03/17
4. Bartels AL, Willemsen AT, Kortekaas R, de Jong BM, de Vries R, de Klerk O, van Oostrom JC, Portman A, Leenders KL (2008) Decreased blood-brain barrier P-glycoprotein function in the progression of Parkinson's disease, PSP and MSA. J Neural Transm 115(7):1001–1009. doi:10.1007/s00702-008-0030-y
5. Bartholomaus I, Kawakami N, Odoardi F, Schlager C, Miljkovic D, Ellwart JW, Klinkert WE, Flugel-Koch C, Issekutz TB, Wekerle H, Flugel A (2009) Effector T cell interactions with meningeal vascular structures in nascent autoimmune CNS lesions. Nature 462(7269):94–98. doi:10.1038/nature08478
6. Bas J, Calopa M, Mestre M, Mollevi DG, Cutillas B, Ambrosio S, Buendia E (2001) Lymphocyte populations in Parkinson's disease and in rat models of parkinsonism. J Neuroimmunol 113(1):146–152
7. Bell RD, Zlokovic BV (2009) Neurovascular mechanisms and blood-brain barrier disorder in Alzheimer's disease. Acta Neuropathol 118(1):103–113. doi:10.1007/s00401-009-0522-3
8. Bjorkqvist M, Wild EJ, Thiele J, Silvestroni A, Andre R, Lahiri N et al (2008) A novel pathogenic pathway of immune activation detectable before clinical onset in Huntington's disease. J Exp Med 205(8):1869–1877. Epub 2008/07/16
9. Blanc EM, Toborek M, Mark RJ, Hennig B, Mattson MP (1997) Amyloid beta-peptide induces cell monolayer albumin permeability, impairs glucose transport, and induces apoptosis in vascular endothelial cells. J Neurochem 68(5):1870–1881
10. Bouchard J, Truong J, Bouchard K, Dunkelberger D, Desrayaud S, Moussaoui S et al (2012) Cannabinoid receptor 2 signaling in peripheral immune cells modulates disease onset and severity in mouse models of Huntington's disease. J Neurosci Off J Soc Neurosci 32(50):18259–18268. Epub 2012/12/15
11. Bowerman M, Vincent T, Scamps F, Perrin FE, Camu W, Raoul C (2013) Neuroimmunity dynamics and the development of therapeutic strategies for amyotrophic lateral sclerosis. Front Cell Neurosci 7:214. doi:10.3389/fncel.2013.00214

12. Bowman GL, Quinn JF (2008) Alzheimer's disease and the blood-brain barrier: past, present and future. Aging Health 4(1):47–55. doi:10.2217/1745509X.4.1.47
13. Britschgi M, Wyss-Coray T (2007) Systemic and acquired immune responses in Alzheimer's disease. Int Rev Neurobiol 82:205–233. doi:10.1016/S0074-7742(07)82011-3
14. Brochard V, Combadiere B, Prigent A, Laouar Y, Perrin A, Beray-Berthat V, Bonduelle O, Alvarez-Fischer D, Callebert J, Launay JM, Duyckaerts C, Flavell RA, Hirsch EC, Hunot S (2009) Infiltration of CD4+ lymphocytes into the brain contributes to neurodegeneration in a mouse model of Parkinson disease. J Clin Invest 119(1):182–192. doi:10.1172/JCI36470
15. Carrano A, Hoozemans JJ, van der Vies SM, van Horssen J, de Vries HE, Rozemuller AJ (2012) Neuroinflammation and blood-brain barrier changes in capillary amyloid angiopathy. Neuro Degener Dis 10(1-4):329–331. doi:10.1159/000334916
16. Carvey PM, Zhao CH, Hendey B, Lum H, Trachtenberg J, Desai BS, Snyder J, Zhu YG, Ling ZD (2005) 6-Hydroxydopamine-induced alterations in blood-brain barrier permeability. Eur J Neurosci 22(5):1158–1168. doi:10.1111/j.1460-9568.2005.04281.x
17. Chao CC, Hu S, Frey WH 2nd, Ala TA, Tourtellotte WW, Peterson PK (1994) Transforming growth factor beta in Alzheimer's disease. Clin Diagn Lab Immunol 1(1):109–110
18. Chen X, Lan X, Roche I, Liu R, Geiger JD (2008) Caffeine protects against MPTP-induced blood-brain barrier dysfunction in mouse striatum. J Neurochem 107(4):1147–1157. doi:10.1111/j.1471-4159.2008.05697.x
19. Compston A, Coles A (2008) Multiple sclerosis. Lancet 372(9648):1502–1517. doi:10.1016/S0140-6736(08)61620-7
20. Cruz-Orengo L, Daniels BP, Dorsey D, Basak SA, Grajales-Reyes JG, McCandless EE, Piccio L, Schmidt RE, Cross AH, Crosby SD, Klein RS (2014) Enhanced sphingosine-1-phosphate receptor 2 expression underlies female CNS autoimmunity susceptibility. J Clin Invest 124(6):2571–2584. doi:10.1172/JCI73408
21. Damiano M, Starkov AA, Petri S, Kipiani K, Kiaei M, Mattiazzi M et al (2006) Neural mitochondrial $Ca2+$ capacity impairment precedes the onset of motor symptoms in G93A Cu/Zn-superoxide dismutase mutant mice. J Neurochem 96(5):1349–1361. Epub 2006/02/16
22. Deane R, Du Yan S, Submamaryan RK, LaRue B, Jovanovic S, Hogg E, Welch D, Manness L, Lin C, Yu J, Zhu H, Ghiso J, Frangione B, Stern A, Schmidt AM, Armstrong DL, Arnold B, Liliensiek B, Nawroth P, Hofman F, Kindy M, Stern D, Zlokovic B (2003) RAGE mediates amyloid-beta peptide transport across the blood-brain barrier and accumulation in brain. Nat Med 9(7):907–913. doi:10.1038/nm890
23. Dietrich JB (2002) The adhesion molecule ICAM-1 and its regulation in relation with the blood-brain barrier. J Neuroimmunol 128(1-2):58–68
24. DiPatre PL, Gelman BB (1997) Microglial cell activation in aging and Alzheimer disease: partial linkage with neurofibrillary tangle burden in the hippocampus. J Neuropathol Exp Neurol 56(2):143–149
25. Donnenfeld H, Kascsak RJ, Bartfeld H (1984) Deposits of IgG and C3 in the spinal cord and motor cortex of ALS patients. J Neuroimmunol 6(1):51–57
26. Drouin-Ouellet J, Sawiak SJ, Cisbani G, Lagace M, Kuan WL, Saint-Pierre M et al (2015) Cerebrovascular and blood-brain barrier impairments in Huntington's disease: potential implications for its pathophysiology. Ann Neurol 78(2):160–177. Epub 2015/04/14
27. Du H, Li P, Wang J, Qing X, Li W (2012) The interaction of amyloid beta and the receptor for advanced glycation endproducts induces matrix metalloproteinase-2 expression in brain endothelial cells. Cell Mol Neurobiol 32(1):141–147. doi:10.1007/s10571-011-9744-8
28. Eikelenboom P, Veerhuis R, Scheper W, Rozemuller AJ, van Gool WA, Hoozemans JJ (2006) The significance of neuroinflammation in understanding Alzheimer's disease. J Neural Transm 113(11):1685–1695. doi:10.1007/s00702-006-0575-6
29. Ellrichmann G, Reick C, Saft C, Linker RA (2013) The role of the immune system in Huntington's disease. Clin Dev Immunol 2013:541259. Epub 2013/08/21
30. Engelhardt B (2008) The blood-central nervous system barriers actively control immune cell entry into the central nervous system. Curr Pharm Des 14(16):1555–1565
31. Engelhardt JI, Appel SH (1990) IgG reactivity in the spinal cord and motor cortex in amyotrophic lateral sclerosis. Arch Neurol 47(11):1210–1216

32. Engelhardt B, Ransohoff RM (2005) The ins and outs of T-lymphocyte trafficking to the CNS: anatomical sites and molecular mechanisms. Trends Immunol 26(9):485–495. Epub 2005/07/26
33. Erickson MA, Banks WA (2013) Blood-brain barrier dysfunction as a cause and consequence of Alzheimer's disease. J Cereb Blood Flow Metab 33(10):1500–1513. doi:10.1038/jcbfm.2013.135
34. Erickson MA, Hartvigson PE, Morofuji Y, Owen JB, Butterfield DA, Banks WA (2012) Lipopolysaccharide impairs amyloid beta efflux from brain: altered vascular sequestration, cerebrospinal fluid reabsorption, peripheral clearance and transporter function at the blood-brain barrier. J Neuroinflammation 9:150. doi:10.1186/1742-2094-9-150
35. Faideau M, Kim J, Cormier K, Gilmore R, Welch M, Auregan G, Dufour N, Guillermier M, Brouillet E, Hantraye P, Deglon N, Ferrante RJ, Bonvento G (2010) In vivo expression of polyglutamine-expanded huntingtin by mouse striatal astrocytes impairs glutamate transport: a correlation with Huntington's disease subjects. Hum Mol Genet 19(15):3053–3067. doi:10.1093/hmg/ddq212
36. Faucheux BA, Bonnet AM, Agid Y, Hirsch EC (1999) Blood vessels change in the mesencephalon of patients with Parkinson's disease. Lancet 353(9157):981–982. doi:10.1016/S0140-6736(99)00641-8
37. Fiala M, Liu QN, Sayre J, Pop V, Brahmandam V, Graves MC, Vinters HV (2002) Cyclooxygenase-2-positive macrophages infiltrate the Alzheimer's disease brain and damage the blood-brain barrier. Eur J Clin Invest 32(5):360–371
38. Filli L, Hofstetter L, Kuster P, Traud S, Mueller-Lenke N, Naegelin Y et al (2012) Spatiotemporal distribution of white matter lesions in relapsing-remitting and secondary progressive multiple sclerosis. Mult Scler 18(11):1577–1584. Epub 2012/04/13
39. Forster C (2008) Tight junctions and the modulation of barrier function in disease. Histochem Cell Biol 130(1):55–70. doi:10.1007/s00418-008-0424-9
40. Frank S, Burbach GJ, Bonin M, Walter M, Streit W, Bechmann I, Deller T (2008) TREM2 is upregulated in amyloid plaque-associated microglia in aged APP23 transgenic mice. Glia 56(13):1438–1447. doi:10.1002/glia.20710
41. Frohman EM, Frohman TC, Gupta S, de Fougerolles A, van den Noort S (1991) Expression of intercellular adhesion molecule 1 (ICAM-1) in Alzheimer's disease. J Neurol Sci 106(1):105–111
42. Fuller S, Munch G, Steele M (2009) Activated astrocytes: a therapeutic target in Alzheimer's disease? Expert Rev Neurother 9(11):1585–1594. doi:10.1586/ern.09.111
43. Garbuzova-Davis S, Haller E, Saporta S, Kolomey I, Nicosia SV, Sanberg PR (2007) Ultrastructure of blood-brain barrier and blood-spinal cord barrier in SOD1 mice modeling ALS. Brain Res 1157:126–137. doi:10.1016/j.brainres.2007.04.044
44. Garbuzova-Davis S, Hernandez-Ontiveros DG, Rodrigues MC, Haller E, Frisina-Deyo A, Mirtyl S, Sallot S, Saporta S, Borlongan CV, Sanberg PR (2012) Impaired blood-brain/spinal cord barrier in ALS patients. Brain Res 1469:114–128. doi:10.1016/j.brainres.2012.05.056
45. Gate D, Rezai-Zadeh K, Jodry D, Rentsendorj A, Town T (2010) Macrophages in Alzheimer's disease: the blood-borne identity. J Neural Transm 117(8):961–970. doi:10.1007/s00702-010-0422-7
46. Gelb DJ, Oliver E, Gilman S (1999) Diagnostic criteria for Parkinson disease. Arch Neurol 56(1):33–39
47. Gerhard A, Pavese N, Hotton G, Turkheimer F, Es M, Hammers A, Eggert K, Oertel W, Banati RB, Brooks DJ (2006) In vivo imaging of microglial activation with [11C](R)-PK11195 PET in idiopathic Parkinson's disease. Neurobiol Dis 21(2):404–412. doi:10.1016/j.nbd.2005.08.002
48. Grammas P (2011) Neurovascular dysfunction, inflammation and endothelial activation: implications for the pathogenesis of Alzheimer's disease. J Neuroinflammation 8:26. doi:10.1186/1742-2094-8-26
49. Grammas P, Ovase R (2001) Inflammatory factors are elevated in brain microvessels in Alzheimer's disease. Neurobiol Aging 22(6):837–842

50. Gray MT, Woulfe JM (2015) Striatal blood-brain barrier permeability in Parkinson's disease. J Cereb Blood Flow Metab 35(5):747–750. doi:10.1038/jcbfm.2015.32
51. Grundke-Iqbal I, Iqbal K, Quinlan M, Tung YC, Zaidi MS, Wisniewski HM (1986) Microtubule-associated protein tau. A component of Alzheimer paired helical filaments. J Biol Chem 261(13):6084–6089
52. Gusella JF, MacDonald ME (1993) Hunting for Huntington's disease. Mol Genet Med 3:139–158. Epub 1993/01/01
53. Hafler DA, Compston A, Sawcer S, Lander ES, Daly MJ, De Jager PL et al (2007) Risk alleles for multiple sclerosis identified by a genomewide study. N Engl J Med 357(9):851–862. Epub 2007/07/31
54. Hardy JA, Higgins GA (1992) Alzheimer's disease: the amyloid cascade hypothesis. Science 256(5054):184–185
55. Harms AS, Cao S, Rowse AL, Thome AD, Li X, Mangieri LR, Cron RQ, Shacka JJ, Raman C, Standaert DG (2013) MHCII is required for alpha-synuclein-induced activation of microglia, CD4 T cell proliferation, and dopaminergic neurodegeneration. J Neurosci 33(23):9592–9600. doi:10.1523/JNEUROSCI.5610-12.2013
56. Heneka MT, Sastre M, Dumitrescu-Ozimek L, Dewachter I, Walter J, Klockgether T, Van Leuven F (2005) Focal glial activation coincides with increased BACE1 activation and precedes amyloid plaque deposition in APP[V717I] transgenic mice. J Neuroinflammation 2:22. doi:10.1186/1742-2094-2-22
57. Henkel JS, Beers DR, Wen S, Bowser R, Appel SH (2009) Decreased mRNA expression of tight junction proteins in lumbar spinal cords of patients with ALS. Neurology 72(18):1614–1616. doi:10.1212/WNL.0b013e3181a41228
58. Hirsch EC, Breidert T, Rousselet E, Hunot S, Hartmann A, Michel PP. (2003) The Role of Glial Reaction and Inflammation in Parkinson's Disease. Annals of the New York Academy of Sciences, 991: 214–228
59. Hodges A, Strand AD, Aragaki AK, Kuhn A, Sengstag T, Hughes G et al (2006) Regional and cellular gene expression changes in human Huntington's disease brain. Hum Mol Genet 15(6):965–977. Epub 2006/02/10
60. Hsiao HY, Chen YC, Huang CH, Chen CC, Hsu YH, Chen HM, Chiu FL, Kuo HC, Chang C, Chern Y (2015) Aberrant astrocytes impair vascular reactivity in Huntington disease. Ann Neurol 78(2):178–192. doi:10.1002/ana.24428
61. Imamura K, Hishikawa N, Sawada M, Nagatsu T, Yoshida M, Hashizume Y (2003) Distribution of major histocompatibility complex class II-positive microglia and cytokine profile of Parkinson's disease brains. Acta Neuropathol 106(6):518–526. doi:10.1007/s00401-003-0766-2
62. Jablonski MR, Jacob DA, Campos C, Miller DS, Maragakis NJ, Pasinelli P, Trotti D (2012) Selective increase of two ABC drug efflux transporters at the blood-spinal cord barrier suggests induced pharmacoresistance in ALS. Neurobiol Dis 47(2):194–200. doi:10.1016/j.nbd.2012.03.040
63. Jaeger LB, Dohgu S, Sultana R, Lynch JL, Owen JB, Erickson MA, Shah GN, Price TO, Fleegal-Demotta MA, Butterfield DA, Banks WA (2009) Lipopolysaccharide alters the blood-brain barrier transport of amyloid beta protein: a mechanism for inflammation in the progression of Alzheimer's disease. Brain Behav Immun 23(4):507–517. doi:10.1016/j.bbi.2009.01.017
64. Jakkula E, Leppa V, Sulonen AM, Varilo T, Kallio S, Kemppinen A et al (2010) Genome-wide association study in a high-risk isolate for multiple sclerosis reveals associated variants in STAT3 gene. Am J Hum Genet 86(2):285–291. Epub 2010/02/18
65. Jangula A, Murphy EJ (2013) Lipopolysaccharide-induced blood brain barrier permeability is enhanced by alpha-synuclein expression. Neurosci Lett 551:23–27. doi:10.1016/j.neulet.2013.06.058
66. Johnson SA, Lampert-Etchells M, Pasinetti GM, Rozovsky I, Finch CE (1992) Complement mRNA in the mammalian brain: responses to Alzheimer's disease and experimental brain lesioning. Neurobiol Aging 13(6):641–648

67. Kim GW, Lewen A, Copin J, Watson BD, Chan PH (2001) The cytosolic antioxidant, copper/zinc superoxide dismutase, attenuates blood-brain barrier disruption and oxidative cellular injury after photothrombotic cortical ischemia in mice. Neuroscience 105(4):1007–1018. Epub 2001/09/01
68. Komine O, Yamanaka K (2015) Neuroinflammation in motor neuron disease. Nagoya J Med Sci 77(4):537–549
69. Kooij G, Mizee MR, van Horssen J, Reijerkerk A, Witte ME, Drexhage JA, van der Pol SM, van Het Hof B, Scheffer G, Scheper R, Dijkstra CD, van der Valk P, de Vries HE (2011) Adenosine triphosphate-binding cassette transporters mediate chemokine (C-C motif) ligand 2 secretion from reactive astrocytes: relevance to multiple sclerosis pathogenesis. Brain 134(Pt 2):555–570. doi:10.1093/brain/awq330
70. Kortekaas R, Leenders KL, van Oostrom JC, Vaalburg W, Bart J, Willemsen AT, Hendrikse NH (2005) Blood-brain barrier dysfunction in parkinsonian midbrain in vivo. Ann Neurol 57(2):176–179. doi:10.1002/ana.20369
71. Kovac A, Zilkova M, Deli MA, Zilka N, Novak M (2009) Human truncated tau is using a different mechanism from amyloid-beta to damage the blood-brain barrier. J Alzheimer's Dis JAD 18(4):897–906. doi:10.3233/JAD-2009-1197
72. Kutzelnigg A, Lassmann H (2005) Cortical lesions and brain atrophy in MS. J Neurol Sci 233(1–2):55–59. Epub 2005/05/17
73. Kwan W, Trager U, Davalos D, Chou A, Bouchard J, Andre R et al (2012) Mutant huntingtin impairs immune cell migration in Huntington disease. J Clin Invest 122(12):4737–4747. Epub 2012/11/20
74. Lampron A, Elali A, Rivest S (2013) Innate immunity in the CNS: redefining the relationship between the CNS and Its environment. Neuron 78(2):214–232. Epub 2013/04/30
75. Larochelle C, Alvarez JI, Prat A (2011) How do immune cells overcome the blood-brain barrier in multiple sclerosis? FEBS Lett 585(23):3770–3780. doi:10.1016/j.febslet.2011.04.066. Epub 2011/05/10
76. Lasiene J, Yamanaka K (2011) Glial cells in amyotrophic lateral sclerosis. Neurol Res Int 2011:718987. doi:10.1155/2011/718987
77. Lee H, Pienaar IS (2014) Disruption of the blood-brain barrier in Parkinson's disease: curse or route to a cure? Front Biosci 19:272–280
78. Li H, Ke H, Ren G, Qiu X, Weng YX, Wang CC (2009) Thermal-induced dissociation and unfolding of homodimeric DsbC revealed by temperature-jump time-resolved infrared spectra. Biophys J 97(10):2811–2819. doi:10.1016/j.bpj.2009.08.049
79. Lin CL, Kong Q, Cuny GD, Glicksman MA (2012) Glutamate transporter EAAT2: a new target for the treatment of neurodegenerative diseases. Future Med Chem 4(13):1689–1700. doi:10.4155/fmc.12.122
80. Liu R, Li JZ, Song JK, Sun JL, Li YJ, Zhou SB, Zhang TT, Du GH (2014) Pinocembrin protects human brain microvascular endothelial cells against fibrillar amyloid-beta(1-40) injury by suppressing the MAPK/NF-kappaB inflammatory pathways. Biomed Res Int 2014:470393. doi:10.1155/2014/470393
81. McGeer PL, McGeer EG (2011) History of innate immunity in neurodegenerative disorders. Front Pharmacol 2:77. doi:10.3389/fphar.2011.00077
82. Minagar A, Alexander JS (2003) Blood-brain barrier disruption in multiple sclerosis. Mult Scler 9(6):540–549
83. Miyazaki K, Ohta Y, Nagai M, Morimoto N, Kurata T, Takehisa Y, Ikeda Y, Matsuura T, Abe K (2011) Disruption of neurovascular unit prior to motor neuron degeneration in amyotrophic lateral sclerosis. J Neurosci Res 89(5):718–728. doi:10.1002/jnr.22594
84. Mogi M, Harada M, Kondo T, Riederer P, Inagaki H, Minami M, Nagatsu T (1994) Interleukin-1 beta, interleukin-6, epidermal growth factor and transforming growth factor-alpha are elevated in the brain from parkinsonian patients. Neurosci Lett 180(2):147–150
85. Mogi M, Harada M, Narabayashi H, Inagaki H, Minami M, Nagatsu T (1996) Interleukin (IL)-1 beta, IL-2, IL-4, IL-6 and transforming growth factor-alpha levels are elevated in ven-

tricular cerebrospinal fluid in juvenile parkinsonism and Parkinson's disease. Neurosci Lett 211(1):13–16
86. Nicaise C, Mitrecic D, Demetter P, De Decker R, Authelet M, Boom A, Pochet R (2009) Impaired blood-brain and blood-spinal cord barriers in mutant SOD1-linked ALS rat. Brain Res 1301:152–162. doi:10.1016/j.brainres.2009.09.018
87. Novak M, Jakes R, Edwards PC, Milstein C, Wischik CM (1991) Difference between the tau protein of Alzheimer paired helical filament core and normal tau revealed by epitope analysis of monoclonal antibodies 423 and 7.51. Proc Natl Acad Sci U S A 88(13):5837–5841
88. Orr CF, Rowe DB, Mizuno Y, Mori H, Halliday GM (2005) A possible role for humoral immunity in the pathogenesis of Parkinson's disease. Brain J Neurol 128(Pt 11):2665–2674. Epub 2005/10/13
89. Palmer AM (2013) Multiple sclerosis and the blood-central nervous system barrier. Cardiovasc Psychiatry Neurol 2013:530356. Epub 2013/02/13
90. Qosa H, LeVine H 3rd, Keller JN, Kaddoumi A (2014) Mixed oligomers and monomeric amyloid-beta disrupts endothelial cells integrity and reduces monomeric amyloid-beta transport across hCMEC/D3 cell line as an in vitro blood-brain barrier model. Biochim Biophys Acta 1842(9):1806–1815. doi:10.1016/j.bbadis.2014.06.029
91. Ray S, Britschgi M, Herbert C, Takeda-Uchimura Y, Boxer A, Blennow K, Friedman LF, Galasko DR, Jutel M, Karydas A, Kaye JA, Leszek J, Miller BL, Minthon L, Quinn JF, Rabinovici GD, Robinson WH, Sabbagh MN, So YT, Sparks DL, Tabaton M, Tinklenberg J, Yesavage JA, Tibshirani R, Wyss-Coray T (2007) Classification and prediction of clinical Alzheimer's diagnosis based on plasma signaling proteins. Nat Med 13(11):1359–1362. doi:10.1038/nm1653
92. Rogers JT, Bush AI, Cho HH, Smith DH, Thomson AM, Friedlich AL, Lahiri DK, Leedman PJ, Huang X, Cahill CM (2008) Iron and the translation of the amyloid precursor protein (APP) and ferritin mRNAs: riboregulation against neural oxidative damage in Alzheimer's disease. Biochem Soc Trans 36(Pt 6):1282–1287. doi:10.1042/BST0361282
93. Rosen DR, Siddique T, Patterson D, Figlewicz DA, Sapp P, Hentati A, Donaldson D, Goto J, O'Regan JP, Deng HX et al (1993) Mutations in Cu/Zn superoxide dismutase gene are associated with familial amyotrophic lateral sclerosis. Nature 362(6415):59–62. doi:10.1038/362059a0
94. Rowland LP, Shneider NA (2001) Amyotrophic lateral sclerosis. N Engl J Med 344(22):1688–1700. doi:10.1056/NEJM200105313442207
95. Sapp E, Kegel KB, Aronin N, Hashikawa T, Uchiyama Y, Tohyama K et al (2001) Early and progressive accumulation of reactive microglia in the Huntington disease brain. J Neuropathol Exp Neurol 60(2):161–172. Epub 2001/03/29
96. Serrano-Pozo A, Frosch MP, Masliah E, Hyman BT (2011) Neuropathological alterations in Alzheimer disease. Cold Spring Harbor Perspect Med 1(1):a006189. doi:10.1101/cshperspect.a006189
97. Sharief MK (1998) Cytokines in multiple sclerosis: pro-inflammation or pro-remyelination? Mult Scler 4(3):169–173
98. Sheffield LG, Marquis JG, Berman NE (2000) Regional distribution of cortical microglia parallels that of neurofibrillary tangles in Alzheimer's disease. Neurosci Lett 285(3):165–168
99. Sheremata WA, Minagar A, Alexander JS, Vollmer T (2005) The role of alpha-4 integrin in the aetiology of multiple sclerosis: current knowledge and therapeutic implications. CNS Drugs 19(11):909–922. Epub 2005/11/05
100. Shin JY, Fang ZH, Yu ZX, Wang CE, Li SH, Li XJ (2005) Expression of mutant huntingtin in glial cells contributes to neuronal excitotoxicity. J Cell Biol 171(6):1001–1012. Epub 2005/12/21
101. Simmons DA, Casale M, Alcon B, Pham N, Narayan N, Lynch G (2007) Ferritin accumulation in dystrophic microglia is an early event in the development of Huntington's disease. Glia 55(10):1074–1084. Epub 2007/06/07

102. Simpson JE, Ince PG, Lace G, Forster G, Shaw PJ, Matthews F, Savva G, Brayne C, Wharton SB, Function MRCC, Ageing Neuropathology Study G (2010) Astrocyte phenotype in relation to Alzheimer-type pathology in the ageing brain. Neurobiol Aging 31(4):578–590. doi:10.1016/j.neurobiolaging.2008.05.015
103. Sospedra M, Martin R (2005) Immunology of multiple sclerosis. Annu Rev Immunol 23:683–747. doi:10.1146/annurev.immunol.23.021704.115707
104. Stanimirovic D, Satoh K (2000) Inflammatory mediators of cerebral endothelium: a role in ischemic brain inflammation. Brain Pathol 10(1):113–126
105. Streit WJ, Mrak RE, Griffin WS (2004) Microglia and neuroinflammation: a pathological perspective. J Neuroinflammation 1(1):14. doi:10.1186/1742-2094-1-14
106. Sui YT, Bullock KM, Erickson MA, Zhang J, Banks WA (2014) Alpha synuclein is transported into and out of the brain by the blood-brain barrier. Peptides 62:197–202. doi:10.1016/j.peptides.2014.09.018
107. Suo Z, Fang C, Crawford F, Mullan M (1997) Superoxide free radical and intracellular calcium mediate A beta(1-42) induced endothelial toxicity. Brain Res 762(1-2):144–152
108. Tai YF, Pavese N, Gerhard A, Tabrizi SJ, Barker RA, Brooks DJ et al (2007) Microglial activation in presymptomatic Huntington's disease gene carriers. Brain J Neurol 130(Pt 7):1759–1766. Epub 2007/04/03
109. Teismann P, Schulz JB (2004) Cellular pathology of Parkinson's disease: astrocytes, microglia and inflammation. Cell Tissue Res 318(1):149–161. Epub 2004/09/01
110. Thomas T, Thomas G, McLendon C, Sutton T, Mullan M (1996) beta-Amyloid-mediated vasoactivity and vascular endothelial damage. Nature 380(6570):168–171. doi:10.1038/380168a0
111. Togo T, Akiyama H, Iseki E, Kondo H, Ikeda K, Kato M, Oda T, Tsuchiya K, Kosaka K (2002) Occurrence of T cells in the brain of Alzheimer's disease and other neurological diseases. J Neuroimmunol 124(1-2):83–92
112. Tooyama I, Kimura H, Akiyama H, McGeer PL (1990) Reactive microglia express class I and class II major histocompatibility complex antigens in Alzheimer's disease. Brain Res 523(2):273–280
113. Trebst C, Ransohoff RM (2001) Investigating chemokines and chemokine receptors in patients with multiple sclerosis: opportunities and challenges. Arch Neurol 58(12):1975–1980
114. van der Wal EA, Gomez-Pinilla F, Cotman CW (1993) Transforming growth factor-beta 1 is in plaques in Alzheimer and Down pathologies. Neuroreport 4(1):69–72
115. Vautier S, Fernandez C (2009) ABCB1: the role in Parkinson's disease and pharmacokinetics of antiparkinsonian drugs. Expert Opin Drug Metab Toxicol 5(11):1349–1358. doi:10.1517/17425250903193079
116. Veszelka S, Toth AE, Walter FR, Datki Z, Mozes E, Fulop L, Bozso Z, Hellinger E, Vastag M, Orsolits B, Kornyei Z, Penke B, Deli MA (2013) Docosahexaenoic acid reduces amyloid-beta induced toxicity in cells of the neurovascular unit. J Alzheimer's Dis JAD 36(3):487–501. doi:10.3233/JAD-120163
117. Wekerle H (2005) Immune pathogenesis of multiple sclerosis. Neurol Sci Off J Ital Neurol Soc Ital Soc Clin Neurophysiol 26(Suppl 1):S1–S2. Epub 2005/05/11
118. Westerlund M, Belin AC, Olson L, Galter D (2008) Expression of multi-drug resistance 1 mRNA in human and rodent tissues: reduced levels in Parkinson patients. Cell Tissue Res 334(2):179–185. doi:10.1007/s00441-008-0686-5
119. Wilhelmus MM, Bol JG, Van Haastert ES, Rozemuller AJ, Bu G, Drukarch B, Hoozemans JJ (2011) Apolipoprotein E and LRP1 increase early in Parkinson's disease pathogenesis. Am J Pathol 179(5):2152–2156. doi:10.1016/j.ajpath.2011.07.021
120. Wong PC, Pardo CA, Borchelt DR, Lee MK, Copeland NG, Jenkins NA et al (1995) An adverse property of a familial ALS-linked SOD1 mutation causes motor neuron disease characterized by vacuolar degeneration of mitochondria. Neuron 14(6):1105–1116. Epub 1995/06/01

121. Wyss-Coray T, Lin C, Sanan DA, Mucke L, Masliah E (2000) Chronic overproduction of transforming growth factor-beta1 by astrocytes promotes Alzheimer's disease-like microvascular degeneration in transgenic mice. Am J Pathol 156(1):139–150
122. Wyss-Coray T, Rogers J (2012) Inflammation in Alzheimer disease-a brief review of the basic science and clinical literature. Cold Spring Harbor Perspect Med 2(1):a006346. doi:10.1101/cshperspect.a006346
123. Yoshiyama Y, Higuchi M, Zhang B, Huang SM, Iwata N, Saido TC, Maeda J, Suhara T, Trojanowski JQ, Lee VM (2007) Synapse loss and microglial activation precede tangles in a P301S tauopathy mouse model. Neuron 53(3):337–351. doi:10.1016/j.neuron.2007.01.010
124. Zhao W, Beers DR, Appel SH (2013) Immune-mediated mechanisms in the pathoprogression of amyotrophic lateral sclerosis. J Neuroimmune Pharmacol 8(4):888–899. doi:10.1007/s11481-013-9489-x
125. Zhong Z, Deane R, Ali Z, Parisi M, Shapovalov Y, O'Banion MK, Stojanovic K, Sagare A, Boillee S, Cleveland DW, Zlokovic BV (2008) ALS-causing SOD1 mutants generate vascular changes prior to motor neuron degeneration. Nat Neurosci 11(4):420–422. doi:10.1038/nn2073
126. Zilka N, Stozicka Z, Kovac A, Pilipcinec E, Bugos O, Novak M (2009) Human misfolded truncated tau protein promotes activation of microglia and leukocyte infiltration in the transgenic rat model of tauopathy. J Neuroimmunol 209(1-2):16–25. doi:10.1016/j.jneuroim.2009.01.013
127. Zlokovic BV (2008) The blood-brain barrier in health and chronic neurodegenerative disorders. Neuron 57(2):178–201. doi:10.1016/j.neuron.2008.01.003
128. Zlokovic BV (2011) Neurovascular pathways to neurodegeneration in Alzheimer's disease and other disorders. Nat Rev Neurosci 12(12):723–738. doi:10.1038/nrn3114

Leakage at Blood-Neural Barriers

Patric Turowski

Abstract Permeability of blood vessels in the brain and retina is usually very low and dominated by the restricting properties of the blood-brain and blood-retinal barriers, respectively. The highly specialised endothelium pervading the brain and the retina displays low permeability due to the nearly complete absence of transcellular transport (with the exception of that of specific nutrients and metabolites) and also to highly differentiated tight junctions. Importantly, the neuroglial cells that are part of cerebral and retinal blood vessels appear to be the main driver for inducing and maintaining these specialised properties of the endothelium. During many traumatic, inflammatory or degenerative neuro- and retinopathologies, this neurovascular unit is compromised leading to reduced vascular endothelial barrier properties and detrimental leakage of blood components into nervous tissue. Importantly, many extracellular permeability-inducing factors such as histamines, kinins, growth factors and lipids can trigger endothelial leakage in varying ways, but in most cases, pathological leakage occurs through consecutive or parallel opening of the paracellular space (characterised by tight junction protein loss) and induction of transcellular vesicles (possibly caveolae). Both pathways are regulated by complex often overlapping protein phosphorylation and GTPase networks, which lends credence to efforts to limit leakage at the BBB and BRB by specific signalling antagonists. Finally, leakage pathways are also exploited to facilitate drug delivery to the brain.

1 Rationale

Leakage within the cardiovascular system is the consequence of endothelial hyperpermeability and can be the cause, or a significant co-morbidity, of a variety of pathologies ranging from cancer and inflammation to metabolic diseases such as diabetes. In the nervous system, blood vessels are highly specialised rendering the endothelium uniquely impermeable. Due to its restricted exchange properties, this

P. Turowski
Cell Biology, UCL Institute of Ophthalmology, 11-43 Bath Street, London EC1V 9EL, UK
e-mail: p.turowski@ucl.ac.uk

R. Lyck, G. Enzmann (eds.), *The Blood Brain Barrier and Inflammation*,
Progress in Inflammation Research, DOI 10.1007/978-3-319-45514-3_5

vascular bed has been called the blood-brain barrier (BBB), with its main function to control the ionic and molecular microenvironment of the nervous system. The inner vasculature of the neuroretina fulfils a similar role and is termed blood-retinal barrier (BRB). In the following BBB will also be used as a generic term covering all blood-neural barriers unless otherwise specified. Whilst neurotoxic components are mostly excluded from the CNS, cellular and molecular crosstalk between the periphery and the brain parenchyma is allowed [1]. The BBB is often compromised in disease and is characterised by uncontrolled passage of leucocytes and macromolecules contributing to oedema and irrevocable CNS damage. Dysfunctional BBB is a pathological hallmark of stroke or trauma [69] but also neuroinflammatory diseases such as multiple sclerosis [2, 4]. In addition, disorders originally considered exclusively neurological in their aetiology, such as Parkinson's and Alzheimer's disease or amyotrophic lateral sclerosis, are accompanied, or possibly even caused, by BBB dysfunction [31, 50, 76]. Similarly, diseases of the neuroretina such as age-related macular degeneration and diabetic retinopathy are associated with a dysfunctional blood-retinal barrier [56]. Unsurprisingly, stemming vascular leakage by targeting molecules in permeability pathways has become a key therapeutic strategy for many of these diseases [75].

2 Measuring Vascular Permeability

The continuous endothelium lining the blood vessels constitutes a regulated size-selective, semipermeable layer that regulates transport of fluid and solutes between the blood and the interstitium. The vascular system needs to be sufficiently permeable to allow vital exchange of small molecules (gases, nutrients and waste products) between the blood and the underlying tissue. This basal permeability differs significantly between vascular beds and is in line with the physiological needs of the tissue in question. Thus, filtration rates are, for instance, high in the vasculature of the kidney but very low in the brain.

Decades of painstaking analyses by physiologists have provided a theoretical framework to vascular permeability [105]. From a cell biologist's perspective, this description of permeability by equations can be a drab and sometimes confusing affair. However, it delivers clear functional parameters, which have to be matched by endothelial structures and macromolecules. In the following we give a simplified view of the physiologists' perspective and highlight some of the main parameters of permeability, which we consider to be important both experimentally and conceptually.

Importantly, physiologists have recognised that the vast capillary network fanning out across each tissue was indeed the area of molecular exchange. To fit experimental data to mathematical equations, the endothelium was viewed as a thin, semipermeable membrane that is dotted by a large number of small pores (of around 4.7 nm) and relatively few larger pores of around 30 nm [97]. In this model the small pores mediate relative unhindered flux of water and very small molecules, whilst the larger pores allowed limited movement of macromolecules. This original model has been refined in many ways, for instance, by modelling the intervascular spaces not as pores but as narrow slits to include knowledge of morphological equivalents such as paracellular clefts between endothelial cells [38]. Compared to the complex underlying biological structures, these models are based on simple assumptions. Nevertheless, they allow the description of fluid transport in single vessel segments with high accuracy [73, 79, 97, 105]. Solute movement across the endothelial wall, if unhindered, can primarily be described by Fick's law of diffusion with concentration gradients being the driving force. However, in reality biological solute flux across the endothelium is dominated by molecular size, charge restrictions and osmotic forces. As a result the more complex Starling equation describes fluid flux across endothelial 'membranes' as the product of (i) the relevant surface area across which exchange can occur, (ii) hydraulic conductivity and (iii) net filtration pressure. Hydraulic conductivity can be measured experimentally and is a coefficient describing the leakiness of the endothelial barrier to water. Net filtration pressure is the sum of the hydrostatic and osmotic forces across the walls of the exchanging vessel. It is defined by a combination of the hydrostatic pressure within the vessel relative to the interstitium, the oncotic pressure difference and the reflection coefficient. The reflection coefficient described the relative sieving properties for a given solute. Lastly, measurements in isolated microvessels can also be heavily influenced by convection, i.e. the solvent drag, as well as local changes in solute concentrations near flow-through pores or slits.

Overall, this means that differences in permeability can be observed during change of (i) the available exchange surface (for instance, the number of perfused capillaries), (ii) hydrostatic pressure (which is influenced by blood pressure and vascular compliance), (iii) macromolecule composition of the blood or (iv) the filtration properties of the endothelial barrier (through regulated opening or altered transcellular transport phenomena).

Cell biologists often measure permeability by a Miles-type assay, where a dye is introduced into the circulation and subsequently its net accumulation is measured in the tissue [74, 94]. Within the strict framework set out by physiologists, a Miles assay does not deliver any parameter of vascular permeability [11, 79]. First, it measures the combination of interstitial and intravascular dye (albeit with the latter making minimal contribution). More crucially, however, it does not take into consideration the removal of dye by vascular reabsorption or lymphatic clearance. In addition, in cases of altered dye flux into the tissue, the Miles assay cannot distinguish between changes in true permeability (i.e. affecting hydraulic conductivity or reflection coefficient) and in hydrostatic pressure (through blood pressure changes or vasorelaxation). Nevertheless, the Miles assay has contributed enormously to our understanding of oedema and permeability-inducing factors such as vascular endothelial growth factor and histamine [79, 134].

In vitro measurements on cultured endothelial cells (optionally in conjunction with mural cells) have also contributed to our understanding of permeability, in particular its underlying signalling and mechanisms of endothelial 'opening'. For this usually the diffusive flux of dyes or ions is measured in either primary endothelial cells or established cell lines [34, 88].

Excellent reviews such as by Dvorak and colleagues [21, 79] describes the physiology and cell biology of basal permeability and acute and chronic leakage of the general vasculature. In the following we will focus on the vasculature pervading the nervous systems.

3 Basal Water and Solute Transport at Blood-Neural Barriers

Basal physiological permeability at vascular blood-neural barriers is very low. Hydraulic conductivity is considerably lower than in muscle and the reflection coefficient to proteins is maximal [38]. Indeed, the endothelium of the BBB is continuous and not fenestrated, thus forcing water and small molecule flux to occur in between or through the cells (paracellularly and transcellularly, respectively) [26]. Furthermore, the presence of complex junctions, which seal the paracellular space and act as molecular sieves, restricts movement of molecules of all sizes. Lastly, fluid-phase transcytosis (also referred to as pinocytosis) is virtually absent at the non-inflamed BBB.

Undoubtedly, the high vascular barrier function of the nervous system must have evolved to preserve the delicate environment of neurons and restrict oedema formation which is particularly damaging in the brain due to its rigid encasement and which can thus easily lead to nerve cell and capillary compression [17].

Nevertheless, the brain and retina are metabolically highly active and require nutrients at high rates as well as adequate waste management. Therefore, a tight vascular BBB and BRB must go hand in hand with sophisticated, specific and selective transport mechanisms for nutrients, immune cells and waste [104]. Lipid-soluble, non-polar molecules as well as oxygen and carbon dioxide can diffuse unrestrictedly along concentration gradients into the brain. In addition, many essential polar nutrients, such as glucose and amino acids, are transported by specific solute carriers. Larger molecular weight proteins and peptides generally enter the CNS via receptor-mediated or closely related adsorptive transcytosis. In turn, ATP-binding cassette (ABC) transporters, such as P-glycoprotein and other multidrug resistance-associated proteins, act as active efflux pumps and prevent entry of lipid-soluble compounds into the CNS. The presence of ABC transporters is the single most important reason why efficient delivery of lipid-soluble therapeutic agents into the brain remains a major challenge. Lastly, leucocyte transmigration through the vascular BBB occurs by harnessing a membrane-intensive, primarily transcellular apparatus [9, 127, 128]. Transport properties of BBB endothelial cells are crucially supported by exceptional apico-basal polarisation of the described transport systems, enabling preferential transport to or from the brain even against concentration gradients [130].

For baseline permeability the architecture and molecular structure of paracellular junctions constitute a molecular equivalent, which is compatible with the physiological measurements, in as much as they constitute a sieving fence. These interendothelial junctions consist of proteins of adherens junctions (AJs) and tight junctions (TJs) which, in endothelial cells and in contrast to epithelial cells, are not organised in distinct bands but are intermingled and functionally interdependent [123]. Endothelial AJs consist of the transmembrane protein vascular endothelial (VE)-cadherin in homotypic interactions with neighbouring VE-cadherin molecules, both in cis and trans [28]. VE-cadherin adhesion is regulated by post-translational modifications (mainly phosphorylation on tyrosines and serines) and association with α-, β-, γ- and p120 catenins, which regulate connection with the actin cytoskeleton and internalisation of AJs. AJ interaction and sensing are essential in the development of cell-cell contacts, the endothelial response to shear stress and the transmission of cytoskeletal force between adjacent cells. AJs also modulate and integrate growth factor signalling with connectivity cues. Disruption of AJs leads to leakage and oedema, at least in the periphery [24]. In turn artificial increase of the adhesiveness of AJs leads to relative impermeability and reduced leucocyte transmigration [107]. The importance of AJs in the development and maintenance of the BBB and BRB lies undoubtedly in their priming the formation of stable TJs [123].

TJs are protein complexes dynamically connecting adjoining epithelial and ECs [119]. By separating luminal from abluminal cellular surfaces, TJs, and to a lesser degree adherens junctions [28], regulate apico-basal diffusion of endothelial membrane components (both lipids and proteins) as well as the passage of solutes and cells through the paracellular space. TJs are composed of transmembrane proteins which regulate cell-cell interactions and the lateral organisation of junction strands. Amongst these claudins (CLDNs) are by far the most dominant found in BBB TJs. In mammals they constitute a family of >20 proteins, which form the structural fabric of TJs through homophilic and heterophilic interactions, occurring both in cis and trans [41]. Depending on its type, a claudin can either seal the paracellular space or form a pore for small molecules, ions or even water. At least three CLDNs with fence-forming properties (3, 5 and 12) have been identified in BBB ECs [43, 123]. It is yet unclear if pore-forming claudins are expressed at the BBB. At the BBB CLDN5 regulates paracellular transport of small (<800 dalton) blood solutes [82]. Occludin is another transmembrane protein present in all epithelial and endothelial TJs, which regulates certain aspects of paracellular diffusion and TJ organisation in vitro [119, 123]. Other transmembrane proteins of TJs include junctional adhesion molecules (JAMs) [51] and, at tricellular junctions, the points where three cells meet, the angulins (with lipolysis-stimulated lipoprotein receptor (LSR) particularly specific for the BBB) [113] which can additionally recruit tricellulin [43]. Enhanced leakage in LSR knockout mice mirrors that of a CLDN5 knockout with enhanced permeability to low but not high molecular weight blood solutes. CLDNs, occludin, junctional adhesion molecules, tricellulin and LSR associate further with cytoplasmic proteins integral to TJ function, such as zona occludens proteins ZO-1, ZO-2 and ZO-3, cingulin, binding partitioning defective proteins PAR-3 and PAR-6, MAGI and MUPP1 and

AF-6 [119, 123]. This results in a densely packed space beneath the cell membrane (zonula occludens) which dynamically links the integral membrane proteins to the cytoskeleton and a multitude of intracellular signalling proteins. It is noteworthy that many of the described molecular features of TJs are not exclusive to the BBB and cannot fully explain the exceptional solute and electrical impermeability of the cerebral vasculature.

The complete lack of fenestrae and nearly undetectable level of fluid-phase transcytosis in healthy BBB endothelium [26] might be due to the lack of a glycoprotein called plasmalemmal vesicle-associated protein-1 (PLVAP or PV-1). PV-1 is a major component of the diaphragm of fenestrated endothelium and is a major structural component of caveolae and fenestrae [45, 117]. Recently, another protein, Mfsda2, was shown to suppress transcytosis in brain EC [12].

Additional factors that influence the filtration behaviour of the BBB and BRB are glycocalyx [62] and the basement membrane (laid down by the endothelium and pericytes) [114]. Taken together, in the case of a healthy BBB, the basal permeability requirements of the brain are fully catered for by absence of fluid-phase transcytosis and fenestrae and the presence of sophisticated interendothelial junctions. This virtually impermeable endothelium is then rendered sufficiently open by specific and selective transport machineries.

4 Pathophysiological Leakage

A pathologically weakened and open blood-neural barrier could provide a better access for leucocytes and thus immune response. Openness may also be beneficial to resolve oedema itself [19]. Nevertheless, it is generally accepted that a disrupted BBB has adverse effects on the CNS, provoking neuronal damage and degeneration. BBB disruption has been observed during traumatic and ischemic brain injury [69] and neuroinflammatory [2] and neurodegenerative diseases [50]. Sight-threatening ischemic diseases of the retina are also often associated with a leaky and dysfunctional BRB [56]. The BBB of some neuropathologies such as multiple sclerosis or its animal model experimental autoimmune encephalomyelitis (EAE) has been extensively studied, and thus much detail is known about the nature and mechanisms of associated barrier breakdown. In others, in particular neurodegenerative diseases with long-term pathogenesis and a paucity of availability of relevant animal models, details of barrier breakdown are still emerging and often unclear. For instance, uptake of a therapeutic antibody in rodent models of AD was minimal and not found to be widespread in the brain [13]. Other important uncertainties relate to the question of whether BBB dysfunction is the cause or consequence of a particular disease [118]. Whilst it is clear that this cannot be generalised – each neurological disease may well have a completely distinct underlying pathogenesis – there is increasing evidence both from epidemiological and experimental studies that BBB dysfunction may constitute the initiating trigger in many cases. It is undisputed that chronic vascular breakdown will contribute significantly to morbidity of these diseases, thus justifying increasing investment into anti-leakage therapies for neuro- and retinopathies.

Just like basal permeability, pathological leakage (be it in the brain or not) is restricted to the microvascular bed [79]. However, it does not only affect capillaries and extends to other areas of the microvasculature in particular postcapillary venules. Importantly, this part of the vascular tree is compliant to inflammatory leucocyte migration [36], illustrating just one aspect of the close relationship between the barrier function to molecules and to cells. In the retina where vascular behaviour can be more easily imaged than in the brain, Barber and Antonetti reported that pathophysiological leakage during diabetes extends to arterioles and postcapillary venules [10].

From a physiological point of view, leakage and oedema can be the consequence of changes to the available exchange surface (the number of capillaries), hydrostatic or osmotic pressure differences and/or the size cut-off of the barrier. The main parametric changes of the endothelium of the diseased BBB are therefore increased hydraulic conductivity and reduced reflection coefficient, thus permitting enhanced flux of water and of solutes of larger sizes. Accordingly, aberrations in both paracellular and transcellular pathways have been observed in the disrupted BBB [116]. In cultured cells these two pathways can be distinguished relatively easily: Paracellular leakage leads to a change in ionic conductance, whilst transcellular leakage will not affect electrical resistance across the endothelium. Most notably though, the debate regarding whether transcellular leakage occurs and if so whether it is of relevance in vivo is still active (see below).

A variety of factors extrinsic to the endothelial cell drive BBB breakdown, and these vary greatly between individual pathologies. Traditional vasogenic agents such as histamines and kinins have been shown to lead to leakage at the BBB experimentally and during neurological disease [30, 48, 49, 101, 102]. Bradykinin analogues have even been explored to open the BBB for pharmaceutical purposes [35]. Growth factors such as vascular endothelial growth factor (VEGF)-A, TYMP and TGF-β play a major role in neuroinflammatory diseases [4, 22, 116]. Inflammatory mediators such as the cytokines IL1β, TNFα and INFγ and the chemokines CCL2 and CXCL8 are also major contributors to BBB or BRB breakdown in particular since most conditions of leakage at blood-neural barriers are accompanied by some form of inflammation [100].

Extracellular proteases in particular the metalloproteinases MMP2 and MMP9 through direct action on junctions (see below) have been found in neuroinflammatory lesions close to sites of enhanced leakage [108]. Concordantly, MMP inhibition can attenuate leakage in modelled pathologies. Thrombin triggering endothelial leakage signalling can also act as a luminal leakage factor, just as free radicals and certain lipid compounds such as prostaglandin, sphingosine-1-phosphate or lysophosphatidic acid [37, 44, 54, 85, 92]. Dysfunctional BBB in Alzheimer's disease may (amongst many other factors) be triggered by binding of amyloid to endothelial cells [52]. Direct binding of infectious agents with BBB endothelial cells also causes barrier disruption and may be critical in the pathogenesis of bacterial and viral meningitis and encephalitis. Viruses and virus components (e.g. HIV, measles virus) or bacteria and their components (e.g. *E. coli*, *Streptococcus*, cholera and pertussis toxins) (reviewed in [116]) can bind to BBB endothelial cells and trigger barrier disruption either by direct signalling or by downregulating barrier determinants such as TJ proteins [53]. MicroRNA secreted, e.g. by metastatic cells, can also modulate barrier properties of cerebral endothelial cells, presumably to facilitate subsequent transendothelial migration [124].

5 The Paracellular Leakage Pathway

Our understanding of the molecular structure and assembly of adherens and tight junctions delivers an excellent explanation for the barrier function of BBB and BRB endothelial cells under baseline conditions, mainly dominated by impermeability and sieving. However, the dynamic rearrangements that occur within the paracellular space during pathological leakage are yet to be defined on the molecular but especially the structural level. Excessive transport is accompanied by the phosphorylation and adhesive properties of TJ and adherens junction complexes, resulting in altered junctional protein interactions and localisation [123]. Concomitant reorganisation of the actin cytoskeleton also contributes to permeability, with contractile forces thought to render the paracellular space more compliant to modification [116]. Such changes give rise to rearrangements in the extracellular adhesive portions of proteins, for instance, different architectures of cis and trans interactions of adherens junction cadherin dimers have been resolved by X-ray crystallography [16]. However, possible consequences of such dimer rearrangements for the overall strand structure and paracellular channels or pores remain purely speculative. For TJ strand modulation, the available ultrastructural information is even sketchier. A clearer mechanism is provided in situations where junction (in particular TJ) protein expression is reduced at the paracellular membrane. Intuitively, the functional outcome is assumed to be fewer junction strands with overall reduced sieving and sealing properties. Indeed, loss of tight junction proteins is a hallmark of many neurovascular diseases. In models of traumatic or neuroinflammatory disease, such as EAE or middle cerebral artery occlusion, the TJ proteins occludin, CLDN3 and CLDN5 and LSR have been shown to be downregulated from vessels in the vicinity of focal points of leakage [55, 65, 89, 113, 129]. In multiple sclerosis pathological examination of brains has revealed significantly reduced immunoreactivity of many TJ and AJ proteins [2]. The loss of occludin from microvessels is also a hallmark of diabetic retinopathy [10, 25].

TJ proteins are lost by one or several mechanisms: Transcriptional downregulation can directly reduce the expression of TJ proteins [53, 120, 133]. Post-translational modification especially phosphorylation has been shown to induce enhanced junction protein endocytosis [77, 86]. Lastly, loss of TJ protein can be the consequence of protease action. For instance, matrix metalloproteinases (MMPs), most notably MMP9, are activated by reactive oxygen species, vascular endothelial growth factor and inflammatory cytokines in many CNS pathologies [26, 116].

MMP activation leads to the degradation of EC basement membrane and enhanced phosphorylation and cleavage of TJ proteins. Subsequent degradation of the inter-endothelial junctions is then thought to lead to weakened TJ strand architecture or in extreme cases to the complete loss of paracellular junctions.

However, definitive proof of a direct relationship of pathologically altered paracellular junctions with endothelial hyperpermeability is missing. Correlative mapping of the region of low TJ expression with areas of hyperpermeability would certainly go a long way towards this goal [89]. The retinal vasculature with its intrinsic accessibility may be particularly well suited for high-resolution imaging in vivo and in situ [106].

Osmotic opening of the BBB, explored in search of strategies to deliver chemotherapy to the brain (see also below), appears to involve disruption of junction strands and opening of gaps between endothelial cells [46, 95]. However, there is no hard evidence that the same happens during pathology. In fact, despite clear reduction of junction protein staining in disease such as stroke, MS, Alzheimer's and diabetic retinopathy or in experimental models thereof, actual disrupted junctions are rarely seen by EM [61]. It is particularly noteworthy that ultrastructural junction changes, if reported in pathogenesis, occur much later than when leakage and oedema can be measured, suggesting that paracellular changes may mechanistically underpin chronic failures of the BBB and BRB [57, 60].

6 The Transcellular Leakage Pathway

Vascular leakage in brain and retinal diseases can occur via a paracellular or transcellular route or indeed through a combination of both. It is somewhat surprising, therefore, that most reviews focus on the paracellular pathway despite there being considerable evidence that the transcellular path may occur. Reasons for this may be that pore-sized channels (often referred to as vesiculo-vascular organelles) [33] do not exist in BBB endothelium. Furthermore, the healthy BBB mostly lacks vesicular structures required for fluid-phase transcytosis [112]. In contrast, at the diseased BBB, fluid-phase transcytosis could occur: Under various pathological conditions, including diabetes, the BBB and BRB are associated with increased number of caveolar vesicles [14, 46, 47]. However, the role of caveolae in endothelial trans-cellular transport has been viewed with scepticism and is still the subject of intense debate [98, 111], ever since their formal description by Palade in 1953 [87]. Physiologists have argued that vesicular transport does not fulfil the dependence of macromolecular flux on hydrostatic pressure and convection. It has also been argued that intraendothelial, tracer-filled caveolar vesicles visualised by EM could have backfilled from the material that has entered the interstitial space through leaky junctions. The observation that caveolin (CAV)1-knockout mice, which lack caveolae altogether, do not as predicted display lower but on the contrary significantly higher overall leakage, appeared to ring the death knell for caveolar transcytosis [64]. However, it should be noted that caveolae can sense mechanical pressure [23] and also form important endothelial signalling platforms, e.g. for endothelial nitric oxide synthase (eNOS) [32], the absence of which undoubtedly changes overall signalling behaviour and presumably the architecture of the paracellular space. Thus, the constitutive CAV1 knockout could conceivably push the endothelial cell phenotype from one with intact gating function by AJ and TJ to another with substantial deficiencies in this area. Indeed, another work in the same CAV1-knockout mice demonstrated that VEGF-A-induced leakage was attenuated, suggesting that whilst basal permeability was not dependent on caveolae, pathological leakage was [21]. Furthermore, caveolae have been shown to be instrumental for lipid transport into the cell [110], whilst Schnitzer and colleagues have shown that antibodies specifically targeting caveolae are pumped rapidly and at substantial rates across wild type but not CAV1-negative endothelium in vivo [84]. Nevertheless, the question remains as to the quantitative contribution of caveolar transcytosis to macromolecular transport across the endothelium, in particular in the brain.

Plasmalemmal vesicles and CAV1-associated processes have been frequently described during early brain oedema, often preceding the disruption of TJs. Thus, early brain oedema has been associated with increased expression of CAV1 [78]. VEGF, undoubtedly a mediator of acute BBB dysfunction, induces pinocytotic vesicles in BRB [47] and blood-tumour barrier endothelium in conjunction with increased expression of caveolin-1 and caveolin-2 [136]. Bradykinin also induces transcellular transport within blood-tumour barrier endothelium with increased expression of caveolin-1 and caveolin-2 [67]. Our laboratory has shown that methamphetamine-induced leakage in isolated cerebral endothelial cells is restricted to a transcellular, presumably caveolar pathway [72]. More recently, in yet unpublished experiments, we have extended these observations to the intact brain, i.e. occurring within a functional neurovascular unit (Chang and Turowski, unpublished observation). Lastly, in a rodent model of stroke, early BBB leakage is associated with increased caveolae and transcytosis (absent in CAV1-knockout mice), whilst the second disruption of the BBB occurring after approximately 2 days is associated with TJ reorganisation [57].

Taken together, whilst transcellular flux appears negligible at the healthy BBB, there is a lot of direct and indirect evidence suggesting that it constitutes a potentially important route of leakage and possibly the most prominent pathway in disease and inflammation. Admittedly, the jury is still out to decide if caveolae are functionally responsible, and further research will always be hampered by caveolae being multifunctional units mediating transport as well as signalling. Experimental separation of these different functions will always confound interpretation and possibly impede future research in this area. Notably, caveolae through CAV1 could also constitute the molecular crossroad where transcellular and paracellular pathways meet as CAV1 can mediate internalisation of the TJ protein occludin [70]. Likewise, nitric oxide signalling might constitute a molecular crossroad between the paracellular and transcellular leakage pathways [39].

7 Endothelial Signalling During BBB or BRB Leakage

The properties of the vascular barrier and its permeability are controlled by a complex system of interacting signalling networks [39, 58]. Naturally, its precise nature depends heavily on the cell-intrinsic and cell-extrinsic stimuli. At the level of the plasma membrane signal, specific receptor protein tyrosine kinases or GPCRs are activated to trigger intracellular changes such as $Ca2^{+}$ transients; activation of various protein kinases, including src family kinases, Akt, and certain PKC isoforms. Subsequent fundamental changes to the assembly and contractility of the actin network are especially important to changes to intercellular junction proteins. Keys for dynamic rearrangements of the cytoskeleton during endothelial hyperpermeability are rhoA GTPAse and its downstream protein kinases such as ROCK or MLCK. In addition, direct modification of AJ and TJ proteins, mostly by reversible phosphorylation, appears to act in concert with the cytoskeletal force changes. These pathways mostly modulate junctions and thus intuitively the paracellular leakage pathway. As discussed before junction modulation often also involves modulation of expression levels, and thus classical nuclear pathways of transcriptional activation and repression are also vitally important during the leakage response. For transcellular leakage, molecules such as eNOS and endocytic regulators (such as SNARES and Rab GTPases) in addition to the aforementioned signalling pathways appear to be required for caveolar assembly, fission and fusion. In addition, transcytosis relies on microtubular traffic rather than actin contractility.

It is commonly assumed that signalling pathways that regulate permeability in the periphery also fulfil key roles at the BBB. However, in view of the unique properties of the endothelial cells in the brain and the retina, it should come as no surprise that many signalling pathways leading towards permeability are significantly different. On the one hand, a fair number of pro-leakage signalling components are clearly the same in the periphery and at the BBB and/or the BRB. For instance, Src family protein kinases, PKC isoforms, various actin regulators such as rhoA as well as nitric oxide and reactive oxygen species, via molecules such eNOS and NADPH oxidase, all have key roles in mediating BBB leakage in various neurological diseases [37, 116]. Moreover, cAMP has a clear role in preserving endothelial barrier function in all vascular beds, with astrocyte-mediated elevation of endothelial cAMP undoubtedly an important determinant specifying BBB impermeability [90]. On the other hand, signalling towards leakage at the BBB and BRB may use certain molecules to completely different effect. For instance, whilst the protein kinase Akt enjoys an undisputed role in inducing permeability in the periphery, in particular in that induced by VEGF, it is not involved at the BBB in vitro or in vivo [48]. Instead, Akt is likely to mediate BBB stabilisation, for instance, in response to Ang-1 [68]. VE-cadherin phosphorylation, which occurs frequently during vascular leakage in the periphery [86], also occurs in the brain where it appears to be mediated by an unusual pathway involving $Ca2^{+}$, AMPK and eNOS [71]. At the same time, the role of AMPK at the BBB is not restricted to a permeability-inducing role: its enhanced activity has also been associated with the opposite effect, such as barrier protection during LPS-induced BBB leakage [121, 131]. Similarly, certain PKC isoforms have been associated with both permeability- and barrier-enhancing activities depending on the neuronal vascular environment [42, 115]. Thus, exact experimental representation of the neural microenvironment and the permeability-inducing factor are clearly critical when attempting to understand leakage at blood-neural barriers and developing anti-leakage therapies. Here, it should also be noted that signalling pathways are mostly explored in cultured endothelial cells where culture conditions and, if used, cell immortalisation procedures may influence signalling processes significantly leading to results with little relevance to in vivo events (Turowski, unpublished observations).

Wnt signalling and associated nuclear signalling of beta-catenin play a pivotal role in the development of the BBB [63, 137], in particular the establishment of the tight junction complex that characterises the mature BBB [7]. Wnt signalling pathways appear to continue to operate to maintain BBB homoeostasis, and thus defective wnt signalling may contribute to pathological BBB breakdown [66].

Taken together, relatively little is known about leakage-specific signalling at the blood-brain barriers. Inference from permeability studies in the periphery as well as generalisation within the BBB should therefore be made with caution.

8 BBB/BRB Function and Dysfunction: Nature and Nurture

Endothelial cells of different vascular beds display very different molecular signatures resulting in extremely divergent phenotypes with glomerular and cerebral endothelial cells on each end of the spectrum [83, 96]. The isolation and culturing in vitro of endothelial cells from various vascular beds have revealed that many of the specific traits of endothelial cells are dependent on their immediate environment in vivo since they are rapidly lost in isolation. However, yet other traits are never lost indicating that the endothelial phenotype is controlled by both nature and nurture. Until recently, the BBB was considered a relatively homogeneous vascular bed. Whilst there are obviously clear differences between endothelial cells at the arterial and venous side of vascular trees, other differences are more surprising and often less well understood on a functional level [89]. For instance, the endothelial barrier antigen (EBA) has long been used to demarcate barrier endothelial cells within the CNS. However, EBA staining reveals striking heterogeneity not only between functionally different vessels within the brain but also within single venules [103]. Neither the environmental cues nor the functional consequences for such heterogeneity on single-cell level are known.

The influence of the environment on the vascular phenotype is particularly well illustrated in circumventricular organs, where junctional protein expression in endothelial cells is low and basal vascular permeability is not in line with other cortical areas [93]. Clearly, the brain vasculature has the capacity to adapt its barrier properties to functional needs, in this case to act as a conduit through which signals flow into and out of the brain. Similarly, albeit not to same degree, the vascular barrier in the hippocampus appears to be comparatively more prone to change during pathological insult [76, 125], which again may indicate a functional need to facilitate communicative exchange with the periphery [122].

Functional differences within the CNS microvasculature exist with respect to barrier function. Generally, postcapillary venules have leakier junctions and thus predicted higher basal permeability [1]. In agreement, leucocyte entry to the brain parenchyma occurs in postcapillary venules but not capillaries [36]. In addition, pathological barrier breakdown does not affect the entire microvascular bed of the retina equally [10]. Clearly, the regulation of endothelial phenotype and consequently vascular permeability at the BBB and BRB relies strongly on environmental cues. Indeed, blood-neural barriers effectively represent the integrated response of a group of cells, which intimately interact and communicate via physical interactions and soluble extracellular factors. Pericytes, astrocytes, perivascular macrophages and microglia are other cells within this so-called neurovascular unit [26].

On the abluminal side, endothelial cells in the brain and the retina are surrounded by pericytes, with which they share a basement membrane. Pericyte coverage at the BBB and BRB is the highest of all vascular beds, with EC to pericyte ratios typically ranging between 1:1 and 3:1 [5]. During embryonic development pericytes are essential for BBB induction [6, 27]. Pericytes are also key in providing the signals that maintain the BBB phenotype of the endothelial cells under their control. A pericyte-deficient BBB features increased macromolecular permeability with increased cytoplasmic vesicles but with intact polarisation and a continued lack of fenestrae. At the healthy BBB, pericytes also suppress plasmalemmal vesicle-associated protein (PLVAP), a marker for fenestrated EC and transcytosis, which is normally only found in non-CNS endothelial cells [26, 109]. Canonical wnt signalling appears to be particularly important since its endothelial loss of function in mice results in reduced CLDN5 and elevated expression of PLVAP [137]. Pericytes also regulate the lipid transporter Mfsda2 which suppresses transcytosis in brain EC [12], indicating that pathological transcellular leakage may be intimately linked to pericyte health. Indeed, pericytes also strongly influence BBB function during disease. Accordingly, dysfunctional pericyte-endothelial interactions appear to drive the pathogenesis of neurodegenerative diseases such as Alzheimer's disease [76]. In addition, pericyte-secreted VEGF may contribute to stroke [8].

Astrocytes ensheath CNS blood vessels further, either directly or via the parenchymal basement membrane. They regulate neuronal function and coordinate multitudinous signals from neurons, the BBB and their microenvironment during such instances as neurovascular coupling that links neural activity to blood flow [91]. Astrocytes also strongly influence both BBB TJs and transport properties [1] with astrocyte-secreted factors increasing TJ expression and transendothelial resistance in cultured endothelial cells, presumably by inducing sustained barrier-protective cAMP signalling in the endothelial cells [90, 99]. In neuroinflammatory diseases, astrocytes are key responders and, by secreting a multitude of cytokines factors such as VEGF-A, induce the dysfunctional BBB phenotype from within the parenchyma [3, 4, 22].

CNS immune cells include microglia, which are highly ramified, phagocytic cells found throughout the CNS. They contribute to innate and adaptive immune responses and neuronal homoeostasis [40]. Perivascular BBB macrophages reside between the basal lamina and astrocytic foot processes, where they are involved in first-line CNS immune surveillance and antigen presentation. Their activation affects the BBB with an increased secretion of cytokines contributing to increased permeability and leucocyte infiltration [29].

BBB properties are not only regulated by the immediate cellular environment but also through peripheral inflammatory cytokines [100, 122]. More surprisingly is the recent discovery by Braniste et al. who showed that gut microbiota influences the embryonic development and permeability of the BBB by regulating TJ expression [15]. The molecular determinant for BBB regulation by gut bacteria in these experiments was proposed to be butyrate, which can also improve the intestinal-epithelial barrier. Undoubtedly, BBB homoeostasis is regulated both on a local and a global level.

9 Opening the BBB for Therapeutic Drug Delivery

The protective nature of the BBB has limited the effectiveness of many systemically administered treatments for severe neurological disease. In particular, glioma patients are disproportionally disadvantaged and fail to benefit from the advances that have been made in the development of cancer drugs [126, 135]. This demonstrates that even where there may be alterations to tumour vascular barrier function, at the invasive edge where the barrier remains intact, there is restriction of drug delivery. In most CNS disease, therefore, the selectiveness of the TJs, the lack of effective fluid-phase transcytosis and in addition the high effectiveness of the many drug efflux pumps at the BBB make it virtually impossible to attain pharmacologically meaningful drug levels in the brain parenchyma.

Ever since the discovery that the BBB could be experimentally disrupted, research has attempted to open the barrier transiently during intravenous drug administration. Various strategies have been tried or proposed with varying success. Osmotic opening has been one of the earliest techniques to be explored and shown to aid drug delivery to the brain [80]. Direct modulation of junctions by targeting CLDN5 by siRNA has also been shown to improve drug delivery to the brain [20]. In addition this opening appears to work both ways since it could also be used to resolve existing oedema during experimental stroke [19]. Similarly, the use of luminally acting permeability-inducing factors such as bradykinin, lipoprotein A or methamphetamine or functional analogues has been proposed or experimentally trialled [35, 85, 125]. However, for all these methods, the destructive effect on the BBB and/or neurons may produce disproportional harm and may thus not be useful as a general strategy for drug delivery, in particular in cases of degenerative and chronic disease. Other techniques are not based on the global opening of the barrier.

Various strategies exploit receptor-mediated transcytosis at the BBB, essentially piggybacking a specific (in most cases the transferrin) receptor with antibody-linked drugs [81, 132]. Nanoparticle-mediated delivery, which is also mostly dependent on endothelial transcytosis, is another much explored mechanism for drug delivery to the brain [59]. However, whilst these latter strategies avoid general opening of the BBB, they would still not restrict specific drug delivery to a specific area within the brain. It is debatable if local specificity of drug delivery is essential for the majority of neurological diseases. Nevertheless, focused ultrasound in combination with microbubbles has been shown to allow transient local opening and enhanced drug delivery to areas of interest [18].

10 Conclusion

Understanding vascular leakage at blood-neural barriers is critically important to deliver solutions for a variety of neurological diseases, notably in order to find ways to seal and normalise the vascular barrier and the underlying neuronal environment, but also in many instances to open this biological obstacle to deliver drugs more effectively to the brain. BBB research has in many respects entered a new age as invaluable knowledge collected during decades of ultrastructural, physiological and pharmacological research is now being combined with integrative cell and development biology and systematic omics to tackle the many remaining challenges posed by this sophisticated structure, which is less of wall and more of a reactive and multi-faceted border.

Acknowledgements The author would like to thank Prof John Greenwood for stimulating discussions over the past 15 years and for critically reviewing this manuscript.

References

1. Abbott NJ, Patabendige AA, Dolman DE, Yusof SR, Begley DJ (2010) Structure and function of the blood-brain barrier. Neurobiol Dis 37:13–25
2. Alvarez JI, Cayrol R, Prat A (2011) Disruption of central nervous system barriers in multiple sclerosis. Biochim Biophys Acta 1812:252–264
3. Alvarez JI, Dodelet-Devillers A, Kebir H, Ifergan I, Fabre PJ, Terouz S, Sabbagh M, Wosik K, Bourbonniere L, Bernard M et al (2011) The Hedgehog pathway promotes blood-brain barrier integrity and CNS immune quiescence. Science 334:1727–1731
4. Argaw AT, Asp L, Zhang J, Navrazhina K, Pham T, Mariani JN, Mahase S, Dutta DJ, Seto J, Kramer EG et al (2012) Astrocyte-derived VEGF-A drives blood-brain barrier disruption in CNS inflammatory disease. J Clin Invest 122:2454–2468
5. Armulik A, Genove G, Betsholtz C (2011) Pericytes: developmental, physiological, and pathological perspectives, problems, and promises. Dev Cell 21:193–215
6. Armulik A, Genove G, Mae M, Nisancioglu MH, Wallgard E, Niaudet C, He L, Norlin J, Lindblom P, Strittmatter K et al (2010) Pericytes regulate the blood-brain barrier. Nature 468:557–561

7. Artus C, Glacial F, Ganeshamoorthy K, Ziegler N, Godet M, Guilbert T, Liebner S, Couraud PO (2014) The Wnt/planar cell polarity signaling pathway contributes to the integrity of tight junctions in brain endothelial cells. J Cereb Blood Flow Metab 34:433–440
8. Bai Y, Zhu X, Chao J, Zhang Y, Qian C, Li P, Liu D, Han B, Zhao L, Zhang J et al (2015) Pericytes contribute to the disruption of the cerebral endothelial barrier via increasing VEGF expression: implications for stroke. PLoS One 10:e0124362
9. Bamforth SD, Lightman SL, Greenwood J (1997) Ultrastructural analysis of interleukin-1 beta-induced leukocyte recruitment to the rat retina. Invest Ophthalmol Vis Sci 38:25–35
10. Barber AJ, Antonetti DA (2003) Mapping the blood vessels with paracellular permeability in the retinas of diabetic rats. Invest Ophthalmol Vis Sci 44:5410–5416
11. Bates DO (2010) Vascular endothelial growth factors and vascular permeability. Cardiovasc Res 87:262–271
12. Ben-Zvi A, Lacoste B, Kur E, Andreone BJ, Mayshar Y, Yan H, Gu C (2014) Mfsd2a is critical for the formation and function of the blood-brain barrier. Nature 509:507–511
13. Bien-Ly N, Boswell CA, Jeet S, Beach TG, Hoyte K, Luk W, Shihadeh V, Ulufatu S, Foreman O, Lu Y et al (2015) Lack of Widespread BBB Disruption in Alzheimer's Disease Models: Focus on Therapeutic Antibodies. Neuron 88:289–297
14. Bouchard P, Ghitescu LD, Bendayan M (2002) Morpho-functional studies of the blood-brain barrier in streptozotocin-induced diabetic rats. Diabetologia 45:1017–1025
15. Braniste V, Al-Asmakh M, Kowal C, Anuar F, Abbaspour A, Toth M, Korecka A, Bakocevic N, Ng LG, Kundu P et al (2014) The gut microbiota influences blood-brain barrier permeability in mice. Sci Transl Med 6:263ra158
16. Brasch J, Harrison OJ, Honig B, Shapiro L (2012) Thinking outside the cell: how cadherins drive adhesion. Trends Cell Biol 22:299–310
17. Bundgaard M, Abbott NJ (2008) All vertebrates started out with a glial blood-brain barrier 4–500 million years ago. Glia 56:699–708
18. Burgess A, Shah K, Hough O, Hynynen K (2015) Focused ultrasound-mediated drug delivery through the blood-brain barrier. Expert Rev Neurother 15:477–491
19. Campbell M, Hanrahan F, Gobbo OL, Kelly ME, Kiang AS, Humphries MM, Nguyen AT, Ozaki E, Keaney J, Blau CW et al (2012) Targeted suppression of claudin-5 decreases cerebral oedema and improves cognitive outcome following traumatic brain injury. Nat Commun 3:849
20. Campbell M, Kiang AS, Kenna PF, Kerskens C, Blau C, O'Dwyer L, Tivnan A, Kelly JA, Brankin B, Farrar GJ et al (2008) RNAi-mediated reversible opening of the blood-brain barrier. J Gene Med 10:930–947
21. Chang SH, Feng D, Nagy JA, Sciuto TE, Dvorak AM, Dvorak HF (2009) Vascular permeability and pathological angiogenesis in caveolin-1-null mice. Am J Pathol 175:1768–1776
22. Chapouly C, Tadesse Argaw A, Horng S, Castro K, Zhang J, Asp L, Loo H, Laitman BM, Mariani JN, Straus Farber R et al (2015) Astrocytic TYMP and VEGFA drive blood-brain barrier opening in inflammatory central nervous system lesions. Brain 138:1548–1567
23. Cheng JP, Mendoza-Topaz C, Howard G, Chadwick J, Shvets E, Cowburn AS, Dunmore BJ, Crosby A, Morrell NW, Nichols BJ (2015) Caveolae protect endothelial cells from membrane rupture during increased cardiac output. J Cell Biol 211:53–61
24. Corada M, Mariotti M, Thurston G, Smith K, Kunkel R, Brockhaus M, Lampugnani MG, Martin-Padura I, Stoppacciaro A, Ruco L et al (1999) Vascular endothelial-cadherin is an important determinant of microvascular integrity in vivo. Proc Natl Acad Sci U S A 96:9815–9820
25. Curtis TM, Gardiner TA, Stitt AW (2009) Microvascular lesions of diabetic retinopathy: clues towards understanding pathogenesis? Eye (Lond) 23:1496–1508
26. Daneman R (2012) The blood-brain barrier in health and disease. Ann Neurol 72:648–672
27. Daneman R, Zhou L, Kebede AA, Barres BA (2010) Pericytes are required for blood-brain barrier integrity during embryogenesis. Nature 468:562–566
28. Dejana E, Orsenigo F, Lampugnani MG (2008) The role of adherens junctions and VE-cadherin in the control of vascular permeability. J Cell Sci 121:2115–2122

29. Denieffe S, Kelly RJ, McDonald C, Lyons A, Lynch MA (2013) Classical activation of microglia in CD200-deficient mice is a consequence of blood brain barrier permeability and infiltration of peripheral cells. Brain Behav Immun 34:86–97
30. Dobrivojevic M, Spiranec K, Sindic A (2015) Involvement of bradykinin in brain edema development after ischemic stroke. Pflugers Arch 467:201–212
31. Drozdzik M, Bialecka M, Mysliwiec K, Honczarenko K, Stankiewicz J, Sych Z (2003) Polymorphism in the P-glycoprotein drug transporter MDR1 gene: a possible link between environmental and genetic factors in Parkinson's disease. Pharmacogenetics 13:259–263
32. Duran WN, Breslin JW, Sanchez FA (2010) The NO cascade, eNOS location, and microvascular permeability. Cardiovasc Res 87:254–261
33. Dvorak AM, Feng D (2001) The vesiculo-vacuolar organelle (VVO). A new endothelial cell permeability organelle. J Histochem Cytochem 49:419–432
34. Eigenmann DE, Xue G, Kim KS, Moses AV, Hamburger M, Oufir M (2013) Comparative study of four immortalized human brain capillary endothelial cell lines, hCMEC/D3, hBMEC, TY10, and BB19, and optimization of culture conditions, for an in vitro blood-brain barrier model for drug permeability studies. Fluids Barriers CNS 10:33
35. Emerich DF, Dean RL, Osborn C, Bartus RT (2001) The development of the bradykinin agonist labradimil as a means to increase the permeability of the blood-brain barrier: from concept to clinical evaluation. Clin Pharmacokinet 40:105–123
36. Engelhardt B, Sorokin L (2009) The blood-brain and the blood-cerebrospinal fluid barriers: function and dysfunction. Semin Immunopathol 31:497–511
37. Fraser PA (2011) The role of free radical generation in increasing cerebrovascular permeability. Free Radic Biol Med 51:967–977
38. Fraser PA, Dallas AD, Davies S (1990) Measurement of filtration coefficient in single cerebral microvessels of the frog. J Physiol 423:343–361
39. Goddard LM, Iruela-Arispe ML (2013) Cellular and molecular regulation of vascular permeability. Thromb Haemost 109:407–415
40. Gomez-Nicola D, Perry VH (2015) Microglial dynamics and role in the healthy and diseased brain: a paradigm of functional plasticity. Neuroscientist 21:169–184
41. Gunzel D, Yu AS (2013) Claudins and the modulation of tight junction permeability. Physiol Rev 93:525–569
42. Harhaj NS, Felinski EA, Wolpert EB, Sundstrom JM, Gardner TW, Antonetti DA (2006) VEGF activation of protein kinase C stimulates occludin phosphorylation and contributes to endothelial permeability. Invest Ophthalmol Vis Sci 47:5106–5115
43. Haseloff RF, Dithmer S, Winkler L, Wolburg H, Blasig IE (2015) Transmembrane proteins of the tight junctions at the blood-brain barrier: structural and functional aspects. Semin Cell Dev Biol 38:16–25
44. Hawkins BT, Gu YH, Izawa Y, del Zoppo GJ (2015) Dabigatran abrogates brain endothelial cell permeability in response to thrombin. J Cereb Blood Flow Metab 35:985–992
45. Herrnberger L, Seitz R, Kuespert S, Bosl MR, Fuchshofer R, Tamm ER (2012) Lack of endothelial diaphragms in fenestrae and caveolae of mutant Plvap-deficient mice. Histochem Cell Biol 138:709–724
46. Hirano A, Kawanami T, Llena JF (1994) Electron microscopy of the blood-brain barrier in disease. Microsc Res Tech 27:543–556
47. Hofman P, Blaauwgeers HG, Tolentino MJ, Adamis AP, Nunes Cardozo BJ, Vrensen GF, Schlingemann RO (2000) VEGF-A induced hyperpermeability of blood-retinal barrier endothelium in vivo is predominantly associated with pinocytotic vesicular transport and not with formation of fenestrations. Vascular endothelial growth factor-A. Curr Eye Res 21:637–645
48. Hudson N, Powner MB, Sarker MH, Burgoyne T, Campbell M, Ockrim ZK, Martinelli R, Futter CE, Grant MB, Fraser PA et al (2014) Differential apicobasal VEGF signaling at vascular blood-neural barriers. Dev Cell 30:541–552
49. Jadidi-Niaragh F, Mirshafiey A (2010) Histamine and histamine receptors in pathogenesis and treatment of multiple sclerosis. Neuropharmacology 59:180–189

50. Jeynes B, Provias J (2011) The case for blood-brain barrier dysfunction in the pathogenesis of Alzheimer's disease. J Neurosci Res 89:22–28
51. Jia W, Martin TA, Zhang G, Jiang WG (2013) Junctional adhesion molecules in cerebral endothelial tight junction and brain metastasis. Anticancer Res 33:2353–2359
52. Keaney J, Walsh DM, O'Malley T, Hudson N, Crosbie DE, Loftus T, Sheehan F, McDaid J, Humphries MM, Callanan JJ et al (2015) Autoregulated paracellular clearance of amyloid-beta across the blood-brain barrier. Sci Adv 1:e1500472
53. Kim BJ, Hancock BM, Bermudez A, Del Cid N, Reyes E, van Sorge NM, Lauth X, Smurthwaite CA, Hilton BJ, Stotland A et al (2015) Bacterial induction of Snail1 contributes to blood-brain barrier disruption. J Clin Invest 125:2473–2483
54. Kim GS, Yang L, Zhang G, Zhao H, Selim M, McCullough LD, Kluk MJ, Sanchez T (2015) Critical role of sphingosine-1-phosphate receptor-2 in the disruption of cerebrovascular integrity in experimental stroke. Nat Commun 6:7893
55. Kim HN, Kim YR, Ahn SM, Lee SK, Shin HK, Choi BT (2015) Protease activated receptor-1 antagonist ameliorates the clinical symptoms of experimental autoimmune encephalomyelitis via inhibiting breakdown of blood-brain barrier. J Neurochem 135:577–588
56. Klaassen I, Van Noorden CJ, Schlingemann RO (2013) Molecular basis of the inner blood-retinal barrier and its breakdown in diabetic macular edema and other pathological conditions. Prog Retin Eye Res 34:19–48
57. Knowland D, Arac A, Sekiguchi KJ, Hsu M, Lutz SE, Perrino J, Steinberg GK, Barres BA, Nimmerjahn A, Agalliu D (2014) Stepwise recruitment of transcellular and paracellular pathways underlies blood-brain barrier breakdown in stroke. Neuron 82:603–617
58. Komarova Y, Malik AB (2010) Regulation of endothelial permeability via paracellular and transcellular transport pathways. Annu Rev Physiol 72:463–493
59. Kreuter J (2014) Drug delivery to the central nervous system by polymeric nanoparticles: what do we know? Adv Drug Deliv Rev 71:2–14
60. Krueger M, Bechmann I, Immig K, Reichenbach A, Hartig W, Michalski D (2015) Blood-brain barrier breakdown involves four distinct stages of vascular damage in various models of experimental focal cerebral ischemia. J Cereb Blood Flow Metab 35:292–303
61. Krueger M, Hartig W, Reichenbach A, Bechmann I, Michalski D (2013) Blood-brain barrier breakdown after embolic stroke in rats occurs without ultrastructural evidence for disrupting tight junctions. PLoS One 8:e56419
62. Li G, Yuan W, Fu BM (2010) A model for the blood-brain barrier permeability to water and small solutes. J Biomech 43:2133–2140
63. Liebner S, Corada M, Bangsow T, Babbage J, Taddei A, Czupalla CJ, Reis M, Felici A, Wolburg H, Fruttiger M et al (2008) Wnt/beta-catenin signaling controls development of the blood-brain barrier. J Cell Biol 183:409–417
64. Lin MI, Yu J, Murata T, Sessa WC (2007) Caveolin-1-deficient mice have increased tumor microvascular permeability, angiogenesis, and growth. Cancer Res 67:2849–2856
65. Liu J, Jin X, Liu KJ, Liu W (2012) Matrix metalloproteinase-2-mediated occludin degradation and caveolin-1-mediated claudin-5 redistribution contribute to blood-brain barrier damage in early ischemic stroke stage. J Neurosci 32:3044–3057
66. Liu L, Wan W, Xia S, Kalionis B, Li Y (2014) Dysfunctional Wnt/beta-catenin signaling contributes to blood-brain barrier breakdown in Alzheimer's disease. Neurochem Int 75:19–25
67. Liu LB, Xue YX, Liu YH (2010) Bradykinin increases the permeability of the blood-tumor barrier by the caveolae-mediated transcellular pathway. J Neurooncol 99:187–194
68. Liu X, Zhou X, Yuan W (2014) The angiopoietin1-Akt pathway regulates barrier function of the cultured spinal cord microvascular endothelial cells through Eps8. Exp Cell Res 328:118–131
69. Lo EH, Dalkara T, Moskowitz MA (2003) Mechanisms, challenges and opportunities in stroke. Nat Rev Neurosci 4:399–415
70. Marchiando AM, Shen L, Graham WV, Weber CR, Schwarz BT, Austin JR 2nd, Raleigh DR, Guan Y, Watson AJ, Montrose MH et al (2010) Caveolin-1-dependent occludin endocytosis is required for TNF-induced tight junction regulation in vivo. J Cell Biol 189:111–126

71. Martinelli R, Gegg M, Longbottom R, Adamson P, Turowski P, Greenwood J (2009) ICAM-1-mediated endothelial nitric oxide synthase activation via calcium and AMP-activated protein kinase is required for transendothelial lymphocyte migration. Mol Biol Cell 20:995–1005
72. Martins T, Burgoyne T, Kenny BA, Hudson N, Futter CE, Ambrosio AF, Silva AP, Greenwood J, Turowski P (2013) Methamphetamine-induced nitric oxide promotes vesicular transport in blood-brain barrier endothelial cells. Neuropharmacology 65:74–82
73. Michel CC, Curry FE (1999) Microvascular permeability. Physiol Rev 79:703–761
74. Miles AA, Miles EM (1952) Vascular reactions to histamine, histamine-liberator and leukotaxine in the skin of guinea-pigs. J Physiol 118:228–257
75. Miller JW, Le CJ, Strauss EC, Ferrara N (2013) Vascular endothelial growth factor a in intraocular vascular disease. Ophthalmology 120:106–114
76. Montagne A, Barnes SR, Sweeney MD, Halliday MR, Sagare AP, Zhao Z, Toga AW, Jacobs RE, Liu CY, Amezcua L et al (2015) Blood-brain barrier breakdown in the aging human hippocampus. Neuron 85:296–302
77. Murakami T, Felinski EA, Antonetti DA (2009) Occludin phosphorylation and ubiquitination regulate tight junction trafficking and vascular endothelial growth factor-induced permeability. J Biol Chem 284:21036–21046
78. Nag S, Venugopalan R, Stewart DJ (2007) Increased caveolin-1 expression precedes decreased expression of occludin and claudin-5 during blood-brain barrier breakdown. Acta Neuropathol 114:459–469
79. Nagy JA, Benjamin L, Zeng H, Dvorak AM, Dvorak HF (2008) Vascular permeability, vascular hyperpermeability and angiogenesis. Angiogenesis 11:109–119
80. Neuwelt EA, Frenkel EP, Diehl J, Vu LH, Rapoport S, Hill S (1980) Reversible osmotic blood-brain barrier disruption in humans: implications for the chemotherapy of malignant brain tumors. Neurosurgery 7:44–52
81. Niewoehner J, Bohrmann B, Collin L, Urich E, Sade H, Maier P, Rueger P, Stracke JO, Lau W, Tissot AC et al (2014) Increased brain penetration and potency of a therapeutic antibody using a monovalent molecular shuttle. Neuron 81:49–60
82. Nitta T, Hata M, Gotoh S, Seo Y, Sasaki H, Hashimoto N, Furuse M, Tsukita S (2003) Size-selective loosening of the blood-brain barrier in claudin-5-deficient mice. J Cell Biol 161:653–660
83. Nolan DJ, Ginsberg M, Israely E, Palikuqi B, Poulos MG, James D, Ding BS, Schachterle W, Liu Y, Rosenwaks Z et al (2013) Molecular signatures of tissue-specific microvascular endothelial cell heterogeneity in organ maintenance and regeneration. Dev Cell 26:204–219
84. Oh P, Borgstrom P, Witkiewicz H, Li Y, Borgstrom BJ, Chrastina A, Iwata K, Zinn KR, Baldwin R, Testa JE et al (2007) Live dynamic imaging of caveolae pumping targeted antibody rapidly and specifically across endothelium in the lung. Nat Biotechnol 25:327–337
85. On NH, Savant S, Toews M, Miller DW (2013) Rapid and reversible enhancement of blood-brain barrier permeability using lysophosphatidic acid. J Cereb Blood Flow Metab 33:1944–1954
86. Orsenigo F, Giampietro C, Ferrari A, Corada M, Galaup A, Sigismund S, Ristagno G, Maddaluno L, Koh GY, Franco D et al (2012) Phosphorylation of VE-cadherin is modulated by haemodynamic forces and contributes to the regulation of vascular permeability in vivo. Nat Commun 3:1208
87. Palade GE (1953) An electron microscope study of the mitochondrial structure. J Histochem Cytochem 1:188–211
88. Patabendige A, Skinner RA, Abbott NJ (2013) Establishment of a simplified in vitro porcine blood-brain barrier model with high transendothelial electrical resistance. Brain Res 1521:1–15
89. Paul D, Cowan AE, Ge S, Pachter JS (2013) Novel 3D analysis of Claudin-5 reveals significant endothelial heterogeneity among CNS microvessels. Microvasc Res 86:1–10
90. Perriere N, Demeuse P, Garcia E, Regina A, Debray M, Andreux JP, Couvreur P, Scherrmann JM, Temsamani J, Couraud PO et al (2005) Puromycin-based purification of rat brain capillary endothelial cell cultures. Effect on the expression of blood-brain barrier-specific properties. J Neurochem 93:279–289

91. Petzold GC, Murthy VN (2011) Role of astrocytes in neurovascular coupling. Neuron 71:782–797
92. Prager B, Spampinato SF, Ransohoff RM (2015) Sphingosine 1-phosphate signaling at the blood-brain barrier. Trends Mol Med 21:354–363
93. Price CJ, Hoyda TD, Ferguson AV (2008) The area postrema: a brain monitor and integrator of systemic autonomic state. Neuroscientist 14:182–194
94. Radu M, Chernoff J (2013). An in vivo assay to test blood vessel permeability. J Vis Exp 73:e50062
95. Rapoport SI, Robinson PJ (1986) Tight-junctional modification as the basis of osmotic opening of the blood-brain barrier. Ann N Y Acad Sci 481:250–267
96. Regan ER, Aird WC (2012) Dynamical systems approach to endothelial heterogeneity. Circ Res 111:110–130
97. Rippe B, Haraldsson B (1994) Transport of macromolecules across microvascular walls: the two-pore theory. Physiol Rev 74:163–219
98. Rippe B, Rosengren BI, Carlsson O, Venturoli D (2002) Transendothelial transport: the vesicle controversy. J Vasc Res 39:375–390
99. Rist RJ, Romero IA, Chan MW, Couraud PO, Roux F, Abbott NJ (1997) F-actin cytoskeleton and sucrose permeability of immortalised rat brain microvascular endothelial cell monolayers: effects of cyclic AMP and astrocytic factors. Brain Res 768:10–18
100. Rochfort KD, Cummins PM (2015) The blood-brain barrier endothelium: a target for proinflammatory cytokines. Biochem Soc Trans 43:702–706
101. Sarker MH, Easton AS, Fraser PA (1998) Regulation of cerebral microvascular permeability by histamine in the anaesthetized rat. J Physiol 507(Pt 3):909–918
102. Sarker MH, Hu DE, Fraser PA (2000) Acute effects of bradykinin on cerebral microvascular permeability in the anaesthetized rat. J Physiol 528(Pt 1):177–187
103. Saubamea B, Cochois-Guegan V, Cisternino S, Scherrmann JM (2012) Heterogeneity in the rat brain vasculature revealed by quantitative confocal analysis of endothelial barrier antigen and P-glycoprotein expression. J Cereb Blood Flow Metab 32:81–92
104. Saunders NR, Daneman R, Dziegielewska KM, Liddelow SA (2013) Transporters of the blood-brain and blood-CSF interfaces in development and in the adult. Mol Aspects Med 34:742–752
105. Scallan J, Huxley VH, Korthuis RJ (2010) Capillary fluid exchange: regulation, functions, and pathology. Morgan & Claypool Life Sciences Publishers, San Rafael
106. Schallek J, Geng Y, Nguyen H, Williams DR (2013) Morphology and topography of retinal pericytes in the living mouse retina using in vivo adaptive optics imaging and ex vivo characterization. Invest Ophthalmol Vis Sci 54:8237–8250
107. Schulte D, Kuppers V, Dartsch N, Broermann A, Li H, Zarbock A, Kamenyeva O, Kiefer F, Khandoga A, Massberg S et al (2011) Stabilizing the VE-cadherin-catenin complex blocks leukocyte extravasation and vascular permeability. EMBO J 30:4157–4170
108. Seo JH, Guo S, Lok J, Navaratna D, Whalen MJ, Kim KW, Lo EH (2012) Neurovascular matrix metalloproteinases and the blood-brain barrier. Curr Pharm Des 18:3645–3648
109. Shue EH, Carson-Walter EB, Liu Y, Winans BN, Ali ZS, Chen J, Walter KA (2008) Plasmalemmal vesicle associated protein-1 (PV-1) is a marker of blood-brain barrier disruption in rodent models. BMC Neurosci 9:29
110. Shvets E, Bitsikas V, Howard G, Hansen CG, Nichols BJ (2015) Dynamic caveolae exclude bulk membrane proteins and are required for sorting of excess glycosphingolipids. Nat Commun 6:6867
111. Simionescu M, Gafencu A, Antohe F (2002) Transcytosis of plasma macromolecules in endothelial cells: a cell biological survey. Microsc Res Tech 57:269–288
112. Simionescu M, Ghinea N, Fixman A, Lasser M, Kukes L, Simionescu N, Palade GE (1988) The cerebral microvasculature of the rat: structure and luminal surface properties during early development. J Submicrosc Cytol Pathol 20:243–261
113. Sohet F, Lin C, Munji RN, Lee SY, Ruderisch N, Soung A, Arnold TD, Derugin N, Vexler ZS, Yen FT et al (2015) LSR/angulin-1 is a tricellular tight junction protein involved in blood-brain barrier formation. J Cell Biol 208:703–711

114. Sorokin L (2010) The impact of the extracellular matrix on inflammation. Nat Rev Immunol 10:712–723
115. Spyridopoulos I, Luedemann C, Chen D, Kearney M, Chen D, Murohara T, Principe N, Isner JM, Losordo DW (2002) Divergence of angiogenic and vascular permeability signaling by VEGF: inhibition of protein kinase C suppresses VEGF-induced angiogenesis, but promotes VEGF-induced, NO-dependent vascular permeability. Arterioscler Thromb Vasc Biol 22:901–906
116. Stamatovic SM, Keep RF, Andjelkovic AV (2008) Brain endothelial cell-cell junctions: how to "open" the blood brain barrier. Curr Neuropharmacol 6:179–192
117. Stan RV, Tse D, Deharvengt SJ, Smits NC, Xu Y, Luciano MR, McGarry CL, Buitendijk M, Nemani KV, Elgueta R et al (2012) The diaphragms of fenestrated endothelia: gatekeepers of vascular permeability and blood composition. Dev Cell 23:1203–1218
118. Stanimirovic DB, Friedman A (2012) Pathophysiology of the neurovascular unit: disease cause or consequence? J Cereb Blood Flow Metab 32:1207–1221
119. Steed E, Balda MS, Matter K (2010) Dynamics and functions of tight junctions. Trends Cell Biol 20:142–149
120. Taddei A, Giampietro C, Conti A, Orsenigo F, Breviario F, Pirazzoli V, Potente M, Daly C, Dimmeler S, Dejana E (2008) Endothelial adherens junctions control tight junctions by VE-cadherin-mediated upregulation of claudin-5. Nat Cell Biol 10:923–934
121. Takata F, Dohgu S, Matsumoto J, Machida T, Kaneshima S, Matsuo M, Sakaguchi S, Takeshige Y, Yamauchi A, Kataoka Y (2013) Metformin induces up-regulation of blood-brain barrier functions by activating AMP-activated protein kinase in rat brain microvascular endothelial cells. Biochem Biophys Res Commun 433:586–590
122. Terrando N, Eriksson LI, Ryu JK, Yang T, Monaco C, Feldmann M, Jonsson Fagerlund M, Charo IF, Akassoglou K, Maze M (2011) Resolving postoperative neuroinflammation and cognitive decline. Ann Neurol 70:986–995
123. Tietz S, Engelhardt B (2015) Brain barriers: crosstalk between complex tight junctions and adherens junctions. J Cell Biol 209:493–506
124. Tominaga N, Kosaka N, Ono M, Katsuda T, Yoshioka Y, Tamura K, Lotvall J, Nakagama H, Ochiya T (2015) Brain metastatic cancer cells release microRNA-181c-containing extracellular vesicles capable of destructing blood-brain barrier. Nat Commun 6:6716
125. Turowski P, Kenny BA (2015) The blood-brain barrier and methamphetamine: open sesame? Front Neurosci 9:156
126. van Tellingen O, Yetkin-Arik B, de Gooijer MC, Wesseling P, Wurdinger T, de Vries HE (2015) Overcoming the blood-brain tumor barrier for effective glioblastoma treatment. Drug Resist Updat 19:1–12
127. von Wedel-Parlow M, Schrot S, Lemmen J, Treeratanapiboon L, Wegener J, Galla HJ (2011) Neutrophils cross the BBB primarily on transcellular pathways: an in vitro study. Brain Res 1367:62–76
128. Wolburg H, Wolburg-Buchholz K, Engelhardt B (2005) Diapedesis of mononuclear cells across cerebral venules during experimental autoimmune encephalomyelitis leaves tight junctions intact. Acta Neuropathol (Berl) 109:181–190
129. Wolburg H, Wolburg-Buchholz K, Kraus J, Rascher-Eggstein G, Liebner S, Hamm S, Duffner F, Grote EH, Risau W, Engelhardt B (2003) Localization of claudin-3 in tight junctions of the blood-brain barrier is selectively lost during experimental autoimmune encephalomyelitis and human glioblastoma multiforme. Acta Neuropathol 105:586–592
130. Worzfeld T, Schwaninger M (2015) Apicobasal polarity of brain endothelial cells. J Cereb Blood Flow Metab 36:340–362
131. Yu HY, Cai YB, Liu Z (2015) Activation of AMPK improves lipopolysaccharide-induced dysfunction of the blood-brain barrier in mice. Brain Inj 29:777–784
132. Yu YJ, Atwal JK, Zhang Y, Tong RK, Wildsmith KR, Tan C, Bien-Ly N, Hersom M, Maloney JA, Meilandt WJ et al (2014) Therapeutic bispecific antibodies cross the blood-brain barrier in nonhuman primates. Sci Transl Med 6:261ra154
133. Yuan L, Le Bras A, Sacharidou A, Itagaki K, Zhan Y, Kondo M, Carman CV, Davis GE, Aird WC, Oettgen P (2012) ETS-related gene (ERG) controls endothelial cell permeability via transcriptional regulation of the claudin 5 (CLDN5) gene. J Biol Chem 287:6582–6591

134. Yuan SY, Rigor RR (2010) Regulation of endothelial barrier function. Morgan & Claypool Life Sciences, San Rafael
135. Zhang F, Xu CL, Liu CM (2015) Drug delivery strategies to enhance the permeability of the blood-brain barrier for treatment of glioma. Drug Des Devel Ther 9:2089–2100
136. Zhao LN, Yang ZH, Liu YH, Ying HQ, Zhang H, Xue YX (2011) Vascular endothelial growth factor increases permeability of the blood-tumor barrier via caveolae-mediated transcellular pathway. J Mol Neurosci 44:122–129
137. Zhou Y, Wang Y, Tischfield M, Williams J, Smallwood PM, Rattner A, Taketo MM, Nathans J (2014) Canonical WNT signaling components in vascular development and barrier formation. J Clin Invest 124:3825–3846

Blood–Brain Barrier Transporters and Neuroinflammation: Partners in Neuroprotection and in Pathology

Victoria Makrides, Elena Dolgodilina, and Daniela Virgintino

Abstract The blood–brain barrier (BBB) controls brain access of molecules and cells, communicates immune status to the central nervous system (CNS), and coordinates responses. Almost all solutes reach the brain from the blood through BBB transendothelial transporters that are often differentially localized on luminal and abluminal endothelial membranes. Therefore, BBB transporters crucially regulate CNS homeostasis and physiological responses to internal and external stimuli. Although the main functions of inflammatory processes are to remove the causes of the insult, to clear necrotic tissue, and to initiate tissue repair, these signals are often associated with further CNS damage in neuroinflammation. The BBB can aid adaptive responses; however, one hallmark of neuroinflammation is a breakdown of the tight endothelial barrier resulting in BBB opening, vascular leakage, and the loss of control over transendothelial transport. Thus, neuroinflammation often leads to pathological changes in CNS functions, such as those associated with the chronic age-related neuropathologies, Alzheimer's disease and Parkinson's disease, mechanical or physical insult and injury to the CNS (traumatic brain injury), infection, oncogenic diseases, and chronic autoimmune diseases, such as, multiple sclerosis and diabetes mellitus. Here, we review responses of selected ATP-binding cassette, solute carrier, receptor, and vesicular BBB endothelial transporters to neuroinflammatory mediators and diseases. Much remains to be learned concerning interactions of age, gender, genetics, microbiome, circadian cycle, diet, exercise,

V. Makrides (✉)
Institute of Physiology, University of Zurich,
Winterthurerstrasse 190, 8057 Zurich, Switzerland
e-mail: makrides@access.uzh.ch

E. Dolgodilina
Institute of Physiology, University of Zurich,
Winterthurerstrasse 190, 8057 Zurich, Switzerland

Zurich Center for Integrative Human Physiology (ZIHP), University of Zurich,
Zurich, Switzerland

D. Virgintino
Department of Basic Medical Sciences, Neurosciences and Sense Organs,
University of Bari, Bari, Italy

R. Lyck, G. Enzmann (eds.), *The Blood Brain Barrier and Inflammation*,
Progress in Inflammation Research, DOI 10.1007/978-3-319-45514-3_6

medication, etc., in neuroinflammatory processes. Crucially, in many regards, it is still unclear what tips the system from providing protective to effecting destructive responses.

1 Introduction

The blood–brain barrier (BBB) has a complex relationship with the immune system. It is simultaneously charged with controlling brain access by and egress of molecules and cells, communicating peripheral status to the central nervous system (CNS), and responding to peripheral and CNS stimuli [1]. The endothelial cells of the BBB form an interface between the blood and the CNS tissue and interstitial fluid (ISF). The barrier properties of endothelial cells are conferred by the expression of contiguous tight junctions (TJs), which virtually seal paracellular capillary pathways, coupled with a lack of endothelial fenestrations. These features effectively minimize solute diffusion and cellular movement between the blood and the brain [2–4]. Although there is a small but detectable paracellular diffusion of water-soluble molecules through the TJs, and some small lipid soluble substances do diffuse transcellularly through the endothelial plasma membranes. However, almost all other solutes are transported through the BBB by active pump, carrier, receptor protein, and vesicular mechanisms that are often differentially localized on the two endothelial plasma membranes (luminal vs abluminal). Therefore, BBB transporters serve as key regulators of CNS homeostasis. For example, comparison of the solute concentrations of brain ISF amino acids with plasma or cerebrospinal fluid (CSF) concentrations reveals steep gradients that are strongly influenced by BBB transporter activity. These data support the crucial roles of BBB transport in controlling physiological responses to internal and external perturbations [5–9].

Neuroinflammation can be broadly characterized as CNS responses to *inflammatory* signals. In general, inflammation is induced by primary stimuli (e.g., diseases, traumas, toxins), which activate cells (i.e., in the CNS the microglia, astroglia, and endothelial cells) to produce cytokines (e.g., interleukins such as IL-1β and IL-6, and tumor necrosis factor [TNF α]), chemokines (e.g., CCL2, CCL5, CXL1), secondary messengers (e.g., nitric oxide [NO], prostaglandins) or reactive oxygen species (ROS) [1]. The primary functions of inflammatory responses are to remove the causes of the insult, to clear necrotic tissue, and to initiate tissue repair. Unfortunately, in neuroinflammation these signals are often associated with further CNS damage [1, 10].

The BBB in general and BBB localized transport proteins specifically are important mediators of CNS–immune system interactions and communication. Microvascular endothelial cells of the brain both produce and transport various inflammatory signals, mediators, and effectors. In this manner, the BBB participates in communicating immune status between the immune system and the CNS [11, 12]. Although the up- or downregulation of BBB transporters as a consequence of neuroinflammatory stimuli can provoke CNS damage, it can also aid adaptive responses by restricting access to and/or removing neurotoxic agents from the CNS, and/or promoting the influx of important components for repair. Adaptive responses to inflammatory mediators include, for example, the induction of *sickness behavior*, defined as physiological and behavioral responses including fever, decreased activity, and social isolation [12, 13]. However, one hallmark effect of neuroinflammation is the breakdown of the tight endothelial barrier, resulting in BBB *opening*, increased paracellular permeability or vascular *leakage*, and the consequent loss of BBB control over transendothelial transport [12, 13]. Therefore, although responses can be protective and promote adaptive physiological outcomes, neuroinflammatory processes often lead to pathological changes in CNS functions, such as are associated with chronic age-related neuropathologies, for example, Alzheimer's disease (AD) and Parkinson's disease (PD), and with CNS mechanical or physical insult, injury (traumatic brain injury [TBI]), infection, or oncogenic diseases, and chronic autoimmune diseases such as multiple sclerosis (MS) and diabetes mellitus (DM).

The focus of this review is on BBB transporters known to be affected by or participate in neuroinflammatory responses. First, the salient features of selected ATP binding cassette (ABC) efflux pumps, solute carrier (SLC) transporters, and receptor-mediated and vesicle-associated transport proteins expressed in BBB are summarized (Table 1). Figures 1 and 2 illustrate their transport mechanisms, substrates, and endothelial membrane localization. Finally, a brief overview of several neuroinflammatory diseases, injuries, or insults is presented, each followed by a summary of what is known about the contributions and regulation of the aforementioned transporters in the context of these inflammatory processes (Figs. 3 and 4).

2 BBB Transporter Proteins

2.1 ATP-Binding Cassette Transporters

The Human Genome Organization classifies the 51 genes encoding the human ABC superfamily as grouped into seven sub-families (ABC-A to ABC-G) [14, 15]. The generic ABC transporter protein structure is a variable number of N terminal transmembrane (TM) domains (TMD) linked to a cytoplasmic nucleotide (ATP) binding domain (NBD); this motif (TMD with NBD) may or may not be repeated to end in a second ATP binding domain (referred to as *full* vs *half* transporters, respectively). The current theory is that ABC transporters evolved as a response to environmental toxins; as such they are potent and efficient efflux pumps that harness energy from ATP hydrolysis to transport drugs, xenobiotics, and metabolites [16] (Fig. 1a). In the brain, ABC pumps are highly expressed in microvascular endothelial cells and to a lesser extent in other neurovascular unit ([NVU]; i.e., astrocytes, pericytes) and immune cells (microglia), where they mediate transport between the brain and the blood stream (Fig. 1, Table 1). Consequently, they contribute to brain detoxification, neuroprotection, neuroregeneration, and overall normal CNS physiology.

Table 1 Selected blood–brain barrier endothelial transporters, substrates, membrane localization, direction of transport, and associated neuroinflammatory diseases and insults

Transporters gene protein	**Substrates**	**Endothelial membrane localization**	**Net transport direction**	**Associated neuroinflammatory disease/insults**
ABC efflux pump transporters				
ABCA1 CERP	Cholesterol	Luminal, abluminal	Bidirectional	AD
ABCB1 MDR1/P-gp	Cationic amphiphilic and lipophilic compounds	Luminal, abluminal	Brain to blood	E, BT, IS, AD MS, PD, ALS, HIV, DM
ABCC1 MRP1	Neutral hydrophobic compounds	Luminal, abluminal	Brain to blood	E, BT, IS, HIV, DM
ABCC4 MRP4	Nucleosides, cyclic nucleotides, prostaglandins, reduced GSH conjugates	Luminal, abluminal	Brain to blood	E, DM
ABCG2 BCRP	Drugs, xenobiotics	Luminal	BMVEC to blood	E, BT, AD, MS, PD, ALS, HIV, DM
Solute carrier transporters				
SLC2A1 GLUT1	Glucose	Luminal, abluminal	Bidirectional	AD, E, IS, DM
SLC7A5 LAT1	Large neutral L-amino acids, leucine, phenylalanine etc.	Luminal, abluminal	Bidirectional	BT, PD, LPS, HE
SLC16A2 MCT8	T3 thyroid hormones	Abluminal	BMVEC to brain	LPS

SLC21A5/SLCO1A4 OATP1A4	Broad: cardiac glycosides, bile acids, steroid conjugates, some cations, some peptides	Luminal, abluminal	Bidirectional	IS, LPS
SLC21A14/SLCO1C1 OATP1C1	Thyroxine	Luminal, abluminal	Bidirectional	LPS
SLC22A8 OAT3	Broad: amphiphilic organic ions, organic anions, organic cations	Abluminal	Brain to BMVEC	LPS
Receptor-mediated and vesicle-associated transport				
LRP1 LRP1	Apolipoproteins, Aβ, ApoE;	Abluminal	Brain to BMVEC	AD, PD, LPS
AGER RAGE	Glycosylated proteins, Aβ,	Luminal	Blood to BMVEC	AD, DM, ALS, LPS
CAV1 CAV1	Large molecular weight proteins, ligand-bound receptor clusters	Luminal	Blood to BMVEC	IS, MS, TBI, HIV, LPS

Refs. [8, 10, 12, 15, 50, 67, 73, 76, 105, 114, 122, 137, 147, 153–155]

AD Alzheimer disease, *Aβ* amyloid beta peptides, *ALS* amyotrophic lateral sclerosis, *ApoE* apolipoprotein E, *BMVEC* brain microvascular endothelial cell, *BT* brain tumors, *DM* diabetes mellitus, *E* epilepsy, *HE* hepatic encephalopathy, *HIV* HIV related brain disorders, *IS* ischemic stroke, *LPS* lipopolysaccharide, *MS* multiple sclerosis, *PD* Parkinson's disease

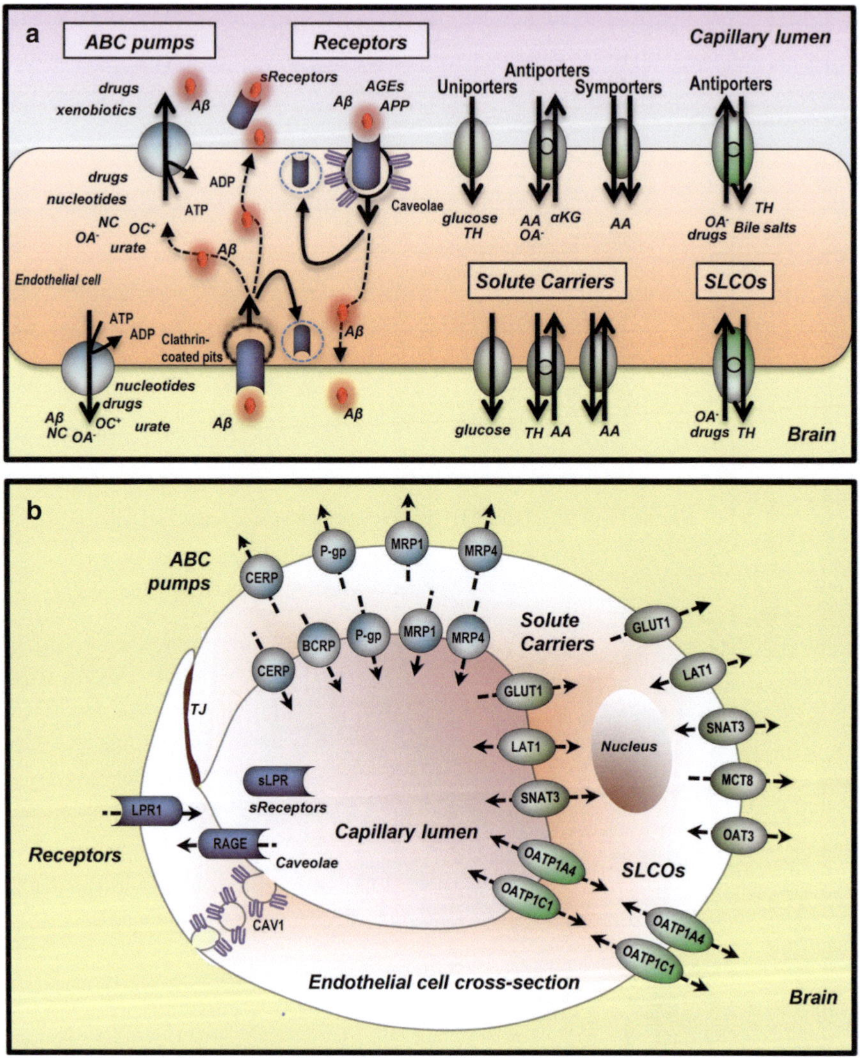
a
ABC pumps
Receptors
Capillary lumen
drugs
xenobiotics
Aβ
sReceptors
AGEs
APP
Antiporters
Uniporters
Symporters
Antiporters
drugs
nucleotides
ADP
ATP
Caveolae
NC
OC+
OA-
urate
glucose
TH
AA
αKG
OA-
AA
OA-
drugs
TH
Bile salts
Endothelial cell
Solute Carriers
SLCOs
ATP
ADP
Clathrin-
coated pits
nucleotides
drugs
Aβ
NC
OC+
OA-
urate
glucose
TH
AA
AA
OA-
drugs
TH
Brain
b
ABC
pumps
CERP
P-gp
MRP1
MRP4
BCRP
Solute
Carriers
GLUT1
LAT1
SNAT3
MCT8
OAT3
Nucleus
TJ
sLPR
sReceptors
LPR1
RAGE
Caveolae
Receptors
Capillary lumen
OATP1A4
OATP1C1
SLCOs
CAV1
Endothelial cell cross-section
Brain

In the BBB, as in other barrier membranes, ABC transporters play a key role in the ADMETox profile of drugs (i.e., drug absorption, distribution, metabolism, excretion, and toxicity). They both control and limit concentrations of substrates with high passive permeability and facilitate the excretion of substrates (and their conjugates) with low passive permeability [17]. The transporters localized to the endothelial luminal membrane actively pump chemicals that have diffused into cells back to the blood, and therefore directly mediate the barrier function of the BBB [18]. Additionally, for abluminal transporters to mediate the net flux of drugs from the periphery to the brain their substrates must first pass the luminal membrane transporters many of which are efflux pumps with overlapping substrate specificity with abluminal transporters. Thus, the extent to which abluminal ABC transporters

Fig. 1 Blood-brain barrier transporters, mechanisms, substrates, and endothelial membrane localization. (**a**). Diagram of the blood-brain barrier (BBB) ATP-binding cassette (ABC pumps), solute carrier (SLC), organic ion transporting peptides (SLCO), receptor-mediated and vesicle-associated transport mechanisms, and selected substrates. *ABC pumps* catalyze the movement of substrates by harnessing energy released from ATP hydrolysis to ADP (indicated by *curved solid lines* with *arrows*) [37, 154]. Direction of transport is indicated by *solid lines* and *arrows* through *circular symbols* indicating transporter proteins. *Solute carrier transporters* operate by three mechanisms, uniport, antiport, and/or symport. Direction of transport is shown as *solid arrows* through *ovals* representing transporters. Uniporters transport substrates along their concentration gradient, antiporters require exchange of substrates between membrane faces, and symporters require co-transport of two or more substrates (e.g., amino acid [AA] and sodium ion [Na^+]). Some SLCs combine antiport with symport transport of substrates, i.e., Na^+ symport with AA coupled with proton antiport [52]. Uniport is indicated by a *single arrow* in the direction of transport. Antiport is indicated by *small circles* between two *oppositely oriented arrows*. Symport is indicated by *two arrows* in the same orientation. *Receptor-mediated transport* of substrates is shown through vesicles (i.e., caveolae and clathrin-coated pits). The scheme shows Aβ internalization and transendothelial transport. Direction of transport is shown by *arrows with dotted lines*. Internalized receptors recycling to membranes are indicated by *solid arrows*. Soluble transporters are labeled sReceptors. Substrate key: *αKG* alpha keto glutarate, *OA*$^-$ organic anions, *OC*$^+$ organic cations, *NC* nonpolar compounds, *Aβ* amyloid beta peptides, *TH* thyroid hormones. (**b**). Simplified scheme showing membrane localization of selected ABC pumps, SLC, RMT, and vesicle-associated proteins of an endothelial cell cross-section through the nuclear region (Fig. 2a–i). The closely associated BBB cell types that with endothelial cells comprise the neurovascular unit (i.e., pericytes, astrocytes, and neurons) and the brain interstitial space are not shown individually, but are jointly referred to as *brain*. The endothelial membranes are shown as separated by the nucleus (nucleus), a transcytotic vesicle (caveolae) is indicated, as are tight junction (TJ) proteins sealing apposing endothelial membranes. Transporters are shown localized to the luminal facing (capillary lumen) and abluminal (brain) membranes. The transport proteins are grouped as: ABC pumps, SLCs, SLCO, and vesicular and receptor associated transport (receptors). Transporter key: ABC pumps are CERP/ABCA1, MRP1/ABCC1, MRP4/ABCC4, P-gp/ABCB1, BCRP/ABCG2; SLC carriers are GLUT1/SLC2A1, LAT1/SLC7A5, MCT8/SLC16A2, SNAT3/SLC38A3, OAT3/SLC22A8, SLCO transporters are OATP1A4/SLCO1A4, and OATP1C1/SLCO1C1. Receptors are *LRP1* low-density lipoprotein receptor-related protein-1, *sLRP1* soluble LRP1, *RAGE* receptor for advanced glycation end products, and the caveolae-associated protein CAV1/caveolin-1. The net direction of substrate transport is indicated by *arrows with broken lines* [8, 10, 12, 15, 50, 67, 73, 76, 105, 114, 122, 137, 153–155]

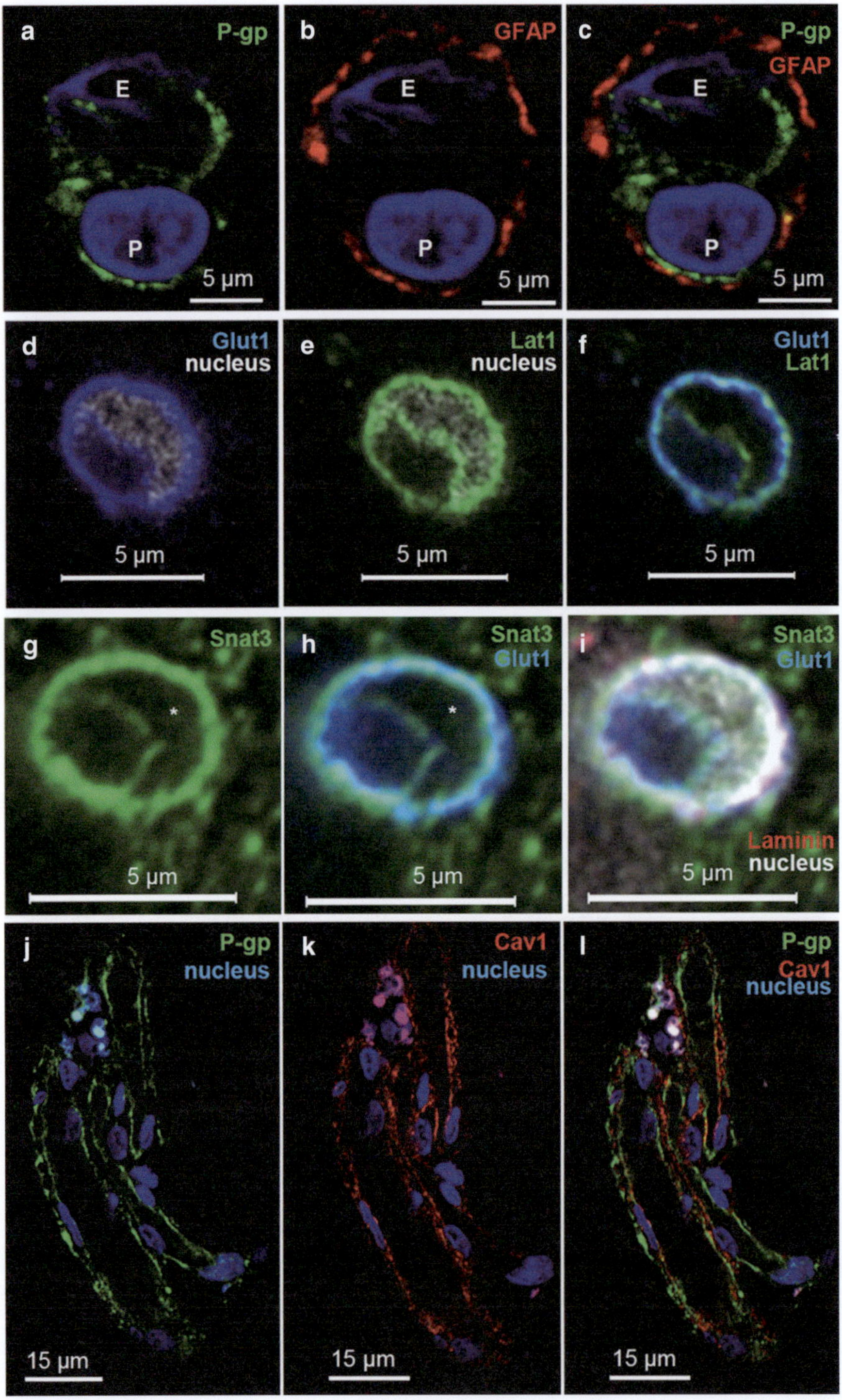
a
P-gp
E
P
5 μm
b
GFAP
E
P
5 μm
c
P-gp
GFAP
E
P
5 μm
d
Glut1
nucleus
5 μm
e
Lat1
nucleus
5 μm
f
Glut1
Lat1
5 μm
g
Snat3
5 μm
h
Snat3
Glut1
5 μm
i
Snat3
Glut1
Laminin
nucleus
5 μm
j
P-gp
nucleus
15 μm
k
Cav1
nucleus
15 μm
l
P-gp
Cav1
nucleus
15 μm

Fig. 2 Brain microvascular expression and localization of selected ABC pumps, SLCs, and RMT and vesicle-associated transport proteins. (**a–c**) Representative adult human parahippocampal cortex tissue sections double immunolabeled for P-glycoprotein/ABCB1 (P-gp) (**a**, **c**, *green*) and glial fibrillary protein (GFAP) (**b**, **c**, *red*), and imaged using confocal microscopy; nuclei were stained with TOPRO-3 (**a–c**, *blue*). Images **a–c** were reprinted with permission from Virgintino et al. [29]. (**d–i**) Representative images of mouse brain tissue sections stained for LAT1 (**e**, **f**, *green*), GLUT1 (**d**, **f**, **h**, **i**, *blue*), SNAT3 (**g–i**, *green*), and laminin (Lam; **i**, *red*), nuclear counter staining was carried out with POPRO-1 (*white*), and sections were imaged by confocal microscopy. **d–f** and **g–i** are of the same sections respectively. All three transporters are localized to both luminal and abluminal membranes. Scale bars are as shown. Images were reprinted with permission from Ruderisch et al. [8]. (**j–p**) Representative images of human cerebral cortex sections double immunolabeled for caveolin-1 (Cav1; **k**, **l**, **m**, **o**, **p**; *red*) and p-glycoprotein (P-gp; **j**, **l**, **m**, **n**, **p**, *green*), with nuclear counterstaining using TOPRO-3 (*blue*). **j–l** Both Cav1 and P-gp are detected on the wall of cortex microvessels and partly co-localize on the luminal membrane; at a higher magnification (**m**) the P-gp signal is localized on the endothelial cells and prevails on the luminal membrane, whereas Cav1 concentrates on the abluminal membrane (unpublished data). (**n–p**) P-gp stain appears concentrated on the luminal compartment of endothelial cells (*arrow*). Cav1 stains the entire cytoplasm, luminal staining shows fine puncta (**n–p**, *arrow*), and abluminal staining shows more intense, large puncta (**o**, *arrowhead*). (**h**) Merged image with co-localization in the endothelial luminal compartment (*arrow*). Note in **m**, **n**, and **p**, P-gp expression on pericytes (pericyte nucleus, P in **n**). Scale bars are as shown. Images shown in **n–p** were reprinted with permission from Virgintino et al. [94] (**a–c** Virgintino et al. [29] – Fig. 3g–i. 7 June 2016: RightsLinks license number 3883041124960; **d–g** Ruderisch et al. [8] – Figs. 1c–f and 3a, b. 7 June 2016: author, have the right to reuse in a book; **n–p** Virgintino et al. [94] – Fig. 2f–h. 7 June 2016: author, have the right to reuse in a book)

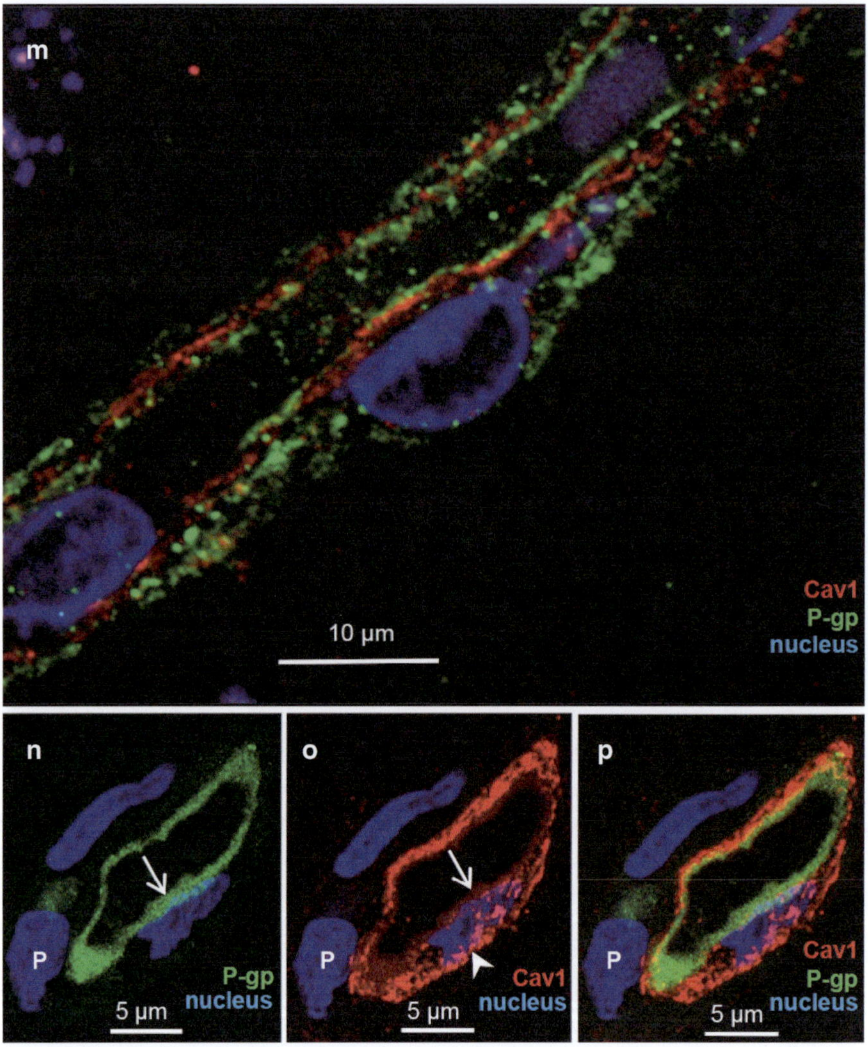

Fig. 2 (continued)

can deliver therapeutics to the CNS is influenced by the substrate affinities and protein expression and activity of the uptake and efflux transporters, and the level of passive substrate permeability [18]. In this way, transporters contribute to the BBB bottleneck for drug delivery to the CNS [19].

There are three identified signaling patterns regulating ABC transporter expression and activity at the BBB: (1) activation of nuclear receptors by endogenous metabolites, nutrients, and xenobiotics, directly increasing levels of multiple ABC transporters; (2) ligand-bound receptors that activate signaling pathways, which activate downstream transcription factors modulating ABC expression; and (3) activation of signaling

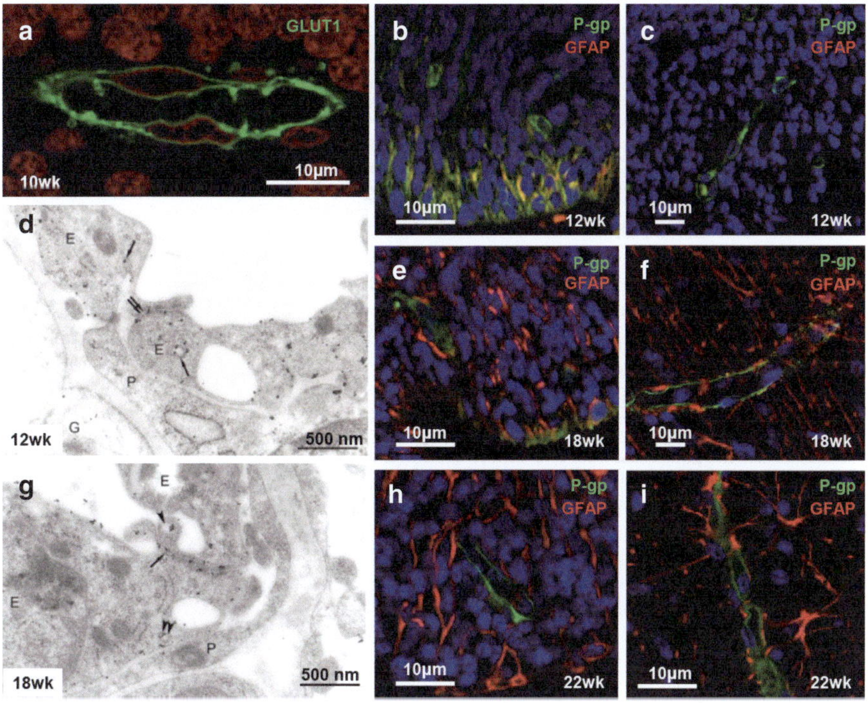

Fig. 3 Neurovascular expression of GLUT1/SLC2A1, P-gp, and GFAP in human telencephalon during early fetal development. (**a**, **d**, **g**) GLUT1 localization in BBB endothelial cells at 10 weeks' (**a**), 12 weeks' (**d**), and 18 weeks' (**g**) gestation. (**a**) Representative vascular field of a section immunostained with anti-GLUT1 antibodies (*green*), nuclei are counterstained with propidium-iodide (*red*), and imaged using confocal microscopy; note the localization of the transporter on both the luminal and abluminal endothelial membranes (unpublished data). **d** and **g** Ultrathin sections immunolabeled for GLUT1, with 5-nm gold particles and silver enhancement, imaged using electron micrscopy; gold particles show the higher density of the transporter on abluminal and lateral endothelial membranes at both 12 and 18 weeks. This is an asymmetrical distribution that already reflects the transporter functional activity as observed in the adult brain. Additionally, in (**d**) a coated vesicle and a caveolae (*arrows*), and the TJs (*double arrows*) are labeled with gold particles; a few particles are also seen within the cytoplasm and the glial end-feet. In (**g**) particles also decorate GLUT1 associated with uncoated (*arrowhead*) and coated (*double arrowheads*) vesicles, in addition to junctional lateral membranes (*arrow*). *E* endothelial cells, *P* pericytes, *G* astroglial cells. Images in **d** and **g** were reprinted with permission from Virgintino et al. [110]. (**b**, **c**, **e**, **f**, **h**, **i**) Telencephalon coronal sections from human fetuses stained with anti-P-gp and anti-GFAP antibodies, and imaged using confocal microscopy. (**b**, **e**, **h**) P-gp/ABCB1 (*green*) and GFAP (*red*) expression in the ventricular zone of the fetal telencephalon at 12, 18, and 22 weeks respectively. Nuclei were counter-stained with TOPRO-3 (*blue*). At the earliest developmental stages both endothelial cells and GFAP-reactive radial glia-like cell bodies express high levels of P-gp; in glial cells, P-gp expression gradually decreases, and at 22 weeks the transporter is only expressed by endothelial cells (unpublished results). (**c**, **f**, **i**) P-gp is precociously expressed by endothelial cells (12 weeks) and is clearly concentrated on the luminal endothelial membrane at 18 and 22 weeks; in nonventricular zone regions neither radial glia cells nor astrocytes (labeled with GFAP) express the P-gp. Scale bars are 10 μm. Image in **c** was reprinted with permission from Virgintino et al. [29]; all other images are previously unpublished (**c** Virgintino et al. [29] – Fig. 5c. 10 June 2016: RightLinks license number: 3885560928199; **d**, **g** Virgintino et al. [156] – Fig. 1a, b. 10 June 2016: RightsLinks license number 3885560560317)

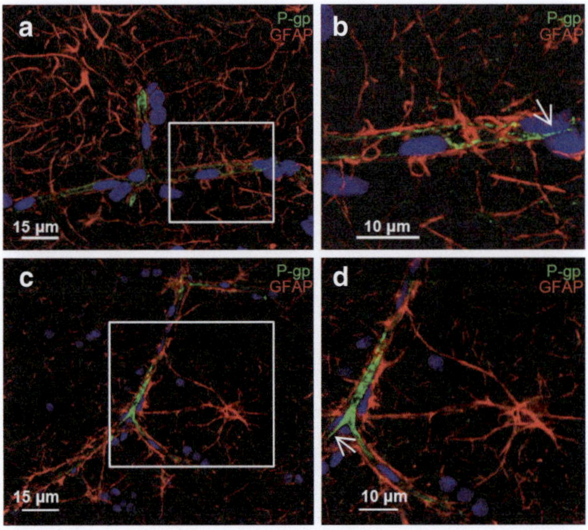

Fig. 4 Neuroinflammatory disease effects on the BBB transporter and TJ proteins. (**a–d**) Representative images of tissue sections from adult human brains. Hippocampal tissue sections from control (**a**, **b**) and mesial temporal lobe epilepsy (MTLE) patients (**c**, **d**) were stained for P-gp/ABCB1 (*green*) and GFAP (*red*) expression; nuclei are counter-stained with TOPTO-3 (*blue*). Stained sections were imaged using confocal microscopy. (**b**, **d**) High magnification images from the boxed regions on **a** and **c** respectively. High expression of P-gp appears restricted to the luminal side of microvascular endothelial cells (**b**, **d**; *arrows*) and is not detectable on GFAP-positive astrocytes in both control (**a**, **b**) and MTLE (**c**, **d**) samples. P-gp expression appears up-regulated in the section from an MTLE patient (**c**, **d**) vs the control section (**a**, **b**) [136]; (unpublished data). (**e–j**) Murine cortical tissue sections from healthy mice (**e–g**) and mice with experimental autoimmune encephalomyelitis (EAE), a murine model for multiple sclerosis (**h–j**). Tissue sections were stained with antibodies raised against caveolin1 (Cav1; *red*) (**e–j**) in double labeled sections with claudin5 (CLDN5; *green*) in (**g**, **j**). For all panels, nuclei were counterstained with TOPRO-3 (*blue*). In healthy mice, endothelial Cav1 forms a regular linear pattern (**e–g**), whereas in mice with EAE, it accumulates within the endothelial cytoplasm (**h–j**). In control microvessels (**g**), CLDN5 staining forms a linear and continuous pattern throughout the endothelial profiles. In EAE mice (**j**), CLDN5 is lost and its pattern is only recognizable as rows of fine puncta, including at the origins of vessel branches. Images shown in panels **e**, **g**, **h**, **i** were reprinted with permission from Errede et al. [96]; all other images are previously unpublished (**e**, **g**, **h**, **j** Errede et al. [96] – respectively Figs. 4a, i, d and 8d. 10 June 10: RightsLinks license number 3885561342159)

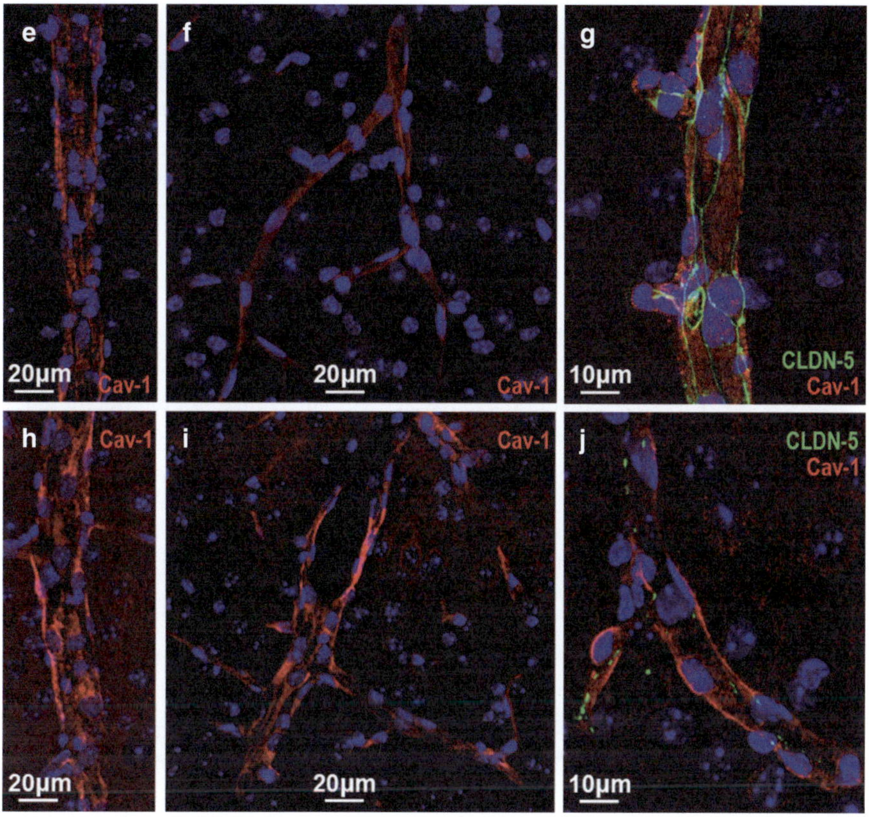

Fig. 4 (continued)

pathways that rapidly and reversibly reduce ABC transport activity [18]. The activation of nuclear receptors, such as pregnane X receptors (PXRs), arylhydrocarbon receptor (AhR), constitutive androstane receptor (CAR), vitamin D receptor (VDR), and the peroxisome proliferator-activated receptor (PPAR), generally upregulates ABC transporter expression. Ligand binding results in the translocation of ligand-receptor complexes to the nucleus for binding to the promoter regions of target genes to facilitate transcriptional complex assembly. In the nucleus, some receptors heterodimerize (e.g., 9-cisretinoic acid receptor [RXR], and CAR) before binding the promoter [18, 20].

Stress signals often trigger activation of nuclear factor-kappa light-chain enhancer of activated B-cells (NF-κB), which is a key transcription factor. For example, inflammation, glutamate, and oxidative stress stimulate NF-κB signaling via p53 or N-methyl-D-aspartate (NMDA) receptor pathways. Additionally, signaling via tumor necrosis factor receptor 1 (TNFR1) activates both NF-κB and activator protein 1 (AP-1)–mediated transcription [15, 18, 21]. Overall, the responsiveness of ABC transporters to endogenous and exogenous stimuli underscores their importance in regulating CNS homeostasis.

2.1.1 Subfamily A Member: CERP (ABCA1)

The ABC transporter sub-family A member 1 gene (*ABCA1*) codes for a protein (cholesterol efflux regulatory protein, CERP/ABCA1) displaying the prototypic full ABC structure with two symmetrical halves each containing TMD and NBD motifs. It is reported to be primarily expressed in the BBB on abluminal endothelial membranes (Fig. 1b, Table 1) [22]. CERP is an efflux pump for cholesterol and phospholipids from cell membranes to the lipid-poor apolipoproteins, apolipoprotein E (ApoE) in the brain, and apolipoprotein AI (ApoA-I) at the periphery [23]. It participates in the first step for the catalysis of high-density lipoproteins (HDL). In that regard it is best known for the genetic defects in *ABCA1* causing the autosomal, co-dominant Tangiers disease, which, when homozygous, leads to the complete absence of HDL. The localization of ABCA1 primarily on abluminal membranes supports a role in brain cholesterol homeostasis through the lipidation of ApoE [22]. The latter has also been implicated in Aβ transport, although this is controversial. One group reported overexpression of ABCA1 in a mouse model by stimulation of liver X receptors (LXR), which increased circulating cholesterol and brain ApoE levels and decreased brain amyloid β (Aβ) [24]. The same authors reported that knockout ABCA1 mice (*Abca1*$^{-/-}$) had lowered cholesterol and ApoE, but demonstrated no change in Aβ [24]. Although a second group reported the knockout to be deficient in HDL overall, they showed decreased ApoE and increased Aβ in the brain [25]. However, it has been recently suggested that rather than transporting Aβ itself, ABCA1 causes ApoE lipidation, which then interacts more efficiently with Aβ for efflux by lipoprotein receptor-related protein 1 (LRP1) and ABCB1 [26].

Substrates In the brain, CERP mediates the efflux of cholesterol and phospholipids to ApoE; however, it has been suggested that this might be an indirect consequence of the formation of lipid-rich microdomains rather than direct ABCA1 binding (Table 1) [27].

BBB Localization ABCA1 is localized mainly on brain capillary endothelial abluminal membranes, but also to a lesser extent on luminal membranes (Fig. 1b, Table 1) [22, 25].

Regulation Stimulation of the activity of the liver (LXR) and peroxisome (PPARα, PPARγ) nuclear receptors increases mRNA expression and both luminal and abluminal BBB protein expression of ABCA1 [24].

Diseases CERP is involved in brain tumors and AD (Table 1) [15, 25, 28].

2.1.2 Subfamily B Member: MDR1/P-gp (ABCB1)

In humans, the *ABCB1* gene codes for the well-known 170-kDa protein, named variously permeability glycoprotein-1 (P-gp), multidrug resistance protein-1 (MDR1), or cluster of differentiation 243 (CD243). Hereafter, it is referred to as either P-gp or ABCB1 in the text and in the figures. It has the prototypic full ABC transporter structure with two associated TMD/NBD domains separated by the substrate binding domain. P-gp is relatively ubiquitously expressed in barrier membranes throughout the body. One of the highest levels of protein expression is in brain microvascular endothelial cells localized to luminal membranes (Figs. 1b and 2j–n) [21, 29]. In the BBB, P-gp regulates the distribution and bioavailability of substrates and removes toxic metabolites and xenobiotics from cells and tissues, including the efflux of compounds from the CNS. It has been suggested that hydrophobic substrates, which diffuse from the blood into endothelial luminal membranes, are directly extracted from the lipid bilayer (the *hydrophobic vacuum cleaner* or *flippase* models). Therefore, lipophilic drugs that are P-gp substrates are transported out of the BBB, without accessing the endothelial interior [18].

There are a number of known polymorphisms affecting the expression or activity of P-gp that are hypothesized to influence drug resistance and the progression of various neuropathologies, for example, epilepsy, AD, schizophrenia, and other illnesses [16]. Currently, 1,630 single-nucleotide polymorphisms (SNPs) have been described, 56 of which are nonsynonymous, i.e., code for a different amino acid. However, even synonymous SNPs, such as C3435T, which codes for an isoleucine in exon 26, can affect the P-gp expression, substrate affinity, and drug resistance of carriers. Ethnicity has been shown be correlated with the genotypic prevalence of SNPs, in addition to P-gp haplotypes (i.e., the inheritance of several linked SNPs). For example, C3435T is linked to G2677T/A (nonsynonymous single nucleotide polymorphism [nsSNP] in exon 21) and G1236T (synonymous SNP in exon 12). The CGC genotype is most common among people of African descent, whereas TTT is prevalent in Asian and Indian ethnicities [16].

Substrates P-gp transports a broad range of cationic amphiphilic and lipophilic compounds [21]. Substrates include a wide variety of drugs (e.g., colchicine, tacrolimus, and quinidine), chemotherapeutic agents (e.g., etoposide, doxorubicin, and vinblastine), cardiac glycosides (e.g., digoxin), immunosuppressive agents, glucocorticoids (e.g., dexamethasone), human immunodeficiency virus (HIV)-type 1 antiretroviral therapeutics (e.g., protease inhibitors, non-nucleoside reverse transcriptase inhibitors), in addition to naturally occurring or endogenous molecules (e.g., lipids, steroids, xenobiotics, peptides, and bilirubin; Table 1). P-gp is also inhibited by a number of antagonists, including atorvastatin, amlodipine, cyclosporin A, dexniguldipine, disulfiram, GF120918, LY475776, LY335979, MS-209, nifedipine, OC144-093, pluronic L61, PSC-833, quinidine, R101933, S9788, VX-710, XR-9576, V-104, and verapamil [30].

BBB Localization Overall, P-gp is highly expressed on luminal membranes of brain capillary endothelial cells (Figs. 1b and 2j–n, Table 1) [21, 29]. It is also expressed on abluminal membranes; however, expression has been shown to be heterogeneous [31]. On some endothelial cells, localization is restricted to luminal endothelial membranes; moreover, on some microvascular tracts it is completely absent [29].

Regulation Several signaling pathways regulate P-gp expression: (1) TNFα signaling through TNFR1 results in endothelin 1 (ET1) release, which signals through nitric oxide synthase (NOS) and protein kinase Cβ (PKCβ) via endothelin B receptors to alter P-gp expression [10]. (2) P-g expression is regulated by the nuclear receptors: PXR, CAR, AhR, PPARα, VDR, and glucocorticoid receptor (GR). A number of drugs, dietary constituents, and nutraceuticals (PXR, CAR, PPARα, VDR, and GR) and environmental contaminants (AhR) are ligands for these receptors [18]. (3) Glutamate signaling by NMDA receptors, cyclooxygenase 2 (COX2), and the prostaglandin E_2 (PGE_2) receptor EP1 upregulates P-gp expression and activity [10]. (4) Astrocyte-derived TGF-β1 has been shown to upregulate P-gp activity and mRNA expression. Recently, TGF-β1 was also shown to regulate P-gp in the developing BBB in guinea pigs [32, 33]. In addition to pathways that upregulate P-gp expression or activity, vascular endothelial growth factor (VEGF) signaling through flk-1 and Src kinase has been shown to rapidly (within minutes), specifically, and reversibly inhibit P-gp activity in isolated brain capillaries. Age-related decline in P-gp protein expression has been suggested to contribute to the progression of neurodegeneration [34]. The intracerebroventricular injection in rats of low doses of VEGF increases the accumulation of P-gp substrates without opening TJs [18].

In cancer cell lines, several microRNAs (miRNAs) (e.g., miR-451) and long noncoding RNAs ([lncRNA]; e.g., MRUL) have been identified that directly control P-gp expression [35, 36].

Diseases MDR1 is involved in brain tumors, ischemic stroke, AD, MS, PD, amyloid lateral sclerosis (ALS), and HIV-related neuropathy (Table 1) [15, 37].

2.1.3 Subfamily C, Members: MRP1 (ABCC1) and MRP4 (ABCC4)

Multidrug resistance associated protein-1 (MRP1) is encoded by the *ABCC1* gene and was the first member of the ABCC family to be cloned. In addition to the canonical P-gp structure with two TMD and two NBDs, MRP1 contains a third N-terminal TMD with five TM helices, an intracellular loop, and an extracellular N-terminus. Overall, it shares only 15 % primary sequence homology with P-gp. In contrast, MRP4, encoded by *ABCC4*, has the standard P-gp-like core structure lacking the additional N-terminal region [38]. Although present on both membranes, MRP1 is predominantly expressed on abluminal BBB membranes [39], whereas MRP4 is localized to the luminal membranes (Fig. 1b) [40]. MRP1 has been shown to efflux many endo- and xenobiotic organic anions. Its preferred endogenous substrates are lipophilic compounds conjugated with glutathione (GSH), glucuronate, or with sulfate (e.g., bile salts) [40, 41]. MRP1 transport of GSH is complex. Some substrates are co-transported with GSH; however, for others, although GSH stimulates MRP1 efflux, there is no reciprocal effect on GSH [42]. Nonetheless, MRP1 may regulate processes, such as oxidative stress, that are influenced by GSH [41]. In addition, by transporting arachidonic acid derivatives, for example, the inflammatory cytokine LTC_4, MRP1 also plays a role in leukotriene-mediated inflammatory responses [42]. Furthermore, data from mouse models for AD crossed with *Abcc1*$^{-/-}$knockout mice were consistent with MRP1-mediated BBB efflux of Aβ [43].

Twenty-three documented nsSNPs have been reported for *ABCC1*, two of which are known to affect substrate binding (e.g., G1299T causes Arg433Ser mutation), or protein expression (G128C results in Cys43Ser substitution) [16]. Protein levels have also been reported to respond to activation of the ApoE receptor, ApoER2, which controls both MRP1 and P-gp levels [37, 44]. Similarly, in a second report by the same group, LXLR was also reported to control protein expression of both ABC transporters, this implies that abluminal MRP1 expression might be coordinated in some way with expression of luminal P-gp. Additionally, preliminary studies suggest that several mi-RNAs downregulate mRNA levels of *ABCC1* [41].

Substrates MRP1 effluxes a wide variety of organic anions, preferentially transporting GSH, glucoronate, or sulfate hydrophobic conjugates (Table 1) [21]. MRP4 transports nucleoside monophosphate analogs, cyclic nucleotides, prostaglandins, reduced GSH conjugates, and some organic anions [21, 45].

BBB Localization MRP1 is localized predominantly on abluminal membranes, although some groups have reported luminal localization (Fig. 1b) [39, 40]. MRP4 is localized on luminal and abluminal brain microvascular endothelial cell membranes (Fig. 1b) [40].

Regulation Several transcription factors have been reported to regulate *ABCC1* gene expression, including Sp1 (through GC elements), c-jun/junD complexes (through putative AP-1 sites), and *MYCN* (through putative E-box elements), the tumor suppressor p53, and Notch1 (through a CBF1 element) [46].

Diseases MRP1 is involved in epilepsy, brain tumors, ischemic stroke, HIV-related neuropathy PD and MRP4 has been shown to be altered in epilepsy and diabetes [15, 21].

2.1.4 Subfamily G, Member BCRP (ABCG2)

The *ABCG2* gene codes for the breast cancer resistance protein (BCRP) transporter, which like all subfamily G members, are structurally half-transporters. Additionally, unlike all other ABC transporter families, the G family transporter NBD is N-terminal to the TMD. BCRP is thought to homo- or heterodimerize, or possibly form larger oligomers, and it has been suggested that oligomerization might control BCRP activity [17, 47, 48]. In humans, BCRP is expressed in the luminal membranes of brain capillary endothelial cells (Fig. 1b, Table 1) and its substrate specificity, which overlaps that of P-gp, includes a broad range of antivirals, anticancer drugs, and antibiotics. BCRP has been shown to associate with a 75 kD transmembrane glycoprotein member of the immunoglobulin super family of receptors, CD147, which is upregulated in the microvascular endothelium and other NVU cells of AD patients. Although 17 naturally occurring nsSNPs have been described in *ABCG2*, only one (i.e., C625A causing Q141K mutation) has an effect on protein activity. It is a loss-of-function mutation that reduces activity by ~50%. Since BCRP effluxes urate, the C625A nsSNP increases plasma uric acid and has been shown in certain populations to be a risk factor for gout and for cardiovascular illnesses, such as hypertension, stroke, diabetes, metabolic syndrome, and coronary heart disease. Drug selection can result in a substitution at Arginine 482 that can change substrate affinity [49].

Substrates BCRP transports a diverse array of substrates, including porphyrin and porphyrin-like compounds, and hydrophobic positively and negatively charged molecules (Table 1) [30, 49].

BBB Localization BCRP is localized on the luminal endothelial membrane (Fig. 1b, Table 1) [50].

Regulation The promotor region contains several putative transcriptional binding sites including those for Sp1 and activator protein-1 (AP-1). However, unlike P-gp, *ABCG2* expression is only minimally affected by PXR or CAR activity. Expression is stimulated by the hypoxia inducible factor 1 (HIF1) binding to a hypoxia response element (HRE) in the promoter [49]. The serine–threonine kinase, RAC-alpha serine/threonine-protein kinase (Akt1/PKB), has been shown to regulate internalization of BCRP from the membrane by an unknown mechanism [49].

Diseases BCRP is involved in epilepsy, brain tumors, AD, MS, PD, ALS, and HIV-related neuropathy (Table 1) [15, 37].

2.2 *Solute Carrier Transporters*

The superfamily of SLC genes is the largest group of ATP-independent transporters and currently includes ~400 members, assigned to 52 families [51, 52]. A common feature of all SLC proteins is the presence of one or more transmembrane domains (TMs), whereas the substrate specificity of SLC transporters is diverse: inorganic cations and anions; organic anions; essential metals; amines; organic, fatty, and amino acids; nucleosides; lipids; bile salts; peptides. Several studies involving microarray and qPCR analysis of mRNA isolated from highly purified brain endothelial cells or in situ hybridization identified a number of BBB-enriched SLC genes [5, 53, 54]. In total ~50 SLC genes were shown to be significantly or strongly expressed. The very high expression of the following SLC members was confirmed by both approaches: glucose transporter (SLC2A1), amino acid transporters (SLC6A6, SLC7A5, SLC38A3, SLC38A5), organic anion transporter (SLC22A8), monocarboxylate transporters (SLC16A1, SLC16A2), and the zinc efflux transporter (SLC30A1).

2.2.1 SLC2: Facilitative Glucose Transporter Family Member: GLUT1 (SLC2A1)

The *SLC2A1* gene codes for a 55-kDa (in BBB endothelial cells) facilitative transporter for glucose called glucose transporter 1 (GLUT1). Structurally GLUT1 contains 12 TM helices, cytosolic N and C termini, and a single *N*-linked oligosaccharide site. GLUT1 transporters function as simple carriers that transport substrates along their chemical gradient [55, 56]. As glucose is the main energy source for neurons, the GLUT1 transporter is critically important for normal brain function [50]. In the adult brain, GLUT1 is predominantly localized on both luminal and abluminal BBB endothelial membranes (Figs. 1b and 2d, f, h, i) [8, 50]. Redistribution between membranes can serve as an important regulatory mechanism for glucose consumption. It has been suggested that the ratio of luminal to abluminal expression of GLUT-1 might control the net BBB influx of glucose. Moreover, ~40 % of total GLUT1 is intracellular and, thus, for certain stimuli, can be rapidly incorporated into membranes [57, 58].

Mechanism and Substrates GLUT1 is a low affinity (~3 mM) sodium-independent facilitative transporter for glucose and other hexoses (Fig. 1a, Table 1; galactose, mannose, glucosamine).

BBB Localization GLUT1 is asymmetrically distributed in the endothelial membranes with abluminal localization normally approximately 3–4 times higher than luminal localization (Figs. 1b and 2d, f, h, i, Table 1) [50, 59].

Regulation Wnt/β-catenin was shown to regulate BBB endothelial GLUT1 expression [10, 54]. Expression is also regulated by HIF-1 [50].

Diseases GLUT1 is involved in AD, epilepsy, stroke, and diabetes. GLUT1 expression increases in ischemic and hypoxic stroke, and decreases in AD and other forms of neurodementia (Table 1) [50, 57, 60–62].

2.2.2 SLC7A-SLC3A2: Neutral Amino Acid Transporter (LAT1-4F2hc)

The *SLC7A5* (LAT1) gene encodes the catalytic subunit of a heterodimeric obligatory sodium-independent amino acid exchanger. The LAT1 catalytic subunit contains 12 TM helices. The *SLC3A2* gene codes for the associated subunit called 4F2hc/CD98. The 4F2hc protein is a type II membrane glycoprotein with a single TM domain, and intracellular N terminus, and an extracellular domain that is homologous with bacterial α-glucosidases. To form the functional transporter (LAT1-4F2hc), the LAT1 protein is linked by a disulfide bond to the associated glycoprotein, 4F2hc [63–67]. LAT1 is highly expressed in the BBB and localized on both luminal and abluminal endothelial membranes, thus supplying the brain with essential amino acids, levodopa (L-DOPA) and thyroid hormones, T3, T4 (Figs. 1b and 2e, f) [8]. LAT1-4F2hc is a high-affinity (micromolar K_m values), obligatory exchanger (1:1) with a broad specificity for branched and aromatic amino acids. As an exchanger, it functions to equilibrate the relative concentrations of substrates across a membrane. However, it can mediate net uptake via tertiary-active transport by exchanging extracellular substrates with intracellular substrates that have been accumulated by, for example, a sodium–amino acid symporter or a facilitative transporter [65, 66, 68]. Recently, LAT1-4F2hc has been shown to be recruited by lysosomal-associated transmembrane protein 4b (LAPTM4b) to the lysosomes where it is required for activation of mammalian target of rapamycin complex 1 (mTORC1), which regulates cell growth, via V-ATPase following essential amino acid or Leu stimulation [69]. 4F2hc also mediates β-integrin signaling, cell fusion, and cell proliferation. It is possible that 4F2hc integrates integrin signaling and amino acid transport [70].

Mechanism and Substrates LAT1/4F2hc is a sodium-independent, obligatory exchanger (1:1 stoichiometry) for large neutral amino acids, including 2-aminobicyclo-(2,2,1)-heptane-2-carboxylic acid (BCH), L-DOPA, and thyroid hormones (Fig. 1, Table 1) [66, 70].

BBB Localization LAT1 localizes on both luminal and abluminal endothelial membranes (Figs. 1b and 2e, f, Table 1) [8].

Regulation Membrane localization is regulated by association with 4F2HC and LAPTM4b [66, 69]. Wnt/β-catenin was shown to regulate LAT1 BBB endothelial expression [10, 54].

Diseases LAT1-4F2hc is involved in brain tumors and systemic inflammation (lipopolysaccharide [LPS]; Table 1) [71, 72].

2.2.3 SLC16A: Monocarboxylate and Thyroid Hormone Transporter Family Member: MCT8 (SLC16A2; aka XPCT, MOT8, MCT7)

The *SLC16A2* gene codes for a protein with 12 TMs, intracellular C and N termini, and a large intracellular loop between TMs 6 and 7. This topology has been confirmed for one family member (MCT1) for which there is also a proposed 3D model. MCT8 contains the consensus sequence for an N-terminal PEST domain, which is generally associated with rapid protein turnover. However, it is unknown if this is true for MCT8 and deletion of the PEST sequence does not reduce substrate binding. MCT8 is a high-affinity thyroid hormone (TH) transporter for iodothyronines T_4, T_3, rT_3, and 3,3′-T_2 (K_m values 2–5 μM), but not sulfated and sulfamated iodothyronines or aromatic amino acids (Phe, Tyr, and Trp). Unlike several other family members neither protons nor sodium are co-transported. MCT8 is ubiquitously expressed, including in brain microvessels where it likely transports T_3 across the BBB. Mutations in the *SLC16A2* gene (located on the X chromosome) can cause severe X-linked psychomotor retardation and impaired TH brain uptake, leading to abnormal normal brain development [73]. Many mutations in the *SLC16A2* gene have been identified [74].

Mechanism and Substrates MCT8 transports T2, T3, rT3, T4 substrates by a facilitative diffusion mechanism in which the chemical gradient provides the energy for transport (Fig. 1a, Table 1) [73].

BBB Localization MCT8 is localized to abluminal BBB endothelial membranes (Fig. 1b, Table 1).

Regulation MCT8 protein expression decreases with age [74].

Diseases MCT8 is regulated by PGE_2 in systemic inflammation (Table 1) [72, 75].

2.2.4 SLC22A: Organic Cation, Anion, and Zwitterion Transporter Family Member: OAT3 (SLC22A8)

The *SCL22A8* gene codes for a 12-TM protein with a large glycosylated extracellular loop (between TMs 1 and 2) that mediates homodimerization, and a large post-translationally modified intracellular loop (between TMs 6 and 7). OAT3 is localized on abluminal BBB endothelial membranes where it may efflux α-ketoglutarate (αKG) for organic anion uptake. Human OAT3 can transport a wide variety of endogenous compounds, including cAMP, cortisol, prostaglandins E_2 and $F_{2\alpha}$. It has also been shown to transport drugs, including benzylpenicillin, tetracycline, valacyclovir, zidovudine, adefovir, cidofovir, tenofovir, and the activated form of oseltamivir, the H_2 receptor antagonists cimetidine, famotidine, ranitidine, and fexofenadine, the diuretics bumetanide, furosemide, and torasemide, the nonsteroidal anti-inflammatory drugs indomethacin, salicylate, ketoprofen, and ibuprofen, the cholesterol-lowering drugs pravastatin and rosuvastatin, the cytostatic methotrexate and topotecan, the antihypertensive drug quinaprilat, the antidiabetic drug sitagliptin, and the neuroprotective drug edaravone sulfate. Additionally, OAT3 transports some xenotoxins [76].

Mechanism and Substrates OAT3 exchanges dicarboxylate against a wide variety of organic anions (Fig 1a, Table 1). It is thought to be involved in the clearance of neurotransmitter metabolites and drugs from the brain [76].

BBB Localization OAT3 is localized on abluminal endothelial membranes (Fig. 1b, Table 1) [76].

Regulation Human OAT3 gene expression has been shown to be up-regulated by hepatocyte nuclear factor HNF-1 and repressed by promoter methylation. The *SLC22A8* promoter contains a cAMP-response element (CRE) that is activated by CRE binding proteins. Human OAT3 activity is stimulated by epidermal growth factor (EGF), PGE_2 and agents that activate protein kinase A PKA or protein kinase C (PKC). Several drugs such as COX 2inhibitors or angiotensin-converting enzyme inhibitors may change the activity or function of OAT3 [76].

Diseases The activity and expression of OAT3 was shown to decrease in an LPS model of systemic inflammation. This is consistent with the observation that elevated PGE_2 decreases OAT3 activity (Table 1) [72, 75, 76].

2.3 *SLCO (formerly SLC21A) Superfamily*

The superfamily of organic anion transporting polypeptides consists of six families (OATP1–6) and each family has subfamilies (e.g., OATP1A, OATP1B, OATP1C) for a total of 11 human, individually identified transporters. By a sodium-independent antiporter mechanism, SLCO transporters mediate uptake of many organic endogenous and exogenous molecules, including bile acids, steroid and thyroid hormones and their conjugates, and numerous drugs and toxins. The two members highlighted below (SLCO1A2/OATP1A2 and SLCO1C1/OATP1C1) have been shown to be expressed in BBB endothelium and to be involved in responses to neuroinflammatory insults [77].

2.3.1 Human Family Member SLCO1A2/OATP1A2 (aka SLC21A3/OATP-A/OATP); Rodent Family Members: Slco1a4/mOatp1a4, rOatp1a4 (aka Slc21a5, Oatp2)

In humans the OATP1A2 transporter coded for by the *SLCO1A2* gene is 670 amino acids, with 12 putative membrane spanning domains and cytoplasmic N and C termini [77, 78]. OATP1A2 has been shown to be expressed in placenta, the small intestine, the liver, and the kidney, and in the brain in the BBB and ciliary epithelium [77–79]. In rats, *Slco1a4* mRNA co-localized with von Willebrand factor ([vWF]; endothelial cell marker) and not with glial fibrillary acidic protein ([GFAP]; astrocyte marker) gene expression. Likewise, in rats, Oatp1a4, and in humans, OATP1A2, protein expression localized to both BBB endothelial membranes, but not the closely situated astrocyte end-feet membranes [79, 80]. OATP1A2, like most OATP super family members, is thought to be an organic ion exchanger [77]. There are data indicating that some OATP transporters may have more than one, possibly interacting, substrate binding sites [77]. OATP1A2 has been shown to mediate the high-affinity transport of substrates such as dioxin and may be responsible for the accumulation of dioxin in the brain observed in P-gp ($Mdr1^{-/-}$) knockout animals [79, 80]. Additionally, OATP1A2 has also been shown to be responsible for the BBB influx of enkephalin (δ_1-opioid receptor agonist) [80].

Substrates OATP1A2 transports bile salts (unconjugated and conjugated bile acids), organic anions and cations, and other amphipathic substrates, including many drugs, drug conjugates, and small peptides, e.g., cardiac glycoside digoxin, δ-opioid peptides, dehydroepiandrosterone sulfate (steroid hormone precursor), thyroxine (T4), and triiodothyronine (T3; Fig. 1a, Table 1) [77, 80].

BBB Localization OATP1A2 is localized on both luminal and abluminal BBB membranes (Fig. 1b, Table 1) [79].

Regulation OATP1A2 protein is upregulated by PXR and the promoter is activated by CAR [77].

Diseases OATP1A2 is involved in stroke, systemic infections (LPS model; Table 1) [72, 75, 81].

2.3.2 SLCO Transporter Member: SLCO1C1/OATP1C1 (aka OATP-F)

The SLCO1C1 gene encodes for a 12 TM protein (with cytoplasmic N and C termini) that mediates the sodium and pH-independent uptake of thyroid hormones in brain tissues. OATP1C1 has narrow substrate specificity and a high affinity for the thyroid hormones T4, rT3, T4 sulfate, but not T3. OATP1C1 is a main thyroid hormone transporter at the BBB. Polymorphisms in the *SLCO1C1* gene have been associated with fatigue and depression in patients suffering from hyperthyroidism. Multiple splice variants are known.

Mechanism and Substrates OATP1C1 is a high-affinity (nM K_m values), pH-independent transporter of bile salts and the thyroid hormones, T4, sT4, and rT3 (Fig. 1a, Table 1) [74, 77]. It has been suggested that OATP1C1 has two T_4 binding sites (high and low affinity) [74]. However, the exact mechanism of transport is poorly understood [77].

BBB Localization OATP1C1 is localized on abluminal endothelial membranes (Fig. 1b, Table 1) [77].

Regulation Protein expression of OATP1C1 is inversely related to thyroid hormone concentration and brain capillary levels increase in hypothyroid rats and decrease in hyperthyroid animals [74].

Diseases OATP1C1 has been shown to be regulated in the LPS model for systemic inflammation (Table 1) [77].

2.4 *Receptor- and Vesicle-Mediated Transport*

Large molecular weight solutes entering the brain via an intact BBB do so by transcytosis. There are three types of endocytic vesicles that occur at the BBB: (1) clathrin-coated pits; (2) caveolae; (3) macropinocytic vesicles. The specific transcytosis of large macromolecules is mediated by receptors or by charge-based absorptive-mediated transcytosis (AMT). Receptor-mediated transport mechanisms involve ligand interactions with receptors at the BBB endothelial apical plasma membrane that cause complexes of ligand-bound receptors to form. The receptor–ligand clusters trigger endocytosis in clathrin-coated pits or caveolae (Fig. 1a). The plasma membrane invaginates to form a vesicle that is released at the opposite facing cell membrane. Ligand dissociation occurs either during transit or upon exocytosis [14, 82]. Alternatively, nonspecific bulk-phase or fluid-phase transcytosis (FMT), where soluble plasma molecules are randomly taken up with bulk plasma, can result in transendothelial transport. However, in healthy BBB, FMT occurs at a low frequency, usually mediated by caveolae, and to an even lesser extent by the more abundant, but negatively charged clathrin-coated pits [83].

2.4.1 Low Density Lipoprotein Receptor-Related Protein 1 (LRP1) Receptors

Low-density LRP1, which is variously known as the α-2-macroglobulin receptor (α2MR), ApoER, or cluster of differentiation 91 (CD91), is an integral plasma membrane receptor involved in receptor-mediated endocytosis. The *LRP1* gene codes for a multimeric protein composed of two noncovalently bound subunits, a

515-kDa extracellular α subunit and an 85-kDa transmembrane and cytoplasmic β subunit. The α subunit has four ligand binding sites containing cysteine-rich complement-type repeats that bind more than 30 ligands, including ApoE and Aβ. Other ligands include extracellular matrix proteins, growth factors, proteases, protease inhibitor complexes, and other proteins involved in lipoprotein metabolism [84–86]. As a result of proteolytic digestion of membrane-bound LRP1, the protein also exists in a soluble form (sLRP1) that circulates in plasma and CSF [85]. The LRP1 cytoplasmic domain contains two NPxY, one YXXL, and two di-leucine motifs that bind cytoplasmic proteins modulating receptor activity. Ligand-bound LRP1 is constitutively endocytosed by a clathrin-dependent mechanism to lysosomes for ligand degradation and LRP1 receptor is recycled back to plasma membranes. Additionally, LRP1 interacts in a phosphorylation-dependent manner with scaffolding and signaling proteins to modulate endocytotic and signaling activity. For example, PKCα phosphorylation of serine (S) and threonine (T) residues on LRP1 allows disabled-1 binding, which decreases endocytosis by 25 %, whereas phosphorylation of S residues by protein kinase A PKA increases endocytosis [85]. Also, it can partner with other plasma membrane proteins that affect LRP1 activity. By these mechanisms, LRP1 participates in regulating lipoprotein metabolism, cell differentiation and motility, and BBB integrity, in addition to the progression of neurodegenerative diseases, cancer, and other pathological conditions [84, 85, 87].

Mechanism and Substrates Bound LRP1 is constitutively endocytosed to liposomes for ligand degradation and recycling of the empty receptors to the plasma membrane [85]. LPR1 binds more than 40 ligands, including aggregated LDL, ApoE, APP, Aβ, α2M, tissue plasminogen activator, proteinase inhibitors, blood coagulation factors, receptor-associated protein, and aprotinin (Fig. 1, Table 1) [50, 84–87].

BBB Localization LRP1 is localized to abluminal endothelial capillary membranes (Fig. 1b, Table 1) [50, 84–86].

Regulation LPR1 transcription is suppressed by the vascular sterol regulatory element binding protein SREBP2 [85]. GLUT1/SLC2A1 deficiency has been shown to increase SREPB2 levels and thereby decrease LRP1 expression. LRP1 endocytotic activity is regulated by phosphorylation on S, T, and tyrosine (Y), which can down- or up-regulate endocytotic activity.

Diseases LRP1 is involved in AD (Table 1) [50, 84–88].

2.4.2 Receptor for Advanced Glycation End Products

The receptor for advanced glycation end products (RAGE) belongs to the major histocompatibility complex class III gene family. Structurally, it is a 35-kD protein composed of an extracellular ligand binding domain, a TMD and a short cytoplasmic signaling domain. It is expressed on a number of cell types, including BBB endothelial cells; however, its expression in endothelial cells is normally low and it is only up-regulated by pathological states or during aging [89]. RAGE has a number of functions depending on the cell type. A main function is as a pattern recognition receptor for the innate immune system. It was named for its ability to bind advance glycation end-products (AGEs). These are glycated lipids or proteins that have formed as a result of exposure to sugars and are prevalent in aging and in the development of inflammatory and degenerative diseases. RAGE also binds a large number of cytokines and regulatory molecules, such as Aβ, transthyretin, etc. [89]. RAGE activation, which increases proinflammatory cytokines and oxidative stress, has been shown to exacerbate many neurodegenerative diseases [89, 90]. Antagonism of RAGE activity using chemical inhibitors or soluble RAGE ([sRAGE]; a naturally occurring isoform that is a RAGE inhibitor) in animal models of neurodegenerative diseases has been shown to attenuate inflammation [89].

Mechanism and Substrates RAGE mediates receptor transcytosis. The substrates are AGEs, S100/calgranulins, HMGB-1 (amphoterin), β-sheet fibrils, and β_2-integrin Mac-1 (Table 1) [90].

BBB Localization RAGE is localized on luminal endothelial membranes (Fig. 1, Table 1) [89].

Regulation Ligand binding and initiates positive feedback, increasing RAGE protein expression [91].

Diseases RAGE is involved in AD, PD, MS, Huntington's disease, ALS, and other neurodegenerative diseases, diabetes, rheumatoid arthritis, inflammatory bowel disease, atherothrombosis, etc. (Table 1) [88–90].

2.4.3 Caveolin-1

The *CAV1* gene encodes for an integral membrane protein, caveolin-1 ([CAV1]; 21–24 kDa), which has cytoplasmic N and C termini. Members of the caveolin family, such as CAV1, are the main scaffold protein components of caveolae. Caveolae, are spherical plasma membrane invaginations composed of lipids (i.e., cholesterol, glycosphingolipids) and lipid-anchored proteins that mediate raft-dependent, clathrin-independent endocytosis. The main role of caveolae is to compartmentalize signaling molecules, and BBB caveolar membranes contain numerous receptors, including RAGE and LRP1 in addition to other signaling molecules [50, 92, 93]. Caveolae regulate the endocytosis, transcytosis, and signaling in the lipid-based endothelial microdomain and thereby control BBB transcellular permeability [50]. High levels of CAV1 are ubiquitously expressed in most tissues, including BBB brain microvascular endothelial cells (Figs. 1b and 2m, o, p) [94]. It has been suggested that CAV1 might coordinate the activities of various BBB cell types [95]. For example, the importance of CAV1 in NO and calcium signaling has been demonstrated in CAV1 knockout mice. CAV1 also influences levels of TJ proteins, with either the loss or increase shown to have different effects in different experimental models (Fig. 4) [96]. The TJ proteins occludin (OCLN) and zonula occludens-1 (ZO1) are associated with CAV1 in the caveolae-related glycated lipid rafts. There is evidence that CAV1 may regulate TJs by controlling degradation by matrix metalloproteinase 9 (MMP-9) [95, 97]. Additionally, both in vitro and in vivo studies have shown that P-gp and CAV1 interactions occur within the BBB caveolae. CAV1 tyrosine phosphorylation is known to regulate P-gp activity and thereby control accumulation of substrates in BBB endothelial cells [29]. CAV1 responds to shear stress by increasing plasma membrane expression and therefore acts as a flow sensor [92].

Mechanism Caveolae transport proteins of high molecular weight and ligand-bound clusters of various transporters (Fig. 1, Table 1). Transendothelial transport is initiated by caveolae detachment from the luminal membrane. This results in formation of endocytic vesicles, which are transported to the basal membranes, where they fuse and release their contents into the interstitial space [98].

BBB Localization In cortical microvessels CAV1 is primarily expressed on luminal membranes in association with caveolae, but expression can also be seen throughout the membranes and in the cytoplasm (Figs. 1 and 2, Table 1) [94].

Regulation CAV1 regulates the plasma membrane formation and detachment of caveolae through a tyrosine phosphorylation-dependent mechanism [29]. Several inflammatory mediators have been shown to regulate CAV1 expression. *CAV1* promoter activity is regulated by TNFα, and IL-15. Vascular endothelial growth factor (VEGF) down-regulates CAV1, as does activation of the endothelial cell CCL receptor, CCR2 [95]

Diseases In BBB, CAV1 dysfunction has been linked to MS (Fig. 4) [96], ischemic stroke, infections, and other inflammatory diseases, in addition to effects on endothelial barrier permeability (Figs. 1, 2, and 4, Table 1) [95, 99, 100, 101].

3 Impacts of Acute and Chronic Neuroinflammatory Diseases and Insults on BBB Transporters

Communication between the immune system and the CNS provide mechanisms for: (1) immune system surveillance of the CNS to remove noxious agents, and to protect and repair brain tissue during and after assault; and (2) brain monitoring of peripheral immune status to coordinate appropriate physiological and behavioral responses [1, 11, 13, 102]. To accomplish this without jeopardizing the brain's security, the NVU establishes a privileged region for immune surveillance between the capillary ablumen and the brain parenchyma proper. Bounded by the astrocyte (glia limitans) and pericyte/endothelial basal membranes, this region has been likened to a "two-walled castle moat" [3]. Low levels of memory/effector T cells continually breach the outer wall (i.e., abluminal capillary barrier), but only trigger immune responses if they encounter specific antigens. Immune cell activation leads to the production of cytokines and proteases that can result in the bridging of the inner wall (i.e., the glia limitans) permitting the influx into the CNS of infiltrating inflammatory cells [1, 3, 11, 102].

Physiological brain homeostasis and therefore, normal neuronal activity, is highly dependent on a functional brain vasculature, which, by extension, critically contributes to the well-being of the whole body. The failure of the BBB functional and morphological integrity required to preserve a normal brain milieu can be a dire consequence of many diseases, one that can rapidly lead to further degeneration [10]. However, before and in addition to BBB failure, there can also be an insidious loss of brain vascular density, surface area, and diameter (thinning). This is a part of normal aging, which can be exacerbated by the disease state [10, 50, 103–105]. Up- and down-stream signaling can lead to abnormal expression of vascular transporters, receptors, etc., and other pathological changes. Together, microvascular morphological and functional abnormalities can diminish neurotoxin clearance and have a negative impact on blood flow and BBB transport, and consequently, the delivery of nutrients, energy substrates, and other necessary substances.

Here, we focus specifically on the impact on and involvement of BBB transport mechanisms for selected diseases and insults with a significant neuroinflammatory etiology, for example, AD, MS, PD, epilepsy, stroke, TBI, diabetes, and cerebral and peripheral infections (HIV). This is a sample of the many diseases in which, for better or worse, the BBB plays a crucial role [10, 12, 72, 106]. Although these diseases all have unique triggers, they share many common mechanisms, including immune system activation, in addition to, for many diseases, the deposition of misfolded or aggregated proteins [89]. This is area of research that expands as it becomes clearer that the etiology of many diseases, including psychiatric illnesses, such as depression and schizophrenia, involve significant neuroinflammatory underpinnings affecting the BBB and vice versa [10, 13, 72, 102].

3.1 Neuroinflammation and Age

A large proportion of research into the neuroinflammatory impacts on the BBB is carried out in young adult human subjects and animal models. However, it is well recognized that many effects are influenced by developmental stage, and additionally, that the immune system responses can be age-dependent [107, 108]. Additionally, brain sensitivity is known to be heightened at various developmental stages, i.e., during pregnancy, infancy, childhood, and advanced age the brain is particularly vulnerable to chemical and infectious challenges. Exposure during early development can have a permanent impact on brain maturation, and likewise, for different reasons, the aged brain is especially vulnerable to sustained neuropathological changes following acute or chronic insults [1, 50, 105, 108].

However, the protective function of the placental barrier is apparently reinforced during pregnancy by an increasingly functional fetal BBB [29]. For example, the hallmark BBB TJ proteins, CLDN5 and OCLN, and the ABC transporter, P-gp, and GLUT1 proteins can be detected as early in fetal development as the telencephalon stage at a gestational age of 12 weeks (Fig. 3) [29, 109]. This is a period that occurs soon after pre-plate formation and vascularization, and which coincides with the beginning of cerebral cortex formation [29]. By mid-gestation, P-gp was shown to co-localize with CAV1 on abluminal membranes suggestive of efflux activity regulated by endothelial caveolae (Fig. 3) [29]. At 12 weeks' gestation, the solute carrier glucose transporter GLUT1 protein is primarily localized on abluminal endothelial membranes. Luminal GLUT1 localization increases in weeks 18 and 22 in human fetal brains (Fig. 3) [110]. The early expression of transporters during brain development and BBB differentiation may thus complement the protective functions of the placental barrier against endo- and xenobiotics and other inflammatory insults [10, 29, 105, 108–111].

In the fetal and perinatal periods the majority of the pathogenic stimuli leading to brain damage arise from: (1) hypoxia–ischemia, which leads to excitotoxicity and oxidative stress; and/or (2) maternal–fetal or post-natal infection/inflammation that stimulates IL-1β, IL-6, and TNFα signaling [108, 112]. Using an LPS model for sustained systemic inflammation during early development in rats, Stolp and co-workers [113] found sustained impacts on BBB barrier tightness, and suggested that this might be due to long-term alterations in TJ distribution. They also observed a reduction in white matter suggestive of a developmental delay, as total white matter volume had recovered by adulthood. In this regard, it was shown TGF-β1, which crosses the placenta and can therefore be derived from blood or brain sources, regulates BBB expression of P-gp in early development [32, 33]. Several inflammatory pathological conditions, such as gestational diabetes and preeclampsia, may result in increased maternal and therefore fetal circulating TGF-β1. In addition, perturbed TGF-β1 levels can be caused by early or delayed gliogenesis, such as can occur in fetal alcohol syndrome and autism, all of which may alter P-gp expression and increase the exposure of the fetus to endo- or xenotoxins, and can contribute to acute and/or potentially long-lasting brain damage [33].

At the other end of the age spectrum, the inflammatory profile of microglia has been shown to increase in humans, nonhuman primates, rodents, canines and other species with advanced age. Normal aging is associated with increased levels of immune mediators and the hyper-activation of microglia in response to immune challenges. For example, it was shown that in aged mice peripheral LPS resulted in prolonged elevation of IL-1β, IL-6, and TNFα. The increased pro- and decreased anti-inflammatory cytokine production translates into a greater risk of developing AD, PD, stroke, brain tumors, etc. [1, 13, 108]. In addition, there is evidence for a neurovascular uncoupling with ageing. The relatively small (~20 %) reductions in cerebral blood flow (CBF) reported for normal aging correlate with lowered protein synthesis in the brain [50]. Severer decreases in CBF that result in disturbances in ISF pH, and water, glutamate, and lactate balance, and which in extreme cases can lead to impaired ATP synthesis and diminished neuronal function, correlate with chronic neurodegenerative diseases [50].

3.2 *Alzheimer's Disease*

Alzheimer's disease is characterized by chronic neuroinflammation and neurovascular dysfunction that initially targets the hippocampal and neocortical regions. AD-related neurodegeneration produces progressive cognitive impairment (dementia) and eventually whole-body organ failure. Secretion of neuroinflammatory mediators from brain microvessels, such as NO, cytokines (e.g., TNFα, TGFβ1, IL-1β, and IL-6), chemokines (e.g., CCL2 and IL-8), prostaglandins, MMPs, and leukocyte adhesion molecules, have been shown to be elevated in AD [114]. Although, the exact molecular mechanisms underlying AD remain controversial, there are two agreed upon hallmarks: (1) an accumulation of extracellular neurotoxic Aβ oligomers in both the soluble form (AβO) and as *senile plaques* on brain blood vessels and in the parenchyma; and, (2) intracellular *neurofibrillary tangles* (NFT) of hyperphosphorylated aggregates of the microtubule-associated protein, tau [50, 104, 115–118]. Aβ peptides (constituting senile plaques) are produced when the initial product from cleavage of the transmembrane amyloid precursor protein (APP) by β-secretase is further digested by the γ-secretase complex forming Aβ40 and Aβ42. There are a large number of intersecting mechanisms that control brain ISF Aβ levels including: (1) Aβ production; (2) BBB receptor-mediated influx and efflux of free Aβ; (3) sequester and/or BBB transport of protein-bound Aβ (e.g., to ApoE); (4) Aβ degradation; (5) CNS removal via ISF–CSF bulk flow; and (6) Aβ oligomer aggregation in the CNS [88]. Oxidative stress due to genetic (e.g., ApoE ε4 allele), lifestyle (e.g., smoking, lack of exercise, high caloric intake), medical (e.g., stroke, hypertension, diabetes, TBI) or risk factors for aging have been proposed to be triggers for AD [118, 119]. The evidence suggests that at least initially, both increased Aβ production and tau hyperphosphorylation, leading to amyloid plague and NFT formation respectively may be protective responses to oxidative stress [104, 118].

Mutations in a number of genes (presenilin1, presenilin2, amyloid precursor protein, ApoE) correlate with an increased risk for developing AD, virtually all of which increase the production and/or accumulation of Aβ. For example, more than 25 deleterious *APP* mutations have been identified, most of which increase cerebral Aβ load and cause autosomal dominant, early onset AD [50, 119, 120]. Recently, an *APP* coding mutation (A673T), which reduces Aβ production by ~40 %, was identified. The *APP* A673T mutation protects against both AD and age-related cognitive decline supporting the theory that reducing Aβ load is neuroprotective [120]. However, for late onset AD there is no increase in Aβ production, and therefore it is thought that Aβ accumulation is primarily due to defects in Aβ clearance across the BBB, or by bulk flow along Virchow–Robin arterial spaces [50, 85]. The major routes for Aβ brain influx and efflux are mediated by BBB-localized RAGE and LRP1 receptors respectively. CNS accumulation of Aβ has been attributed to the synergistic effects of RAGE up-regulation and the down-regulation of LRP1 and increased sLRP1 shedding that occur with age and Aβ exposure [85]. Some ABC transporters, including CERP and P-gp, have also been suggested to influence Aβ levels [15, 28]. Reduced Aβ brain clearance gives rise to Aβ plaque deposition on blood vessels and attendant cerebral amyloid angiopathy (i.e., cerebral vessel wall fragility and recurrent lobar intracerebral micro-hemorrhages) [28, 50].

3.2.1 AD and BBB Transporters

ABC Transporter, CERP In humans, a loss-of-function variant of *ABCA1* occurs in the general population at a rate of 1:500 people. This variant is associated with low levels of circulating ApoE and a high risk of AD and cerebrovascular disease [28]. The authors suggest that because low ApoE is associated with increased AD that the *ABCA1* loss-of-function variant might increase AD risk, primarily by decreasing cerebral ApoE levels and consequently increasing the Aβ burden. The authors further noted that the increased risk for AD and cerebrovascular disease are independent of the increased incidence of atherosclerosis, supporting the hypothesis that BBB dysfunction, cerebrovascular disease, and AD share common mechanisms [28].

ABC Transporter, P-gp A number studies of both AD and non-AD subjects have shown an inverse correlation between P-gp levels and Aβ plagues (in addition to disease progression in AD) [15]. Recently a positron emission tomography (PET) study in AD patients demonstrated a reduction in the transport of the P-gp substrate (*R*)-^{11}C verapamil in several cortical brain regions, compared with healthy control individuals, providing direct evidence for reduced P-gp function in AD patients [15]. This finding was mirrored in a mouse model for AD over-expressing human APP (Tg2576 mice) in which P-gp activity was decreased 70 % and P-gp protein decreased 60 % [15]. The mechanism underlying these decreases has been recently shown to involve the Aβ 40 peptide-mediated ubiquitination, internalization, and proteasomal degradation of P-gp [121]. In vitro studies using isolated mouse capillaries have demonstrated that P-gp can mediate Aβ efflux [15, 122]. Inhibition of P-gp in a mouse model increases Aβ levels within hours [15]. Mice that lack P-gp at the BBB (knockouts for *mdr1a* and *mdr1b* genes) have reduced clearance of Aβ from the CNS and lower levels of LRP1 in brain capillaries. Crossing *mdr1a*/*mdr1b* null mice with APP-overexpressing mice accelerates the accumulation of Aβ and amyloid deposition. This supports the model that Aβ clearance through the BBB occurs through P-gp efflux from BBB in concert with abluminal LRP1 influx from the brain [15, 50, 105].

Solute Carrier Transporter, GLUT1 Reduced BBB GLUT1 expression has been reported for individuals with AD and other forms of neurodementia [50, 123, 124]. PET studies have demonstrated that individuals diagnosed with age-associated cognitive decline have significantly reduced glucose uptake before conversion to AD. These deficits are more pronounced in AD patients and involve the frontal cortices. Reductions in glucose utilization in the hippocampus during normal aging can predict cognitive decline years in advance of a clinical diagnosis. Consistent with this concept, pre-symptomatic early-onset autosomal dominant familial AD individuals carrying mutations in the presenilin-1 gene show widespread AD-like reductions in glucose utilization in the absence of structural brain atrophy [50]. Furthermore, AD animal models that are additionally heterozygous for the GLUT1 gene (*Glut1*$^{+/-}$) exhibit the early onset of degenerative structural changes of the BBB, and reductions in cerebral vascularization and blood flow. These animals exhibit increased Aβ load, neuronal loss and dysfunction, and cognitive impairment. Taken together, reduced GLUT1 expression has a strong impact on AD cerebrovascular and neuronal degeneration [61].

Receptor-Mediated Transporter, LRP1 LRP1 is a major efflux transporter for Aβ across the BBB. Radiolabeled Aβ bound on the abluminal endothelial membrane was shown to be rapidly effluxed to the blood [85]. Also, from 70–90 % of blood Aβ can be sequestered by circulating sLRP1. This creates a brain to blood gradient that further drives brain efflux [85]. LRP1 has been shown to mediate internalization of several ligands, including ApoE and Kunitz protease inhibitor containing APP precursors, which are components of Aβ senile plaques, are genetically associated with AD, and may influence Aβ efflux [84, 125]. During AD, LRP-1 transport of Aβ out of the CNS is affected by a number of factors, including oxidative stress, in addition to transcriptional factor and other associated protein levels. By stimulating oxidative stress, senile plaques set up a positive feedback loop in which oxidized LRP1 and sLRP1, which have significantly lower Aβ binding affinities, increase the Aβ load [85]. A number of mechanisms contribute to decreased LRP1 expression and activity during normal aging and in AD, and therefore increased Aβ load. For example, AD and age decrease GLUT1 expression, and AD increases in serum response factor (SRF) and myocardin (MYOC), all of which increase SREB2 (GLUT1 directly and SRF/MYOC transactivates SREB2). SREB2 suppresses LRP1 transcription and consequently surface expression [85, 125]. Along with age, Aβ exposure increases sLRP1 formation, thereby reducing abluminal membrane levels of LRP1. AD correlates with the down-regulation of vascular mesenchyme homeobox gene 2, (MEOX2) which promotes LRP1 proteosomal degradation, decreasing plasma membrane LRP1. AD leads to decreased Aβ-LRP1 internalization. For endocytosis, the C-terminal domain of the Aβ-LRP1 complex binds to the phosphatidylinositol-binding clathrin assembly protein (PICALM), which, presumably due to protein conformational changes, rapidly triggers clathrin-mediated endocytosis of Aβ-LRP1. In AD, PICALM expression is down-regulated, reducing clearance of Aβ-LRP1 and increasing the Aβ load [85, 105].

Receptor-Mediated Transporter, RAGE RAGE contributes to AD by mediating the influx of Aβ across the BBB, and by stimulating inflammatory and procoagulant responses, specifically in endothelium. Furthermore, a correlation between low plasma sRAGE and the incidence of neurodegenerative disorders, including AD and non-AD forms of dementia, has been demonstrated [89]. In the aging brain, increases in RAGE promote an influx of plasma Aβ through BBB transcytosis, NF-kB activation of endothelial and other cells, and proinflammatory (e.g., TNF-α, IL-6) cytokine secretion [125]. Additionally, RAGE through NF kB-dependent mechanisms stimulates apoptosis [88]. RAGE itself is up-regulated by NF-kB, producing positive feed-back, leading to chronic inflammation [89].

3.3 Parkinson's Disease

Parkinson's disease is the most common neurodegenerative movement disorder, with symptoms including resting tremor, slowed movement, postural instability, and muscle rigidity, in addition to nonmotor symptoms. It is an age-dependent disorder characterized by the progressive degeneration of dopaminergic neurons in the substantia nigra (SN). Inclusions within SN neurons called *Lewy bodies* (LBs), composed of α-synuclein (α-syn) and ubiquitinated proteins, are a hallmark pathological marker of PD. Studies have shown that normally low CSF levels of total α-syn are further depressed in PD, whereas levels of CSF oligomerized α-syn are increased. These observations have led to the suggestion that defective BBB α-syn transport might contribute to PD [126]. It is unclear whether LBs directly cause neuronal death and/or function to protect neuronal viability by sequestering smaller neurotoxic protein aggregates [102]. There is an apparent circadian dysfunction to PD as several nonmotor symptoms exhibit a diurnal cycle. In addition, L-DOPA medication used to treat PD is typically administered during the day, because absorption is higher than at night [127]. Continued administration of L-DOPA can lead to dyskinesia as a side effect. This has been shown to be due to L-DOPA stimulating increased angiogenesis, which also results in BBB leakage, as illustrated by increased albumin extravasation, and increases the rapid fluctuations in CNS L-DOPA concentrations that can be observed in PD [128]. Neuroinflammation is ubiquitous in PD patients and experimental models of PD. Phagocyte activation, increased synthesis, and release of proinflammatory cytokines, complement activation, activation of microglia, and release of ROS have all been described [50].

3.3.1 PD and BBB Transporters

ABC Transporter, P-gp The synonymous C345T polymorphism correlates with altered P-gp function and expression (e.g., shown to reduce duodenal P-gp by 65-fold). Changes in expression may be due to linkage with a nonsynonymous missense mutation, G2677(A/T). The C345TT genotype correlates with a twofold higher early onset of PD than late onset PD. Pesticide exposure significantly exacerbates the risk of PD in both heterozygous (threefold increased risk in carriers) and homozygous (fivefold increased risk) C345T carriers [129].

Solute Carrier Transporter, LAT1 Parkinson's disease is treated with the naturally occurring dopamine precursor L-DOPA, which unlike dopamine, is transported by system L transporters, such as LAT2/SLC7A8 in the intestine and LAT1 in BBB endothelium [50, 127]. LAT1 has also been suggested to transport the pesticide paraquat, which is used in animal (rodent) models to produce PD symptoms. However, the significance in humans has not yet been proven [50].

Receptor-Mediated Transporter, LRP1 α-Syn is bidirectionally transported across the BBB by a saturable mechanism that was proposed to involve LRP1 but not P-gp, as efflux was not inhibited by the P-gp antagonist, cyclosporine. Brain α-syn efflux rates were very rapid and measured as ~eightfold higher than Aβ efflux (known to be transported by LRP1) and ~fourfold higher than α-syn influx from the blood [126].

3.4 Multiple Sclerosis

Multiple sclerosis results in the chronic progressive or relapsing and remitting destruction of the myelin sheaths surrounding CNS axons. It is thought to be an autoimmune disease. It has been shown that during MS, and in *experimental autoimmune encephalomyelitis* (EAE), a widely used animal model for MS, cytotoxic T cells are responsible for neuronal cell death, most likely infiltrating the brain via the BBB [50, 96, 130, 131]. Proinflammatory signals stimulate the BBB to up-regulate adhesion molecule expression promoting leukocyte diapedesis [50, 96, 130, 132]. The chemokine CCL2, which chemoattracts monocytes, memory T-cells, dendritic cells, natural killer cells, and microglia, has been shown to be present in the blood, CSF, and brain of individuals with MS and EAE experimental animal models [133]. By causing TJ cytoskeletal dissociation and decreased expression, CCL2 plays a central role in the co-morbid BBB opening observed in MS and EAE. These CCL2-mediated effects on TJ proteins are lessened by CAV1 knockout [97].

3.4.1 MS and BBB Transporters

Vesicular-Associated, CAV1 CAV1 has been shown to be up-regulated in EAE, especially in demyelinated lesions (Fig. 4) [96, 132]. Reduction of CAV1 was shown to be protective in an EAE model. EAE-inoculated knockout mice (*Cav1*$^{-/-}$) had significant reductions in inflammatory infiltration, reduced lesions and demyelination, and preservation of the myelin sheath [132]. The expression of intercellular adhesion protein 1/CD54) and vascular adhesion protein 1/CD106, which are crucial for T-cell extravasation, has been shown to be strongly up-regulated in wild-type EAE mice and to co-localize with CAV1, particularly at infiltration sites within lesions. *Cav1*$^{-/-}$ knockout mice show no such up-regulation [132]. Furthermore, although the number of activated peripheral T lymphocytes following inoculation in *Cav*$^{-/-}$ mice was comparable with wild-type EAE mice, there was a significant decrease in the numbers of CNS-infiltrating T-cells indicating that CAV1 plays a critical role in EAE pathology [132]. There was a significant up-regulation of circulating CAV1 in plasma, possibly as a result of caveolae disruption, that correlated with EAE disease progression. The authors suggested that secreted CAV1 serum levels might serve as a biomarker for clinical severity in EAE and MS [132].

3.5 Epilepsy

Recurrent spontaneous seizures, known as *epilepsy*, constitute one of the most common neurological pathological conditions, and almost a third of those stricken exhibit *drug-resistant* or *refractory epilepsy*. The underlying causes of epilepsy are thought to be either unknown or resulting from genetic/developmental, or structural/metabolic (i.e., TBI, stroke, brain tumors, infections, autoimmune) sources [134]. Congenital or genetic factors tend to be associated with early (childhood) development, whereas infections and other types of mechanical brain trauma can lead to epilepsy at any age [134]. Infections and infestations are the most common risk factors for seizures and acquired epilepsy [134]. In epilepsy triggered by infections, although the mechanisms generating acute seizures are multifactorial, a common underlying factor is the release of inflammatory cytokines [134]. Likewise, so-called sterile inflammation, a homeostatic repair response to injury triggered by *danger signals*, also activates inflammatory pathways. In either case, so-called pattern recognition receptors are activated, resulting in the transcriptional activation of NFkB-sensitive inflammatory genes. Chronic stimulation of the inflammatory signals can lead to BBB damage, neuronal cell death, and persistent neuronal hyperexcitability [134, 135].

3.5.1 Epilepsy and BBB Transporters

ABC Transporters A screen of adults with mesial temporal lobe epilepsy (MTLE) and hippocampal sclerosis found increased mRNA and protein levels for P-gp and MRP1-5 in endothelial cells of the BBB [60, 136]. However, whether this increase is constitutive or induced by chronic seizures and/or drug treatments is still controversial [60]. Animal models of MTLE indicated transient P-gp overexpression following seizures, supporting the notion that seizures themselves regulate P-gp. Conversely, intrinsically high levels of P-gp have been shown to correlate with cortical developmental abnormalities in humans. A number of animal models for epilepsy have also been shown to overexpress P-gp, MRP1, MRP2, and BCRP. Some studies have found increased P-gp levels in nonresponders vs responders in animal models [60]. Furthermore, functional polymorphisms in P-gp transporters may also result in pharmacoresistance [137]. Finally, drug treatment has been shown to upregulate P-gp in BBB via a signaling pathway initiated by excess extracellular glutamate that activates NF-kB through NMDA receptors [15].

Solute Carrier, GLUT1 GLUT1 deficiency syndrome due to mutations in GLUT1 leads to impaired BBB glucose transport, which alters brain metabolism and if untreated results in seizures [60]. PET studies with a glucose analog, fluoro-2-deoxy-d-glucose (FDG) have shown that in the epileptogenic temporal cortex of patients with complex partial seizures, both blood flow and glucose metabolism are altered. According to FDG-PET data, the zone of reduced metabolism aligns with a region of decreased BBB endothelial membrane glucose transporter activity. However, no down-regulation in glucose metabolism or transport was observed in neuropil. In addition, endothelial GLUT 1 protein expression was also down-regulated around the seizure focus [60].

3.6 Stroke

Stroke refers to an interruption in cerebral blood flow. Hemorrhagic stroke is due to intracranial bleeding and an ischemic stroke occurs when the cerebral blood flow is interrupted by either thrombosis or embolism. Cerebral blood flow can be blocked by either clot formation within cerebral large or small arteries (referred to as *large-vessel thrombosis* and *lacunar stroke* respectively). Stroke can result in sudden paralysis, impaired speech, vision loss, and death [138]. Cerebral ischemia produces a progression of events. Initially, within hours, the BBB endothelium undergoes noticeable cell death (e.g., microvacuoles, an eosinophilic cytoplasm, and pyknotic nuclei) and consequent BBB barrier compromise. Additionally, during this time frame, leukocyte infiltration begins. This is followed by a prolonged period (weeks) marked by the appearance of macrophages and astrocytic proliferation. Finally, during the chronic stage, a pseudocyst forms [131]. Many of the risk factors associated with stroke result from reduced CBF. Among the risk factors are aging, hypertension, diabetes, and hypercholesterolemia. Some risk factors increase the susceptibility of neurons and other brain cells to injury (aging, diabetes), whereas others additionally impair protective mechanisms that stabilize CBF during reperfusion post-ischemic events (hypertension, diabetes), making recovery more fragile. Induction of ROS and systemic and cerebral inflammatory signals are common mechanisms by which risk factors contribute to the vulnerability to ischemic events [138]. Additionally, cerebral ischemia–reperfusion injury is often mediated by free radical triggers that lead to increased BBB permeability, brain edema, hemorrhage, and inflammation [95].

3.6.1 Stroke and BBB Transporters

Efflux ABC and SLCO Transporters, P-gp, BCRP, OATP1A4 A rat model of stroke using a transient occlusion of the middle cerebral artery (tMCAO) resulted in a delayed upregulation (14d post infarct) of *Mdr1* and *Slco1a4*/*Oatp1a4* (*Oatp2*) mRNA and protein in the ischemic penumbra [81]. Contralateral staining for P-gp showed the expected co-localization with vWF. P-gp protein was mainly observed in the newly formed blood vessels. As expected, OAT1A4 protein expression was not observed in the controls. In ischemic animals, similar to P-gp, the observable OATP1A4 protein expression coincided with vasculogenesis (at d14) co-localized with endothelial markers. *Bcrp* mRNA and endothelial protein expression was significantly up-regulated before that of *P-gp* and *Slco1a4/Oatp1a4. Bcrp* gene expression (mRNA and protein) in the peri-infarcted region was increased by 3 days and remained elevated at 14 days, returning to baseline by 28 days post-infarct [81]. BCRP has been suggested to be important for hematopoietic growth and division [81].

Solute Carrier Transporters, GLUT1 Ischemia and hypoxia have been shown to increase GLUT1 expression and consequently glucose uptake in brain endothelial cells both *in vitro* and *in vivo* conditions [139, 140].

Vesicular-Associated, CAV1 During stroke positive feedback between CAV1 and NO promotes BBB breakdown during cerebral reperfusion injury. Normally, CAV1 binds and inhibits the endothelial NOS (eNOS). However, during stroke, CAV1 is down-regulated and eNOS produces high levels of toxic NO that further down-regulates CAV1. Additionally, NO activates MMPs, degrades TJs, and further affects BBB functions [95].

3.7 *Traumatic Brain Injury (TBI)*

Traumatic brain injury (TBI) is defined as biochemical injury of brain tissue resulting from either focal (i.e., penetrating) or diffuse (i.e., injury due to nonpenetrating accelerating or rotational forces) trauma. Neuroinflammation in TBI is mediated by activated microglial cells [1]. Secondary injury, which can cause a long-lasting and progressive pathology following an acute event, is caused by various imbalances such as excess neuroexcitatory neurotransmitters, electrolyte imbalances, injuries such as ischemia, and mitochondrial dysfunction [1]. TBI activates microglia and astrocytes, promoting both neuroinflammatory and neuro-repair processes [1]. In animal models for TBI such as *cold injury*, BBB breakdown is rapidly induced within hours following acute TBI. Initially, it involves arterioles and venules at the lesion edges and is accompanied by spreading edema. Once repair is initiated, BBB breakdown accompanies angiogenesis at the lesion site [92].

3.7.1 TBI and BBB Transporters

Vesicular-Associated Transport Protein, CAV1 There is a heterogeneous response of CAV1 to TBI. TBI rapidly induces the up-regulation of CAV1 in arterioles in the lesion in which the BBB is compromised. Additionally, in a later study, the authors showed a significant increase in phosphorylated CAV1 in these vessels, which promoted caveolae transcytosis and BBB breakdown. The increase was sustained for several days before returning to basal levels. The authors noted that immediately following TBI fewer capillaries showed BBB breakdown and in those cells CAV1 was reduced [92, 93].

3.8 *Human Immunodeficiency Virus-1 (HIV1) Neurodementia*

Human immunodeficiency virus-1 (HIV1) belongs to the *Lentivirus* genus of the Retroviridae virus family. HIV1, like all lentiviruses, is a single-stranded positive-sense enveloped RNA virus that can infect nondividing cells. The viral strand is reverse transcribed into double-stranded DNA, and integrated into cellular DNA, where it can be *latent* (dormant) for long periods before being transcribed, packaged, and released [141]. In about half of individuals with HIV1, infection causes encephalitis and neurobehavioral impairments [142, 143]. HIV1 neuroinvasion, which can take place as early as 10 days post-infection, can occur through a so-called *Trojan horse* strategy in which HIV bypasses the BBB by infecting circulating peripheral monocytes. In infected monocytes the HIV viruses remain dormant until after the monocytes enter the CNS during routine immune surveillance and differentiate into macrophages. Once infected monocytes and perivascular macrophages extravasate, they secrete proinflammatory mediators such as TNFα and CCL2 [142, 144]. Furthermore, release of the HIV1 regulatory protein, transactivator of transcription (Tat), and the viral envelope protein, gp120, has been shown to induce oxidative stress and affect BBB permeability [142–144].

3.8.1 HIV Neurodementia and BBB Transporters

ABC Transporters, P-gp, MRP1, and BCRP HIV1 regulates ABC transporters via cytokine secretion, lipid rafts, Rho/ROCK signaling, NF kB, and other mechanisms. Additionally, drugs used to treat HIV1 are ABC pump inhibitors and inducers, in addition to substrates. The impacts of HIV1 on ABC pump expressions are variable and differences have been seen in various cell types and species. In human post-mortem brains P-gp was down-regulated; however, it is not clear if this resulted from exposure to HIV1 viral proteins or anti-viral therapy effects [143]. In contrast, brain microvascular in vivo animal and in vitro cell model data show that P-gp mRNA and protein expression is up-regulated by Tat and gp120 exposure [143]. Treatment of primary rat brain microvascular cells with Tat was also found to increase BCRP/MRP1 mRNA and protein levels [143].

Vesicular Associated, CAV1 Transport A number of HIV1 effects on BBB TJs are mediated by CAV1 and silencing CAV1 has been shown to protect TJ integrity [95]. The HIV1 protein, Tat, induces redox-activated signals in caveolae, which, via Ras and RhoA, up-regulate CAV1 and decrease TJ expression. Furthermore, HIV1-infected monocytes have been shown to increase CAV1 by inducing ERK1/2 and

Akt. Subsequent CAV1-mediated activation of MMP-9 leads to TJ degradation [95]. Silencing of CAV1 has also been shown to be protective against the observed HIV1-associated increases in cerebral Aβ levels [105].

Receptor-Mediated, VEGFR2 Tat interacts with VEGFR2 and, via signaling through MEK1/2, down-regulates the TJ claudin5 (CLDN5) [143].

3.9 Systemic Inflammation/LPS Models

Peripheral challenges, such as those that occur in bacterial and viral infections and chronic inflammatory diseases, can result in system-wide inflammation that affects the CNS in part through the BBB [145]. There are a number of models for assessing effects [72]. One widely used method to study systemic inflammation in animal models is LPS administration. LPS is an immunogenic proteoglycan found in the outer membrane of some gram-negative bacteria [72]. Intraperitoneal administration of LPS induces the systemic and cerebral increase of inflammatory cytokines and the systemic inflammatory response syndrome, including changes in temperature and activity [146]. The effects of LPS on the BBB are largely mediated by prostanoids (e.g., PGE_2, and NO synthesis by endothelial cells) [72, 146, 147]. Additionally, in other NVU cells, such as astrocytes, LPS broadly induces pro-inflammatory and cytotoxic pathways, resulting in the production of a number of substances that affect the BBB, including IL-1β, IL-6, TNFα, and prostaglandins [72].

3.9.1 Systemic Inflammation (LPS Model) and BBB Transporters

ABC Transporter, P-gp Lipopolysaccharide challenge was shown to cause the rapid release of TNFα that activated signaling through the TNF-R1 receptor, resulting in ET-1 release and ET_B receptor signaling, strongly down-regulating P-gp [15, 72, 148]. In another study LPS did not significantly change the expression of brain microvascular Mrp4 protein [146].

Solute Carrier Transporter, LAT1 Gene expression of *Lat1* in endothelial capillaries was rapidly (within 4 h) and strongly depressed following LPS dosing. This was followed by a marked increase in Lat1 during the recovery from endotoxemia, which by 48 h was approximately three times the control level. These changes in *Lat1* mRNA were only moderately reflected in protein levels, which remained nearly stable at 24 and 48 h post-LPS dosing [71, 72].

SLC and SLCO Transporters, MCT8, OATP1A4, OATP1C1, OAT3
Injection of LPS in either mice or rats resulted in the early down-regulation of mRNAs for both *OAT1C1* and *SLC16A2* thyroid hormone transporters in endothelial cells, and mRNA levels rebounded to higher than control levels within 24 h. OAT1C1 but not

MCT8 protein in BBB endothelial cells was significantly up-regulated at 24 h after LPS administration. The authors suggested that systemic inflammation might down-regulate brain TH uptake [72, 75]. In another study at 24 h post-LPS dosing, the protein expression of the PGE_2 transporters, OATP1A4 (decreased 39 %) and OAT3 (decreased 29 %) was significantly decreased [15, 72, 75, 146].

Receptor-Mediated Transporter LRP1, RAGE Repeated dosing with LPS was found to decrease LRP1 mRNA levels, but not the protein expression of LRP1 or RAGE [72, 149]. However, LPS was found to stimulate a net increase in brain Aβ levels by other cytokine-mediated and BBB-independent mechanisms [149].

Other Transporters Lipopolysaccharide has been reported to increase expression of transporters for PGE_2, TNFα, leptin, insulin, monamines, lysosomal enzymes, leukemia inhibitory factor, and gp120 [15, 72]. TNFα was shown to induce an ATP-dependent transporter of electrophile–glutathione conjugates, RLIP76 (also called RALBP1).

3.10 Diabetes Mellitus

Diabetes mellitus refers to a number of metabolic diseases that involve chronically elevated blood glucose levels. If untreated in the short term it can cause diabetic ketoacidosis and nonketotic hyperosmolar coma and in the long run increases the risk for cardiovascular disease, vascular dementia, AD, stroke, kidney failure, damage to eyes, ulceration of the feet, etc. [150]. There is strong evidence supporting the proinflammatory and oxidative impacts of DM on the BBB [57, 118]. Hyperglycemia can damage endothelial cells and lead to endothelial dysfunction, reduced perfusion and proliferation, and increased permeability [32]. In both type 1 DM, in which the pancreas fails to produce sufficient insulin, and type 2 DM, in which cells become insulin resistant, glucotoxicity activates the pancreatic β-cell inflammasome, releasing IL-1β. By binding to its cognate IL1R receptors in islet cells, IL-1β activates NF-kB transcriptional synthesis of cytokines and chemokines, triggering the destruction of insulin-secreting β-cells [102]. Many, if not most, diabetic individuals face repetitive cycles of hypoglycemia that can lead to severe CNS impairment. This is in part due to treatment to reduce blood glucose levels, along with an impairment of the counterregulatory adrenomedullary epinephrine responses that would normally decrease insulin and increase glucagon, coupled with hypoglycemic unawareness [57].

3.10.1 DM and BBB Transporters

ABC Transporters, P-gp, BCRP, MRP1, and ssMRP4 The impact on the ABC transporters, P-gp and BCRP, is somewhat controversial and several studies of BBB expression using rat models have reported contradictory results. Reichel and co-workers [150] reported that 14 days of streptozotocin (STZ)-induced diabetes induced slight increases in gene expression of *Abcb1* and *Abcg2* mRNA, but no significant increase in P-gp or BCRP protein levels. However, several analogous studies by Liu and co-workers [118] reported decreased levels in BBB-expressed P-gp and BCRP protein. They further reported that insulin treatment restored normal P-gp levels [118]. They concluded that insulin activates PKC/NF-kB pathways that directly regulate P-gp expression and function. An in vitro study using a human brain microvascular model (hCMEC/D3) demonstrated that acute hypoglycemia can up-regulate P-gp, BCRP, MRP1, and MRP4 mRNA and protein expression. The authors further reported that repeated hyperglycemic episodes up-regulate P-gp expression [151].

Solute Carrier Transporter, GLUT1 Results concerning the impact of hyperglycemia and diabetes on BBB GLUT1 expression and activity are mixed. Several studies using STZ-induced rat models for diabetes reported significant decreases in GLUT1 activity and protein expression [57]. The local consumption of glucose was increased parallel to the moderate but significant decreases in microvessel GLUT1 expression for chronic but not acute hyperglycemia [57]. In humans, acute hyperglycemia in nondiabetic subjects also failed to result in statistically significant differences in glucose transport [62]. The consensus from mouse and rat models of hypoglycemia (both acute and chronic) is that it results in an increase in brain glucose, with some studies also finding a corollary decrease in brain glycogen. Additionally, increased expression of GLUT1 mRNA and protein was reported. GLUT1 protein localization was redistributed luminal membranes, suggesting that compensatory mechanisms for increasing brain glucose might be triggered by hypoglycemia [57].

Receptor-Mediated Transporter, RAGE RAGE is activated on cerebral vasculature and, as elsewhere, ligand binding increases the release of ROS, proinflammatory cytokines, chemokines, MMPs, etc. The inflammatory and oxidative environment mediated by RAGE hyperactivity enhances the generation of RAGE ligands and therefore RAGE receptor expression; thereby promoting a positive feedback loop [57, 90, 91].

4 Concluding Remarks

The interplay between the immune and the neurovascular systems is complex [1]. Immune mediators, for better or worse, have an impact on nearly every aspect of neurophysiology, from development to learning, psychological well-being, to pathological conditions and aging [13]. However, studying the NVU is technically challenging because of the physiology and morphology of the component cells, in addition to their close apposition and functional interdependence. This has motivated development of elegant strategies and novel in vitro and in vivo models for examining and testing individual cell types and their transporters, their functions, roles, and co-operations. However, for all the progress made, our understanding of the system clearly remains incomplete. For example, we still do not understand what tips the system from providing protective to destructive responses; or how to tip it back to health. Nor have the impacts of age, gender, genetics, microbiome, circadian cycle, diet, exercise, medication, and their interactions been fully investigated [1, 13, 57, 102, 105, 138]. Beyond the research needed, more powerful analytics are required. The existing informatics and computational resources (i.e. systems biology) could be employed to integrate and synthesize information from large datasets and from studies carried out by different groups. These tools can be used to facilitate the rational design and analytical interpretation of the multifactorial studies that are vital for understanding the immune–NVU interactions, and for successfully applying that knowledge to the clinic [152].

References

1. DiSabato DJ, Quan N, Godbout JP (2016) Neuroinflammation: the devil is in the details. J Neurochem. 2016; 139(*2*):136–153. doi:10.1111/jnc.13607
2. Abbott NJ, Ronnback L, Hansson E (2006) Astrocyte-endothelial interactions at the blood–brain barrier. Nat Rev Neurosci 7(1):41–53. doi:10.1038/nrn1824
3. Coisne C, Engelhardt B (2011) Tight junctions in brain barriers during central nervous system inflammation. Antioxid Redox Signal 15(5):1285–1303. doi:10.1089/ars.2011.3929
4. Chen Y, Liu L (2012) Modern methods for delivery of drugs across the blood–brain barrier. Adv Drug Deliv Rev 64(7):640–665, http://dx.doi.org/10.1016/j.addr.2011.11.010
5. Lyck R, Ruderisch N, Moll AG, Steiner O, Cohen CD, Engelhardt B, Makrides V, Verrey F (2009) Culture-induced changes in blood–brain barrier transcriptome: implications for amino-acid transporters in vivo. J Cereb Blood Flow Metab 29(9):1491–1502. doi:10.1038/jcbfm.2009.72
6. Hawkins RA, O'Kane RL, Simpson IA, Vina JR (2006) Structure of the blood–brain barrier and its role in the transport of amino acids. J Nutr 136(1 Suppl):218s–226s
7. Al Ahmad A, Taboada CB, Gassmann M, Ogunshola OO (2011) Astrocytes and pericytes differentially modulate blood–brain barrier characteristics during development and hypoxic insult. J Cereb Blood Flow Metab 31(2):693–705. doi:10.1038/jcbfm.2010.148
8. Ruderisch N, Virgintino D, Makrides V, Verrey F (2011) Differential axial localization along the mouse brain vascular tree of luminal sodium-dependent glutamine transporters Snat1 and Snat3. J Cereb Blood Flow Metab 31(7):1637–1647. doi:10.1038/jcbfm.2011.21
9. Dolgodilina E, Imobersteg S, Laczko E, Welt T, Verrey F, Makrides V (2015) Brain interstitial fluid glutamine homeostasis is controlled by blood–brain barrier SLC7A5/LAT1 amino acid transporter. J Cereb Blood Flow Metab. 2016; 36(*11*):1929–1941. doi:10.1177/0271678x15609331
10. Neuwelt EA, Bauer B, Fahlke C, Fricker G, Iadecola C, Janigro D, Leybaert L, Molnar Z, O'Donnell ME, Povlishock JT, Saunders NR, Sharp F, Stanimirovic D, Watts RJ, Drewes LR (2011) Engaging neuroscience to advance translational research in brain barrier biology. Nat Rev Neurosci 12(3):169–182. doi:10.1038/nrn2995
11. Ransohoff RM, Engelhardt B (2012) The anatomical and cellular basis of immune surveillance in the central nervous system. Nat Rev Immunol 12(9):623–635. doi:10.1038/nri3265
12. Erickson MA, Dohi K, Banks WA (2012) Neuroinflammation: a common pathway in CNS diseases as mediated at the blood–brain barrier. Neuroimmunomodulation 19(2):121–130. doi:10.1159/000330247
13. Dantzer R, O'Connor JC, Freund GG, Johnson RW, Kelley KW (2008) From inflammation to sickness and depression: when the immune system subjugates the brain. Nat Rev Neurosci 9(1):46–56. doi:10.1038/nrn2297

14. Abbott NJ, Patabendige AAK, Dolman DEM, Yusof SR, Begley DJ (2010) Structure and function of the blood–brain barrier. Neurobiol Dis 37(1):13–25, http://dx.doi.org/10.1016/j.nbd.2009.07.030
15. Mahringer A, Fricker G (2016) ABC transporters at the blood–brain barrier. Expert Opin Drug Metab Toxicol 12(5):499–508. doi:10.1517/17425255.2016.1168804
16. Silverton L, Dean M, Moitra K (2011) Variation and evolution of the ABC transporter genes ABCB1, ABCC1, ABCG2, ABCG5 and ABCG8: implication for pharmacogenetics and disease. Drug Metabol Drug Interact 26(4):169–179. doi:10.1515/DMDI.2011.027
17. Jani M, Ambrus C, Magnan R, Jakab KT, Beéry E, Zolnerciks JK, Krajcsi P (2014) Structure and function of BCRP, a broad specificity transporter of xenobiotics and endobiotics. Arch Toxicol 88(6):1205–1248. doi:10.1007/s00204-014-1224-8
18. Miller DS (2015) Regulation of ABC transporters at the blood–brain barrier. Clin Pharmacol Ther 97(4):395–403. doi:10.1002/cpt.64
19. Pardridge WM (2005) The blood–brain barrier: bottleneck in brain drug development. NeuroRx 2(1):3–14
20. Li H, Wang H (2010) Activation of xenobiotic receptors: driving into the nucleus. Expert Opin Drug Metab Toxicol 6(4):409–426. doi:10.1517/17425251003598886
21. Bernstein H-G, Hölzl G, Dobrowolny H, Hildebrandt J, Trübner K, Krohn M, Bogerts B, Pahnke J (2014) Vascular and extravascular distribution of the ATP-binding cassette transporters ABCB1 and ABCC1 in aged human brain and pituitary. Mech Ageing Dev 141–142:12–21, http://dx.doi.org/10.1016/j.mad.2014.08.003
22. Kubo Y, Ohtsuki S, Uchida Y, Terasaki T (2015) Quantitative determination of luminal and abluminal membrane distributions of transporters in porcine brain capillaries by plasma membrane fractionation and quantitative targeted proteomics. J Pharm Sci 104(9):3060–3068. doi:10.1002/jps.24398
23. Wahrle SE, Jiang H, Parsadanian M, Legleiter J, Han X, Fryer JD, Kowalewski T, Holtzman DM (2004) ABCA1 is required for normal central nervous system ApoE levels and for lipidation of astrocyte-secreted apoE. J Biol Chem 279(39):40987–40993. doi:10.1074/jbc.M407963200
24. Burns MP, Vardanian L, Pajoohesh-Ganji A, Wang L, Cooper M, Harris DC, Duff K, Rebeck GW (2006) The effects of ABCA1 on cholesterol efflux and Aβ levels in vitro and in vivo. J Neurochem 98(3):792–800. doi:10.1111/j.1471-4159.2006.03925.x
25. Do TM, Dodacki A, Alata W, Calon F, Nicolic S, Scherrmann JM, Farinotti R, Bourasset F (2015) Age-dependent regulation of the blood–brain barrier influx/efflux equilibrium of amyloid-beta peptide in a mouse model of Alzheimer's disease (3xTg-AD). J Alzheimers Dis 49(2):287–300. doi:10.3233/jad-150350
26. ElAli A, Rivest S (2013) The role of ABCB1 and ABCA1 in beta-amyloid clearance at the neurovascular unit in Alzheimer's disease. Front Physiol 4:45. doi:10.3389/fphys.2013.00045
27. Reboul E, Dyka FM, Quazi F, Molday RS (2013) Cholesterol transport via ABCA1: new insights from solid-phase binding assay. Biochimie 95(4):957–961. doi:10.1016/j.biochi.2012.11.009
28. Nordestgaard LT, Tybjaerg-Hansen A, Nordestgaard BG, Frikke-Schmidt R (2015) Loss-of-function mutation in ABCA1 and risk of Alzheimer's disease and cerebrovascular disease. Alzheimers Dement 11(12):1430–1438. doi:10.1016/j.jalz.2015.04.006
29. Virgintino D, Errede M, Girolamo F, Capobianco C, Robertson D, Vimercati A, Serio G, Di Benedetto A, Yonekawa Y, Frei K (2008) Fetal blood–brain barrier P-glycoprotein contributes to brain protection during human development. J Neuropathol Exp Neurol 67(1):50–61
30. Chen Z, Shi T, Zhang L, Zhu P, Deng M, Huang C, Hu T, Jiang L, Li J (2016) Mammalian drug efflux transporters of the ATP binding cassette (ABC) family in multidrug resistance: a review of the past decade. Cancer Lett 370(1):153–164. doi:10.1016/j.canlet.2015.10.010
31. Bendayan R, Ronaldson PT, Gingras D, Bendayan M (2006) In situ localization of P-glycoprotein (ABCB1) in human and rat brain. J Histochem Cytochem 54(10):1159–1167. doi:10.1369/jhc.5A6870.2006

32. Liu L, Liu X-D (2014) Alterations in function and expression of ABC transporters at blood–brain barrier under diabetes and the clinical significances. Front Pharmacol 5:273. doi:10.3389/fphar.2014.00273
33. Baello S, Iqbal M, Bloise E, Javam M, Gibb W, Matthews SG (2014) TGF-beta1 regulation of multidrug resistance P-glycoprotein in the developing male blood–brain barrier. Endocrinology 155(2):475–484. doi:10.1210/en.2013-1472
34. Bartels AL, Kortekaas R, Bart J, Willemsen ATM, de Klerk OL, de Vries JJ, van Oostrom JCH, Leenders KL (2009) Blood–brain barrier P-glycoprotein function decreases in specific brain regions with aging: a possible role in progressive neurodegeneration. Neurobiol Aging 30(11):1818–1824, http://dx.doi.org/10.1016/j.neurobiolaging.2008.02.002
35. Wang Y, Zhang D, Wu K, Zhao Q, Nie Y, Fan D (2014) Long noncoding RNA MRUL promotes ABCB1 expression in multidrug-resistant gastric cancer cell sublines. Mol Cell Biol 34(17):3182–3193. doi:10.1128/MCB.01580-13
36. Chen Z, Ma T, Huang C, Zhang L, Lv X, Xu T, Hu T, Li J (2013) MiR-27a modulates the MDR1/P-glycoprotein expression by inhibiting FZD7/beta-catenin pathway in hepatocellular carcinoma cells. Cell Signal 25(12):2693–2701. doi:10.1016/j.cellsig.2013.08.032
37. ElAli A, Hermann DM (2012) Liver X receptor activation enhances blood–brain barrier integrity in the ischemic brain and increases the abundance of ATP-binding cassette transporters ABCB1 and ABCC1 on brain capillary cells. Brain Pathol 22(2):175–187. doi:10.1111/j.1750-3639.2011.00517.x
38. Kruh GD, Belinsky MG (2003) The MRP family of drug efflux pumps. Oncogene 22(47):7537–7552
39. Soontornmalai A, Vlaming M, Fritschy J-M (2006) Differential, strain-specific cellular and subcellular distribution of multidrug transporters in murine choroid plexus and blood–brain barrier. Neuroscience 138(1):159–169
40. Nies A, Jedlitschky G, König J, Herold-Mende C, Steiner H, Schmitt H-P, Keppler D (2004) Expression and immunolocalization of the multidrug resistance proteins, MRP1–MRP6 (ABCC1–ABCC6), in human brain. Neuroscience 129(2):349–360
41. Cole SP (2014) Targeting multidrug resistance protein 1 (MRP1, ABCC1): past, present, and future. Annu Rev Pharmacol Toxicol 54:95–117. doi:10.1146/annurev-pharmtox-011613-135959
42. Bakos É, Homolya L (2007) Portrait of multifaceted transporter, the multidrug resistance-associated protein 1 (MRP1/ABCC1). Pflügers Arch 453(5):621–641
43. Krohn M, Lange C, Hofrichter J, Scheffler K, Stenzel J, Steffen J, Schumacher T, Brüning T, Plath A-S, Alfen F, Schmidt A, Winter F, Rateitschak K, Wree A, Gsponer J, Walker LC, Pahnke J (2011) Cerebral amyloid-β proteostasis is regulated by the membrane transport protein ABCC1 in mice. J Clin Invest 121(10):3924–3931. doi:10.1172/JCI57867
44. ElAli A, Hermann DM (2010) Apolipoprotein E controls ATP-binding cassette transporters in the ischemic brain. Sci Signal 3(142):ra72. doi:10.1126/scisignal.2001213
45. Hermann DM, Bassetti CL (2007) Implications of ATP-binding cassette transporters for brain pharmacotherapies. Trends Pharmacol Sci 28(3):128–134
46. Cole SP (2014) Multidrug resistance protein 1 (MRP1, ABCC1), a "multitasking" ATP-binding cassette (ABC) transporter. J Biol Chem 289(45):30880–30888. doi:10.1074/jbc.R114.609248
47. Nahalkova J, Volkmann I, Aoki M, Winblad B, Bogdanovic N, Tjernberg LO, Behbahani H (2010) CD147, a γ-secretase associated protein is upregulated in Alzheimer's disease brain and its cellular trafficking is affected by presenilin-2. Neurochem Int 56(1):67–76
48. Moitra K, Lou H, Dean M (2011) Multidrug efflux pumps and cancer stem cells: insights into multidrug resistance and therapeutic development. Clin Pharmacol Ther 89(4):491
49. Krishnamurthy P, Schuetz JD (2005) The ABC transporter Abcg2/Bcrp: role in hypoxia mediated survival. Biometals 18(4):349–358
50. Zlokovic BV (2008) The blood–brain barrier in health and chronic neurodegenerative disorders. Neuron 57(2):178–201

51. Hediger MA, Romero MF, Peng JB, Rolfs A, Takanaga H, Bruford EA (2004) The ABCs of solute carriers: physiological, pathological and therapeutic implications of human membrane transport proteins. Introduction. Pflugers Arch 447(5):465–468. doi:10.1007/s00424-003-1192-y
52. Hediger MA, Clemencon B, Burrier RE, Bruford EA (2013) The ABCs of membrane transporters in health and disease (SLC series): introduction. Mol Aspects Med 34(2–3):95–107. doi:10.1016/j.mam.2012.12.009
53. Dahlin A, Royall J, Hohmann JG, Wang J (2009) Expression profiling of the solute carrier gene family in the mouse brain. J Pharmacol Exp Ther 329(2):558–570. doi:10.1124/jpet.108.149831
54. Daneman R, Zhou L, Agalliu D, Cahoy JD, Kaushal A, Barres BA (2010) The mouse blood–brain barrier transcriptome: a new resource for understanding the development and function of brain endothelial cells. PLoS One 5(10):e13741. doi:10.1371/journal.pone.0013741
55. Uldry M, Thorens B (2004) The SLC2 family of facilitated hexose and polyol transporters. Pflugers Arch 447(5):480–489. doi:10.1007/s00424-003-1085-0
56. Mueckler M, Thorens B (2013) The SLC2 (GLUT) family of membrane transporters. Mol Aspects Med 34:121–138. doi:10.1016/j.mam.2012.07.001
57. Prasad S, Sajja RK, Naik P, Cucullo L (2014) Diabetes mellitus and blood–brain barrier dysfunction: an overview. J Pharmacovigil 2(2):125. doi:10.4172/2329-6887.1000125
58. Leybaert L (2005) Neurobarrier coupling in the brain: a partner of neurovascular and neurometabolic coupling? J Cereb Blood Flow Metab 25(1):2–16. doi:10.1038/sj.jcbfm.9600001
59. Farrell CL, Pardridge WM (1991) Blood–brain barrier glucose transporter is asymmetrically distributed on brain capillary endothelial lumenal and ablumenal membranes: an electron microscopic immunogold study. Proc Natl Acad Sci U S A 88(13):5779–5783
60. Oby E, Janigro D (2006) The blood–brain barrier and epilepsy. Epilepsia 47(11):1761–1774. doi:10.1111/j.1528-1167.2006.00817.x
61. Winkler EA, Nishida Y, Sagare AP, Rege SV, Bell RD, Perlmutter D, Sengillo JD, Hillman S, Kong P, Nelson AR, Sullivan JS, Zhao Z, Meiselman HJ, Wenby RB, Soto J, Abel ED, Makshanoff J, Zuniga E, De Vivo DC, Zlokovic BV (2015) GLUT1 reductions exacerbate Alzheimer's disease vasculo-neuronal dysfunction and degeneration. Nat Neurosci 18(4):521–530. doi:10.1038/nn.3966
62. Patching SG (2016) Glucose transporters at the blood–brain barrier: function, regulation and gateways for drug delivery. Mol Neurobiol. 2016; 22:1–32. doi:10.1007/s12035-015-9672-6
63. Mastroberardino L, Spindler B, Pfeiffer R, Skelly PJ, Loffing J, Shoemaker CB, Verrey F (1998) Amino-acid transport by heterodimers of 4F2hc/CD98 and members of a permease family. Nature 395(6699):288–291. doi:10.1038/26246
64. Verrey F, Jack DL, Paulsen IT, Saier MH Jr, Pfeiffer R (1999) New glycoprotein-associated amino acid transporters. J Membr Biol 172(3):181–192
65. Verrey F (2003) System L: heteromeric exchangers of large, neutral amino acids involved in directional transport. Pflugers Arch 445(5):529–533. doi:10.1007/s00424-002-0973-z
66. Verrey F, Closs EI, Wagner CA, Palacin M, Endou H, Kanai Y (2004) CATs and HATs: the SLC7 family of amino acid transporters. Pflugers Arch 447(5):532–542. doi:10.1007/s00424-003-1086-z
67. Verrey F, Meier C, Rossier G, Kuhn LC (2000) Glycoprotein-associated amino acid exchangers: broadening the range of transport specificity. Pflugers Arch 440(4):503–512
68. Ramadan T, Camargo SM, Herzog B, Bordin M, Pos KM, Verrey F (2007) Recycling of aromatic amino acids via TAT1 allows efflux of neutral amino acids via LAT2-4F2hc exchanger. Pflugers Arch 454(3):507–516. doi:10.1007/s00424-007-0209-3
69. Milkereit R, Persaud A, Vanoaica L, Guetg A, Verrey F, Rotin D (2015) LAPTM4b recruits the LAT1-4F2hc Leu transporter to lysosomes and promotes mTORC1 activation. Nat Commun 6:7250. doi:10.1038/ncomms8250
70. Fotiadis D, Kanai Y, Palacin M (2013) The SLC3 and SLC7 families of amino acid transporters. Mol Aspects Med 34(2–3):139–158. doi:10.1016/j.mam.2012.10.007

71. Wittmann G, Mohacsik P, Balkhi MY, Gereben B, Lechan RM (2015) Endotoxin-induced inflammation down-regulates L-type amino acid transporter 1 (LAT1) expression at the blood–brain barrier of male rats and mice. Fluids Barriers CNS 12:21. doi:10.1186/s12987-015-0016-8
72. Varatharaj A, Galea I (2016) The blood–brain barrier in systemic inflammation. Brain Behav Immun. 2016; 16. pii: S0889–1591(*16*):30055–1. doi:10.1016/j.bbi.2016.03.010
73. Halestrap AP (2013) Monocarboxylic acid transport. Compr Physiol 3(4):1611–1643. doi:10.1002/cphy.c130008
74. Visser WE, Friesema EC, Visser TJ (2011) Minireview: thyroid hormone transporters: the knowns and the unknowns. Mol Endocrinol 25(1):1–14. doi:10.1210/me.2010-0095
75. Wittmann G, Szabon J, Mohacsik P, Nouriel SS, Gereben B, Fekete C, Lechan RM (2015) Parallel regulation of thyroid hormone transporters OATP1c1 and MCT8 during and after endotoxemia at the blood–brain barrier of male rodents. Endocrinology 156(4):1552–1564. doi:10.1210/en.2014-1830
76. Koepsell H (2013) The SLC22 family with transporters of organic cations, anions and zwitterions. Mol Aspects Med 34(2–3):413–435. doi:10.1016/j.mam.2012.10.010
77. Hagenbuch B, Stieger B (2013) The SLCO (former SLC21) superfamily of transporters. Mol Aspects Med 34(2–3):396–412. doi:10.1016/j.mam.2012.10.009
78. Meier PJ, Eckhardt U, Schroeder A, Hagenbuch B, Stieger B (1997) Substrate specificity of sinusoidal bile acid and organic anion uptake systems in rat and human liver. Hepatology 26(6):1667–1677. doi:10.1002/hep.510260641
79. Gao B, Stieger B, Noe B, Fritschy JM, Meier PJ (1999) Localization of the organic anion transporting polypeptide 2 (Oatp2) in capillary endothelium and choroid plexus epithelium of rat brain. J Histochem Cytochem 47(10):1255–1264
80. Gao B, Hagenbuch B, Kullak-Ublick GA, Benke D, Aguzzi A, Meier PJ (2000) Organic anion-transporting polypeptides mediate transport of opioid peptides across blood–brain barrier. J Pharmacol Exp Ther 294(1):73–79
81. Dazert P, Suofu Y, Grube M, Popa-Wagner A, Kroemer HK, Jedlitschky G, Kessler C (2006) Differential regulation of transport proteins in the periinfarct region following reversible middle cerebral artery occlusion in rats. Neuroscience 142(4):1071–1079. doi:10.1016/j.neuroscience.2006.07.056
82. Jones AR, Shusta EV (2007) Blood–brain barrier transport of therapeutics via receptor-mediation. Pharm Res 24(9):1759–1771. doi:10.1007/s11095-007-9379-0
83. Hervé F, Ghinea N, Scherrmann J-M (2008) CNS delivery via adsorptive transcytosis. AAPS J 10(3):455–472. doi:10.1208/s12248-008-9055-2
84. Kang NH, Lee WK, Yi BR, Park MA, Lee HR, Park SK, Hwang KA, Park HK, Choi KC (2012) Modulation of lipid metabolism by mixtures of protamine and chitooligosaccharide through pancreatic lipase inhibitory activity in a rat model. Lab Anim Res 28(1):31–38. doi:10.5625/lar.2012.28.1.31
85. Ramanathan A, Nelson AR, Sagare AP, Zlokovic BV (2015) Impaired vascular-mediated clearance of brain amyloid beta in Alzheimer's disease: the role, regulation and restoration of LRP1. Front Aging Neurosci 7:136. doi:10.3389/fnagi.2015.00136
86. Kang DE, Pietrzik CU, Baum L, Chevallier N, Merriam DE, Kounnas MZ, Wagner SL, Troncoso JC, Kawas CH, Katzman R, Koo EH (2000) Modulation of amyloid beta-protein clearance and Alzheimer's disease susceptibility by the LDL receptor-related protein pathway. J Clin Invest 106(9):1159–1166. doi:10.1172/jci11013
87. Lillis AP, Mikhailenko I, Strickland DK (2005) Beyond endocytosis: LRP function in cell migration, proliferation and vascular permeability. J Thromb Haemost 3(8):1884–1893. doi:10.1111/j.1538-7836.2005.01371.x
88. Deane R, Wu Z, Zlokovic BV (2004) RAGE (yin) versus LRP (yang) balance regulates Alzheimer amyloid beta-peptide clearance through transport across the blood–brain barrier. Stroke 35(11 Suppl 1):2628–2631. doi:10.1161/01.STR.0000143452.85382.d1

89. Ray R, Juranek JK, Rai V (2016) RAGE axis in neuroinflammation, neurodegeneration and its emerging role in the pathogenesis of amyotrophic lateral sclerosis. Neurosci Biobehav Rev 62:48–55. doi:10.1016/j.neubiorev.2015.12.006
90. Vazzana N, Santilli F, Cuccurullo C, Davi G (2009) Soluble forms of RAGE in internal medicine. Intern Emerg Med 4(5):389–401. doi:10.1007/s11739-009-0300-1
91. Schmidt AM, Yan SD, Wautier JL, Stern D (1999) Activation of receptor for advanced glycation end products: a mechanism for chronic vascular dysfunction in diabetic vasculopathy and atherosclerosis. Circ Res 84(5):489–497
92. Nag S, Venugopalan R, Stewart DJ (2007) Increased caveolin-1 expression precedes decreased expression of occludin and claudin-5 during blood–brain barrier breakdown. Acta Neuropathol 114(5):459–469. doi:10.1007/s00401-007-0274-x
93. Nag S, Manias JL, Stewart DJ (2009) Expression of endothelial phosphorylated caveolin-1 is increased in brain injury. Neuropathol Appl Neurobiol 35(4):417–426. doi:10.1111/j.1365-2990.2008.01009.x
94. Virgintino D, Robertson D, Errede M, Benagiano V, Girolamo F, Maiorano E, Roncali L, Bertossi M (2002) Expression of P-glycoprotein in human cerebral cortex microvessels. J Histochem Cytochem 50(12):1671–1676
95. Zhao YL, Song JN, Zhang M (2014) Role of caveolin-1 in the biology of the blood–brain barrier. Rev Neurosci 25(2):247–254. doi:10.1515/revneuro-2013-0039
96. Errede M, Girolamo F, Ferrara G, Strippoli M, Morando S, Boldrin V, Rizzi M, Uccelli A, Perris R, Bendotti C, Salmona M, Roncali L, Virgintino D (2012) Blood–brain barrier alterations in the cerebral cortex in experimental autoimmune encephalomyelitis. J Neuropathol Exp Neurol 71(10):840–854. doi:10.1097/NEN.0b013e31826ac110
97. Song L, Ge S, Pachter JS (2007) Caveolin-1 regulates expression of junction-associated proteins in brain microvascular endothelial cells. Blood 109(4):1515–1523. doi:10.1182/blood-2006-07-034009
98. Kim MP, Park SI, Kopetz S, Gallick GE (2009) Src family kinases as mediators of endothelial permeability: effects on inflammation and metastasis. Cell Tissue Res 335(1):249–259. doi:10.1007/s00441-008-0682-9
99. Schubert W, Frank PG, Woodman SE, Hyogo H, Cohen DE, Chow CW, Lisanti MP (2002) Microvascular hyperpermeability in caveolin-1 (−/−) knock-out mice. Treatment with a specific nitric-oxide synthase inhibitor, L-NAME, restores normal microvascular permeability in Cav-1 null mice. J Biol Chem 277(42):40091–40098. doi:10.1074/jbc.M205948200
100. Gratton JP, Lin MI, Yu J, Weiss ED, Jiang ZL, Fairchild TA, Iwakiri Y, Groszmann R, Claffey KP, Cheng YC, Sessa WC (2003) Selective inhibition of tumor microvascular permeability by cavtratin blocks tumor progression in mice. Cancer Cell 4(1):31–39
101. Bauer PM, Yu J, Chen Y, Hickey R, Bernatchez PN, Looft-Wilson R, Huang Y, Giordano F, Stan RV, Sessa WC (2005) Endothelial-specific expression of caveolin-1 impairs microvascular permeability and angiogenesis. Proc Natl Acad Sci U S A 102(1):204–209. doi:10.1073/pnas.0406092102
102. Turner MD, Nedjai B, Hurst T, Pennington DJ (2014) Cytokines and chemokines: at the crossroads of cell signalling and inflammatory disease. Biochim Biophys Acta 1843(11):2563–2582. doi:10.1016/j.bbamcr.2014.05.014
103. Bailey TL, Rivara CB, Rocher AB, Hof PR (2004) The nature and effects of cortical microvascular pathology in aging and Alzheimer's disease. Neurol Res 26(5):573–578. doi:10.1179/016164104225016272
104. Ogunshola OO, Antoniou X (2009) Contribution of hypoxia to Alzheimer's disease: is HIF-1alpha a mediator of neurodegeneration? Cell Mol Life Sci 66(22):3555–3563. doi:10.1007/s00018-009-0141-0
105. Zhao Z, Nelson Amy R, Betsholtz C, Zlokovic Berislav V (2015) Establishment and dysfunction of the blood–brain barrier. Cell 163(5):1064–1078, http://dx.doi.org/10.1016/j.cell.2015.10.067
106. de Vries HE, Kuiper J, de Boer AG, Van Berkel TJ, Breimer DD (1997) The blood–brain barrier in neuroinflammatory diseases. Pharmacol Rev 49(2):143–155

107. Rothaug M, Becker-Pauly C, Rose-John S (2016) The role of interleukin-6 signaling in nervous tissue. Biochim Biophys Acta 1863(6 Pt A):1218–1227. doi:10.1016/j.bbamcr.2016.03.018
108. Strazielle N, Ghersi-Egea JF (2013) Physiology of blood–brain interfaces in relation to brain disposition of small compounds and macromolecules. Mol Pharm 10(5):1473–1491. doi:10.1021/mp300518e
109. Virgintino D, Robertson D, Benagiano V, Errede M, Bertossi M, Ambrosi G, Roncali L (2000) Immunogold cytochemistry of the blood–brain barrier glucose transporter GLUT1 and endogenous albumin in the developing human brain. Brain Res Dev Brain Res 123(1):95–101
110. Virgintino D, Maiorano E, Errede M, Vimercati A, Greco P, Selvaggi L, Roncali L, Bertossi M (1998) Astroglia-microvessel relationship in the developing human telencephalon. Int J Dev Biol 42(8):1165–1168
111. Engelhardt B, Liebner S (2014) Novel insights into the development and maintenance of the blood–brain barrier. Cell Tissue Res 355(3):687–699. doi:10.1007/s00441-014-1811-2
112. Aden U, Favrais G, Plaisant F, Winerdal M, Felderhoff-Mueser U, Lampa J, Lelievre V, Gressens P (2010) Systemic inflammation sensitizes the neonatal brain to excitotoxicity through a pro-/anti-inflammatory imbalance: key role of TNFalpha pathway and protection by etanercept. Brain Behav Immun 24(5):747–758. doi:10.1016/j.bbi.2009.10.010
113. Stolp HB, Dziegielewska KM, Ek CJ, Potter AM, Saunders NR (2005) Long-term changes in blood–brain barrier permeability and white matter following prolonged systemic inflammation in early development in the rat. Eur J Neurosci 22(11):2805–2816. doi:10.1111/j.1460-9568.2005.04483.x
114. Zlokovic BV (2011) Neurovascular pathways to neurodegeneration in Alzheimer's disease and other disorders. Nat Rev Neurosci 12(12):723–738. doi:10.1038/nrn3114
115. Makrides V, Shen TE, Bhatia R, Smith BL, Thimm J, Lal R, Feinstein SC (2003) Microtubule-dependent oligomerization of tau. Implications for physiological tau function and tauopathies. J Biol Chem 278(35):33298–33304. doi:10.1074/jbc.M305207200
116. Makrides V, Massie MR, Feinstein SC, Lew J (2004) Evidence for two distinct binding sites for tau on microtubules. Proc Natl Acad Sci U S A 101(17):6746–6751. doi:10.1073/pnas.0400992101
117. Feinstein SC, Wilson L (2005) Inability of tau to properly regulate neuronal microtubule dynamics: a loss-of-function mechanism by which tau might mediate neuronal cell death. Biochim Biophys Acta 1739(2–3):268–279. doi:10.1016/j.bbadis.2004.07.002
118. Liu Z, Li T, Li P, Wei N, Zhao Z, Liang H, Ji X, Chen W, Xue M, Wei J (2015) The ambiguous relationship of oxidative stress, Tau hyperphosphorylation, and autophagy dysfunction in Alzheimer's disease. Oxid Med Cell Longev 2015:352723. doi:10.1155/2015/352723
119. Nelson AR, Sweeney MD, Sagare AP, Zlokovic BV (2016) Neurovascular dysfunction and neurodegeneration in dementia and Alzheimer's disease. Biochim Biophys Acta 1862(5):887–900. doi:10.1016/j.bbadis.2015.12.016
120. Jonsson T, Atwal JK, Steinberg S, Snaedal J, Jonsson PV, Bjornsson S, Stefansson H, Sulem P, Gudbjartsson D, Maloney J, Hoyte K, Gustafson A, Liu Y, Lu Y, Bhangale T, Graham RR, Huttenlocher J, Bjornsdottir G, Andreassen OA, Jonsson EG, Palotie A, Behrens TW, Magnusson OT, Kong A, Thorsteinsdottir U, Watts RJ, Stefansson K (2012) A mutation in APP protects against Alzheimer's disease and age-related cognitive decline. Nature 488(7409):96–99. doi:10.1038/nature11283
121. Hartz AMS, Zhong Y, Wolf A, LeVine H, Miller DS, Bauer B (2016) Aβ40 reduces P-glycoprotein at the blood–brain barrier through the ubiquitin–proteasome pathway. J Neurosci 36(6):1930–1941. doi:10.1523/jneurosci.0350-15.2016
122. Wolf A, Bauer B, Hartz AM (2012) ABC transporters and the Alzheimer's disease enigma. Front Psychiatry 3:54. doi:10.3389/fpsyt.2012.00054
123. Simpson IA, Chundu KR, Davies-Hill T, Honer WG, Davies P (1994) Decreased concentrations of GLUT1 and GLUT3 glucose transporters in the brains of patients with Alzheimer's disease. Ann Neurol 35(5):546–551. doi:10.1002/ana.410350507

124. Mooradian AD, Chung HC, Shah GN (1997) GLUT-1 expression in the cerebra of patients with Alzheimer's disease. Neurobiol Aging 18(5):469–474
125. Shibata M, Yamada S, Kumar SR, Calero M, Bading J, Frangione B, Holtzman DM, Miller CA, Strickland DK, Ghiso J, Zlokovic BV (2000) Clearance of Alzheimer's amyloid-β(1–40) peptide from brain by LDL receptor–related protein-1 at the blood–brain barrier. J Clin Invest 106(12):1489–1499
126. Sui YT, Bullock KM, Erickson MA, Zhang J, Banks WA (2014) Alpha synuclein is transported into and out of the brain by the blood–brain barrier. Peptides 62:197–202. doi:10.1016/j.peptides.2014.09.018
127. Camargo SM, Vuille-dit-Bille RN, Mariotta L, Ramadan T, Huggel K, Singer D, Gotze O, Verrey F (2014) The molecular mechanism of intestinal levodopa absorption and its possible implications for the treatment of Parkinson's disease. J Pharmacol Exp Ther 351(1):114–123. doi:10.1124/jpet.114.216317
128. Westin JE, Lindgren HS, Gardi J, Nyengaard JR, Brundin P, Mohapel P, Cenci MA (2006) Endothelial proliferation and increased blood–brain barrier permeability in the basal ganglia in a rat model of 3,4-dihydroxyphenyl-l-alanine-induced dyskinesia. J Neurosci 26(37):9448–9461. doi:10.1523/jneurosci.0944-06.2006
129. Drozdzik M, Bialecka M, Mysliwiec K, Honczarenko K, Stankiewicz J, Sych Z (2003) Polymorphism in the P-glycoprotein drug transporter MDR1 gene: a possible link between environmental and genetic factors in Parkinson's disease. Pharmacogenetics 13(5):259–263. doi:10.1097/01.fpc.0000054087.48725.d9
130. Engelhardt B, Ransohoff RM (2005) The ins and outs of T-lymphocyte trafficking to the CNS: anatomical sites and molecular mechanisms. Trends Immunol 26(9):485–495. doi:10.1016/j.it.2005.07.004
131. Lopes Pinheiro MA, Kooij G, Mizee MR, Kamermans A, Enzmann G, Lyck R, Schwaninger M, Engelhardt B, de Vries HE (2016) Immune cell trafficking across the barriers of the central nervous system in multiple sclerosis and stroke. Biochim Biophys Acta 1862(3):461–471. doi:10.1016/j.bbadis.2015.10.018
132. Wu H, Deng R, Chen X, Wong WC, Chen H, Gao L, Nie Y, Wu W, Shen J (2016) Caveolin-1 is critical for lymphocyte trafficking into central nervous system during experimental autoimmune encephalomyelitis. J Neurosci 36(19):5193–5199. doi:10.1523/jneurosci.3734-15.2016
133. Mahad DJ, Ransohoff RM (2003) The role of MCP-1 (CCL2) and CCR2 in multiple sclerosis and experimental autoimmune encephalomyelitis (EAE). Semin Immunol 15(1):23–32
134. Vezzani A, Fujinami RS, White HS, Preux PM, Blumcke I, Sander JW, Loscher W (2016) Infections, inflammation and epilepsy. Acta Neuropathol 131(2):211–234. doi:10.1007/s00401-015-1481-5
135. Bozzi Y, Caleo M (2016) Epilepsy, seizures, and inflammation: role of the C-C motif ligand 2 chemokine. DNA Cell Biol 35(6):257–260. doi:10.1089/dna.2016.3345
136. Lachos J, Zattoni M, Wieser H-G, Fritschy J-M, Langmann T, Schmitz G, Errede M, Virgintino D, Yonekawa Y, Frei K (2011) Characterization of the gene expression profile of human hippocampus in mesial temporal lobe epilepsy with hippocampal sclerosis. Epilepsy Res Treat 2011:11. doi:10.1155/2011/758407
137. Loscher W, Potschka H (2005) Drug resistance in brain diseases and the role of drug efflux transporters. Nat Rev Neurosci 6(8):591–602. doi:10.1038/nrn1728
138. Moskowitz MA, Lo EH, Iadecola C (2010) The science of stroke: mechanisms in search of treatments. Neuron 67(2):181–198. doi:10.1016/j.neuron.2010.07.002
139. Vannucci SJ, Reinhart R, Maher F, Bondy CA, Lee WH, Vannucci RC, Simpson IA (1998) Alterations in GLUT1 and GLUT3 glucose transporter gene expression following unilateral hypoxia-ischemia in the immature rat brain. Brain Res Dev Brain Res 107(2):255–264
140. Yeh WL, Lin CJ, Fu WM (2008) Enhancement of glucose transporter expression of brain endothelial cells by vascular endothelial growth factor derived from glioma exposed to hypoxia. Mol Pharmacol 73(1):170–177. doi:10.1124/mol.107.038851
141. Weiss RA (1993) How does HIV cause AIDS? Science 260(5112):1273–1279

142. Toborek M, Lee YW, Flora G, Pu H, Andras IE, Wylegala E, Hennig B, Nath A (2005) Mechanisms of the blood–brain barrier disruption in HIV-1 infection. Cell Mol Neurobiol 25(1):181–199
143. McRae M (2016) HIV and viral protein effects on the blood brain barrier. Tissue Barriers 4(1):e1143543. doi:10.1080/21688370.2016.1143543
144. Ivey NS, MacLean AG, Lackner AA (2009) AIDS and the blood–brain barrier. J Neurovirol 15(2):111–122. doi:10.1080/13550280902769764
145. Bennani-Baiti B, Toegel S, Viernstein H, Urban E, Noe CR, Bennani-Baiti IM (2015) Inflammation modulates RLIP76/RALBP1 electrophile-glutathione conjugate transporter and housekeeping genes in human blood–brain barrier endothelial cells. PLoS One 10(9):e0139101. doi:10.1371/journal.pone.0139101
146. Akanuma S, Uchida Y, Ohtsuki S, Tachikawa M, Terasaki T, Hosoya K (2011) Attenuation of prostaglandin E2 elimination across the mouse blood–brain barrier in lipopolysaccharide-induced inflammation and additive inhibitory effect of cefmetazole. Fluids Barriers CNS 8:24. doi:10.1186/2045-8118-8-24
147. Banks WA, Erickson MA (2010) The blood–brain barrier and immune function and dysfunction. Neurobiol Dis 37(1):26–32. doi:10.1016/j.nbd.2009.07.031
148. Hartz AM, Bauer B, Fricker G, Miller DS (2006) Rapid modulation of P-glycoprotein-mediated transport at the blood–brain barrier by tumor necrosis factor-alpha and lipopolysaccharide. Mol Pharmacol 69(2):462–470. doi:10.1124/mol.105.017954
149. Jaeger LB, Dohgu S, Hwang MC, Farr SA, Murphy MP, Fleegal-DeMotta MA, Lynch JL, Robinson SM, Niehoff ML, Johnson SN, Kumar VB, Banks WA (2009) Testing the neurovascular hypothesis of Alzheimer's disease: LRP-1 antisense reduces blood–brain barrier clearance, increases brain levels of amyloid-beta protein, and impairs cognition. J Alzheimers Dis 17(3):553–570. doi:10.3233/jad-2009-1074
150. Reichel V, Burghard S, John I, Huber O (2011) P-glycoprotein and breast cancer resistance protein expression and function at the blood–brain barrier and blood-cerebrospinal fluid barrier (choroid plexus) in streptozotocin-induced diabetes in rats. Brain Res 1370:238–245. doi:10.1016/j.brainres.2010.11.012
151. Sajja RK, Rahman S, Cucullo L (2016) Drugs of abuse and blood–brain barrier endothelial dysfunction: a focus on the role of oxidative stress. J Cereb Blood Flow Metab 36(3):539–554. doi:10.1177/0271678x15616978
152. Margineanu DG (2016) Neuropharmacology beyond reductionism – a likely prospect. Biosystems 141:1–9. doi:10.1016/j.biosystems.2015.11.010
153. Panzenboeck U, Kratzer I, Sovic A, Wintersperger A, Bernhart E, Hammer A, Malle E, Sattler W (2006) Regulatory effects of synthetic liver X receptor- and peroxisome-proliferator activated receptor agonists on sterol transport pathways in polarized cerebrovascular endothelial cells. Int J Biochem Cell Biol 38(8):1314–1329. doi:10.1016/j.biocel.2006.01.013
154. ElAli A, Hermann DM (2011) ATP-binding cassette transporters and their roles in protecting the brain. Neuroscientist 17(4):423–436. doi:10.1177/1073858410391270
155. Strazielle N, Ghersi-Egea J-F (2015) Efflux transporters in blood–brain interfaces of the developing brain. Front Neurosci 9:21. doi:10.3389/fnins.2015.00021
156. Virgintino D, Robertson D, Monaghan P, Errede M, Ambrosi G, Roncali L, Bertossi M (1998) Glucose transporter GLUT1 localization in human foetus telencephalon. Neurosci Lett 256:147–150

microRNAs in Brain Endothelium and Inflammation

D. Roig-Carles, C. Cerutti, M.A. Lopez-Ramirez, D. Wu, David K. Male, H.E. de Vries, and I.A. Romero

Abstract Blood-brain barrier (BBB) dysfunction, characterised by increased permeability across brain endothelium and/or leukocyte extravasation into CNS tissue, is associated with changes in the gene expression profile of brain endothelium and is therefore potentially controlled by epigenetic factors. MicroRNAs (miRNAs) are single-stranded, short, non-coding RNA molecules that mediate post-transcriptional gene silencing. Recent studies have demonstrated alterations of miRNA expression at the BBB in experimental autoimmune encephalomyelitis, in multiple sclerosis, and in cytokine-stimulated human brain endothelial cells. These results suggest that brain endothelial miRNAs play a role in the pathophysiology of the BBB in inflammation. In this chapter, we will first give an overview of miRNA biology and then review the current knowledge of the role of miRNAs in regulating BBB function, particularly in the context of multiple sclerosis.

1 Introduction

Blood-brain barrier (BBB) dysfunction is a major hallmark of many central nervous system (CNS) disorders such as multiple sclerosis and even non-neurological conditions (e.g. acute systemic infection) and is characterised by

$Author contributed equally with all other contributors

D. Roig-Carles (✉) • C. Cerutti • M.A. Lopez-Ramirez • D. Wu • D.K. Male • I.A. Romero
Department of Life, Health and Chemical Sciences, The Open University,
Walton Hall, Milton Keynes MK7 6BJ, UK
e-mail: David.roig-carles@open.ac.uk

H.E. de Vries
Department of Molecular Cell Biology and Immunology, MS center Amsterdam, VU University medical center, Amsterdam, The Netherlands

R. Lyck, G. Enzmann (eds.), *The Blood Brain Barrier and Inflammation*,
Progress in Inflammation Research, DOI 10.1007/978-3-319-45514-3_7

increased permeability across the endothelium, both para- and trans-cellular, alteration in the expression of cell-surface receptors and/or transporters and activation of endothelial cells to support leucocyte extravasation into the CNS parenchyma [1, 65]. We are just beginning to understand the molecular factors controlling BBB permeability (for a recent review, see [3]) and the mechanisms of leucocyte extravasation across brain endothelium [74] under pathological conditions, but many of these pathophysiological effects are underpinned by overt acute or chronic changes in gene expression in cerebral endothelial cells. As an example, activation of cultured human brain endothelial cells (BEC) with the pro-inflammatory mediators TNFα and IFNγ leads to changes in the expression of approximately 5 % of all genes identified using transcriptome microarray technology [53].

In pathological conditions, changes in the brain endothelial pattern of gene expression may occur at the transcriptional level, for example, by regulating the activity and/or expression of transcriptional regulators or, alternatively, following epigenetic changes involving DNA methylation and/or chromatin remodelling. In addition, regulation of gene expression has been shown to occur at the post-transcriptional level, specifically by a class of gene expression regulators termed microRNAs (miRs or miRNAs for short). In this chapter, we will first briefly review the biology of miRNAs as their role in regulating BBB function is a relatively novel concept. Then, we will review our current understanding of the role of brain endothelial miRNAs in inflammation, particularly in the context of multiple sclerosis. Further research has been carried out on this topic in the context of stroke, and this is reviewed in [69] and [93].

2 microRNAs

miRNAs are a class of endogenous, 20–25 nucleotide-long highly conserved, single-stranded, small non-coding RNA molecules that mediate post-transcriptional gene silencing [14].

It was in the early 1990s when two independent groups identified the first miRNA, lin-4, in the nematode *Caenorhabditis elegans* [47, 87]. A second miRNA, let-7, was discovered in 2000 [62, 68], followed by three different groups that published their discovery of new miRNAs in *Science* contributing to the concept of miRNAs as an abundant and distinct group of gene regulators [44–46]. These findings established a new field of research that has grown exponentially since then. Indeed, miRNAs have been defined as critical regulators for many biological processes such as differentiation, cell cycle, development, apoptosis and disease in various model organisms and in humans [26] demonstrating astonishing evolutionary conservation between different vertebrate and invertebrate species [27].

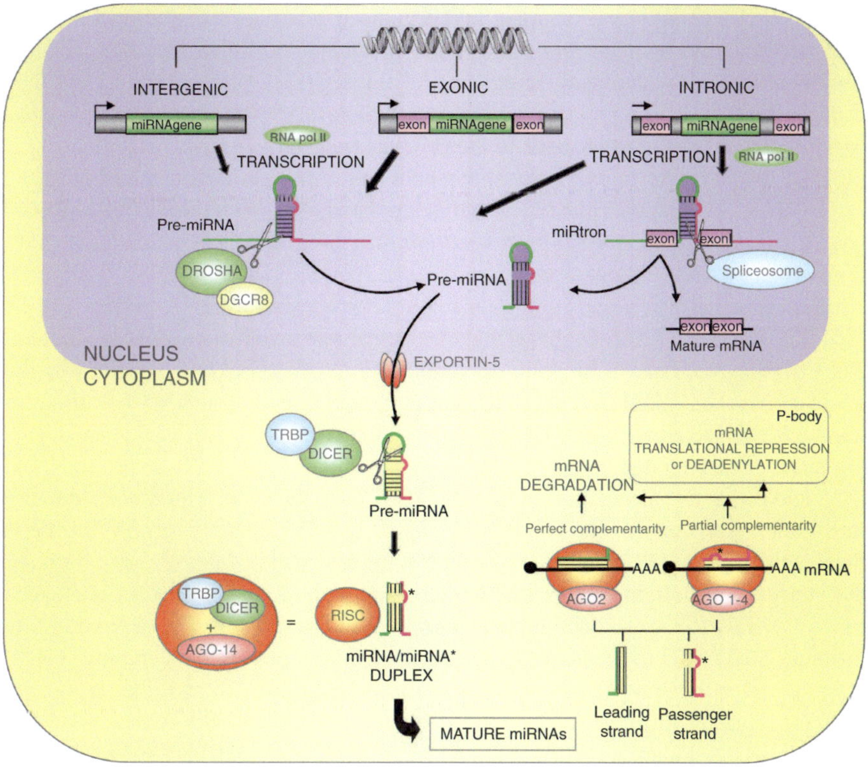

Fig. 1 Canonical pathway for the biogenesis of miRNAs. miRNA biogenesis takes place in the nucleus (*violet*) and in the cytoplasm (*yellow*) of the cell. Intergenic, exonic and intronic miRNA genes are transcribed by RNA polymerase II (RNA pol II) to generate pri-miRNAs or miRtrons, which are further cleaved by a microprocessor formed by DROSHA and DGCR8 or by the spliceosome, respectively, to originate the 60–70 nucleotide precursor miRNA (pre-miRNA). pre-miRNA is exported from the nucleus to the cytoplasm by exportin-5 and further processed by DICER associated with TRBP in a miRNA/miRNA* duplex. This duplex is cleaved by the RISC complex (formed by DICER, TRBP and AGO proteins) to originate two mature miRNA species: miRNA (leading) and miRNA* (passenger). These are incorporated into the RISC complex by AGO2 or AGO1-4, depending on sequence complementarity, and directed to the target mRNA (*black line*) [29, 43, 57, 76]

2.1 *Biogenesis*

More than a thousand miRNA sequences have been identified in the human genome, which are located in different regions of the genome. In humans, miRNA genes have been defined as intergenic (between two genes) and within genes, the latter being either intronic or exonic [71] (Fig. 1). Currently, it is estimated that 1 % of human genes encodes for miRNAs [6].

miRNA biogenesis is a complex and organised process that is compartmentalised between the nucleus and the cytoplasm and is classified into two classes, canonical and non-canonical, based on the transcription process used [29]. The canonical pathway is mediated by RNA polymerase II (Pol II) and involves the transcription of the miRNA gene to the primary miRNA transcript, pri-miR or pri-miRNA, with similar transcriptional mechanisms to protein-coding genes (Fig. 1). pri-miRNAs are capped, spliced and polyadenylated [48]. Structurally, pri-miRNAs have a unique imperfect stem-loop structure of approximately 70 nucleotides with flanking 5′- and 3′-single-stranded ends [95]. Exceptionally, miRNAs containing Alu repeat elements on their promoter might be processed by RNA polymerase III [10]. The pri-miRNA is recognised and cleaved, at one strand of the double-stranded RNA towards the base of the loop, by a microprocessor complex comprised by the nuclear enzyme DROSHA (RNase III type endonuclease) and an RNA-binding protein DGCR8 (DiGeorge syndrome critical region gene 8) [30, 32]. The cleaved hairpin sequence of 70 nucleotides is the precursor miRNA (pre-miR or pre-miRNA) [32], and it is actively transported from the nucleus to the cytoplasm associated with its cofactor Ran coupled to GTP by the nuclear export receptor exportin-5. GTP is then replaced by GDP inducing exportin-5 to release its pre-miRNA cargo into the cytoplasm [9, 55]. In the cytoplasm, the terminal loop of the pre-miRNA is cleaved off by DICER (RNase III endonuclease enzyme) associated with TAR RNA-binding protein (TRBP) [7] to originate a mature double-stranded miRNA duplex. TRBP alters the cleaved pre-miRNA and physically bridges DICER with the Argonaute proteins (Ago) to form the RNA-induced silencing complex (RISC).

The miRNA duplex is composed by a mature miRNA strand, also known as leading or guide strand, and an miRNA*, also known as passenger or star strand. Mature miRNAs are incorporated into RISC complexes that together with proteins of the GW182 family lead the post-transcriptional gene silencing process. Furthermore, this complex has been found sometimes to be enriched within p-bodies that are generated as a consequence of the gene silencing process [23]. Traditionally, the passenger miRNA* strand was believed to be present at a much lower frequency than the mature miRNA [18, 50] because it appeared to be degraded and removed from the RISC complex [57]. However, recent evidence shows that the star forms are functional and they can also mediate post-transcriptional gene regulation. Currently, it is not understood how one of the miRNA strands is incorporated into the RISC complex, while the other is degraded. Several studies have observed that most pre-miRNAs generate a ratio between miRNA and miRNA*, but the loading of one of the two strands will depend on the cellular type, developmental stage, environmental factors, availability of their mRNA targets and other factors [17, 43, 70]. As a result of the functional role of both strands, the nomenclature of miRNA/miRNA* is being replaced by indicating the strand of the pre-miRNA where the mature miRNA originates, for example, miR-155-5p and miR-155-3p,

The non-canonical miRNAs are structurally and functionally similar to canonical miRNAs. However, they follow a different maturation pathway, which normally skips one or several steps of the traditional canonical pathway. For instance, non-canonical miRNAs have different origins that could include introns, small nucleolar

RNAs and endogenous small hairpin RNAs or tRNAs (for more information, see [2]). Furthermore, some non-canonical miRNAs are synthesised independently from Dicer but mediated by other enzymes such as Ago1 [29], whereas others are Drosha and DGCR8 independent. Additionally, another miRNA biogenetic pathway such as that involving the let-7 family of miRNAs is dependent on the activity of TUTases, terminal uridylyl transferases, that enhance the efficiency of Dicer-mediated processing [31].

2.2 *Function of microRNAs*

miRNAs usually repress their target genes at the post-transcriptional level [5] by either inducing mRNA target degradation or inhibiting its translation [64, 84]. One single miRNA can potentially regulate hundreds or thousands of different target mRNAs, and a specific mRNA can potentially be a target for many different miRNAs. It has been predicted that more than 1000 miRNAs regulate over 60 % of all protein-coding genes in mammals [26].

Typically, a mature miRNA recognises target mRNAs by Watson-Crick base pairing, where the miRNA 5′ region, also known as seed region (2–7 nucleotides of the 22–25 nt that formed the miRNA), hybridises with the 3' UTR (3' untranslated region) of the target mRNA [35]. Depending on the complementarity between the seed (miRNA) and the match (mRNA) regions, different mechanisms of gene silencing occur resulting in either mRNA degradation by deadenylation (perfect match base pairing) or mRNA translation repression (mismatch base pairing) [5, 25] (Fig. 2). The mechanism of miRNA repression of gene expression depends on the specific features of the mRNA targets and on their abundance. In addition, mammalian cells present an mRNA turnover, which is highly variable and can range from minutes to days making miRNA gene targets difficult to predict [73].

Alongside the well-known gene repression function of miRNAs, recent studies have highlighted the possibility that some miRNAs bind to their target mRNA in regions different from the 3′ UTR, promoting the expression of the gene instead of repressing it (for more information, see [79]).

2.3 *Identification of microRNA-mRNA Interactions*

2.3.1 Bioinformatic Approach

Computational and experimental approaches to predict miRNA targets based on miRNA-mRNA pairs have been developed in the absence of high-throughput biological methods. Several computational approaches that assign putative mRNA targets to specific miRNAs include databases such as miRanda, PicTar, TargetScan, mirSV, RNA22, RNAhybrid, PITA and DIANA-microT. These databases were

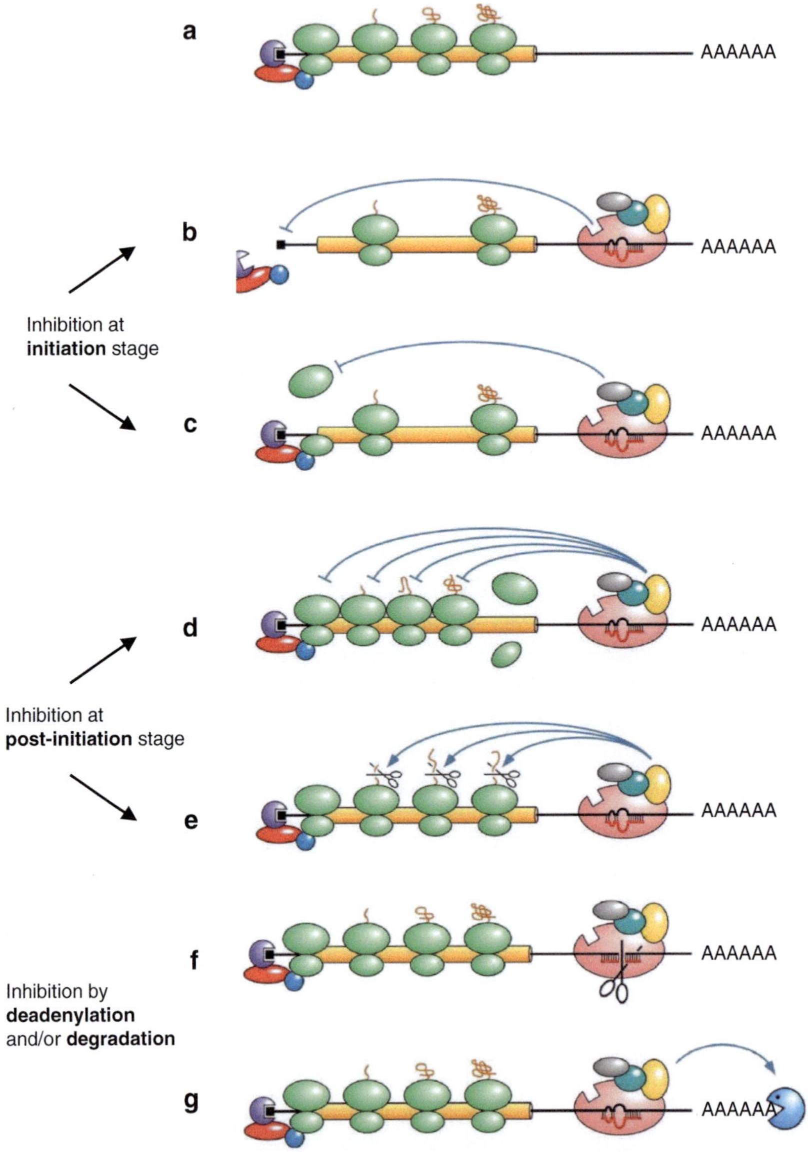

Fig. 2 Mechanisms of mRNA target translational repression and degradation by miRNAs. (**a**) mRNA undergoing translation in the absence of a bound miRNA. (**b**) Inhibition of translation initiation by competition between RISC and eIF4E for cap binding. (**c**) Inhibition of translation initiation at a step after cap recognition, such as by impeding the association of the small and large ribosomal subunits. (**d**) Inhibition of translation elongation coupled to premature termination. (**e**) Co-translational degradation of nascent polypeptides. (**f**) mRNA undergoing endonucleolytic cleavage by Ago2, as guided by a fully complementary miRNA. (**g**) mRNA undergoing poly(A) removal by the Ccr4/Not deadenylase (Pac-Man) complex, as directed by a partially complementary miRNA. *Legend*: Black square, m7G 5'cap; amber cylinder, protein-coding region; and AAAAAA, poly(A) tail. Ribosomes are coloured green, nascent polypeptides are brown and the eIF4E subunit of the cap-binding complex is violet; RISC is depicted as a ribonucleoprotein complex comprising an miRNA (*red*), Ago (*pink*) and other protein subunits (Adapted from [89])

developed combining algorithms for predicting mRNA target for miRNAs with the limited empirical evidence from a few validated target sites for some miRNAs [8, 39, 42, 49, 59]. Overall, algorithms used by databases for predicted targets rely on cross-species conservation as a requirement to reduce the number of false positives, but the complex characteristics of miRNAs make computational approaches a difficult challenge for miRNA target prediction.

For this reason, bioinformatic analyses are just the first step for the validation of miRNA-mRNA interactions. Once a candidate pair has been selected, an experimental validation of the interaction is usually performed. In addition to target prediction databases, other in silico target prediction tools may involve Chip-seq, microarray databases and databases that profile the expression pattern of miRNAs and of mRNAs.

2.3.2 Experimental Validation of microRNA Targets

Experimental validation of mRNA-miRNA interactions is a complicated and not a standardised process that requires the combination of different approaches in order to make results as robust as possible. Typically, experimental validation approaches include manipulation of cellular miRNA levels, reporter assays and mutation analysis [81].

Manipulation of miRNA Levels

Chemically synthesised double-stranded miRNA precursors may be transfected into cells in vitro to mimic upregulation of endogenous miRNAs [33, 66]. However, the miRNA precursor is transiently expressed with this approach and might not be sufficient for longer-term studies. Therefore, constitutive miRNA expression has also been used following transfection of DNA plasmids into cells or transduction approaches using adenoviruses, adeno-associated viruses, and lentiviruses as delivery systems [34]. Then, the functional consequences in the expression of the predicted target gene of interest can be assessed by different methods at the mRNA or protein level, using either RT^2-qPCR or Western blotting and/or ELISA techniques, respectively. Furthermore, new approaches are being developed based on pull-down assays using transfection of labelled miRNA precursors (e.g. biotin) followed by isolation of RNA complexes using beads (e.g. magnetic beads coated with streptavidin) and RT^2-qPCR or microarray techniques for identification of pulled mRNA targets [85].

Gene knockout and antisense technologies are well-established methodologies to repress miRNA function. Knockout of the miRNA processing enzyme Dicer1 can affect the expression of many mature miRNAs, whereas knockout of the miRNA (either transient or constitutive) itself or site-directed mutagenesis in the miRNA-encoding sites might inhibit miRNA-mediated gene regulation [81]. A widespread technique to inhibit miRNA expression and assess loss-of-function effects is transient transfection of modified oligoribonucleotides complementary to the mature miRNA sequence, also known as antagomirs [36].

Reporter Gene Assay and Mutation Analysis

Both reducing and overexpressing miRNA levels are important methods to verify miRNA as expression regulators of possible mRNA targets. However, these techniques do not demonstrate whether the changes observed in the levels of the candidate mRNA target are due to direct binding of the miRNA to its predicted 3′ UTR site. For this reason, reporter gene assays constitute a widely used approach for the validation of miRNA functions [83]. Reporter genes for these assays may include EGFP or lacZ, but, due to its short mRNA half-life in mammalian cells, luciferase is most commonly used [78]. This experimental approach is based on cloning the 3′ UTR of the target mRNA of interest immediately downstream of the reporter gene open reading frame (ORF) sequence. Then, both plasmid and miRNA are transfected into cells. Thereafter, the reporter gene is detected 24–48 h after transfection, and the miRNA-mRNA relationship is determined by the activity of the reporter gene [38]. In addition, mutagenesis of the seed region in the 3′ UTR of the predicted mRNA target can be applied to this experimental approach to determine specific binding to the miRNA and assess interactions between them [83].

3 Role of Cerebrovascular microRNAs in Neuroinflammation

Many studies have pointed out a role for miRNAs in the regulation of the immune system [75], and, for example, they appear particularly important in B and T cell homeostasis and immunological function [20, 37, 91]. This pivotal role suggests their possible involvement in the development of inflammatory and/or autoimmune diseases [15]. In the context of the CNS, miRNAs have been shown to be altered in cells of the immune system and/or in plasma of multiple sclerosis (MS) patients and of mice with experimental autoimmune encephalomyelitis (EAE), one animal model of MS (reviewed by [56]) as well as in CNS infections (for an overview, see [94, 96]). Not only do their levels change in immune cells but they also appear to regulate immune cell function [56]. For example, miR-326 was one of the first miRNAs shown to regulate immune function in MS by suppressing differentiation of Th17 cells [21]. Comparatively fewer studies have been published demonstrating a role for miRNAs in regulating the function of CNS-resident cells in neuroinflammation (again, reviewed in [56]). This field was pioneered by Junker et al. [40], as they were the first to demonstrate specific changes in the levels of many miRNAs in acute MS lesions compared to normal-appearing white matter and chronic silent lesions, isolated by laser-capture microdissection, and identified several miRNAs that targeted the regulatory protein CD47 in astrocytes [40].

In the context of the BBB, and specifically of brain endothelial cells, their miRNA profile has been shown to be altered following cytokine treatment [67]. Cytokine-induced changes in miRNA expression are time dependent suggesting that miRNAs are expressed at different stages to either promote or inhibit the inflammatory process (Fig. 3). Only a few miRNAs including miR-155-5p and miR-155-3p, miR-19b-5p, miR-21-3p, miR-23a-5p and miR-29b-1-5p were rapidly upregulated at 2 h and/or 6 h after cytokine treatment of the human cerebral endothelial cell line, hCMEC/D3. However, when hCMEC/D3 cells were stimulated with TNFα/IFNγ for 24 h, changes in the levels of many miRNAs were identified (Table 1). Indeed, 32 miRNAs were upregulated including miRNAs previously shown to be involved in immune responses such as miR-155 and miR-146a and miR-146b [61, 82], whereas 117 miRNAs were downregulated including miRNAs previously identified to be abundant in endothelium such as miR-126 and the let-7 family.

Little is known about miRNA regulation under inflammatory conditions in other cells of the neurovascular unit such as pericytes, and only a few studies have focused on astrocytes and perivascular microglia/macrophages. For the rest of this chapter, we will focus on the few studies that have investigated miRNAs in brain endothelium under inflammatory conditions and only comment briefly on studies on other cells of the neurovascular unit when applicable (reviewed in [3]).

3.1 Pro- and Anti-inflammatory Cerebrovascular microRNAs: inflammiRs

In this section, we will review the miRNAs upregulated by inflammation, which are commonly known as inflammation-associated miRNAs or inflammiRs.

3.1.1 microRNA-155

miR-155 is a well-known inflammiR, which is highly conserved among species and, in humans, is originated from an exon of a non-coding sequence of the *BIC* gene located on chromosome 21q21. Furthermore, miR-155 is related to a wide range of biological functions including cancer, immunity and haematopoiesis (for a review, see [24]).

A recent study identified miR-155 as a crucial miRNA in neuroinflammation at the BBB [54]. It was reported that miR-155-5p was expressed at the neurovascular unit of both MS patients and mice with EAE. Moreover, miR-155-5p levels rapidly increased by the effect of inflammatory cytokines in cultured human brain endothelial cells [54]. Functionally, loss of miR-155 in EAE mice and in mice exposed to acute systemic inflammation induced by lipopolysaccharide (LPS) reduced

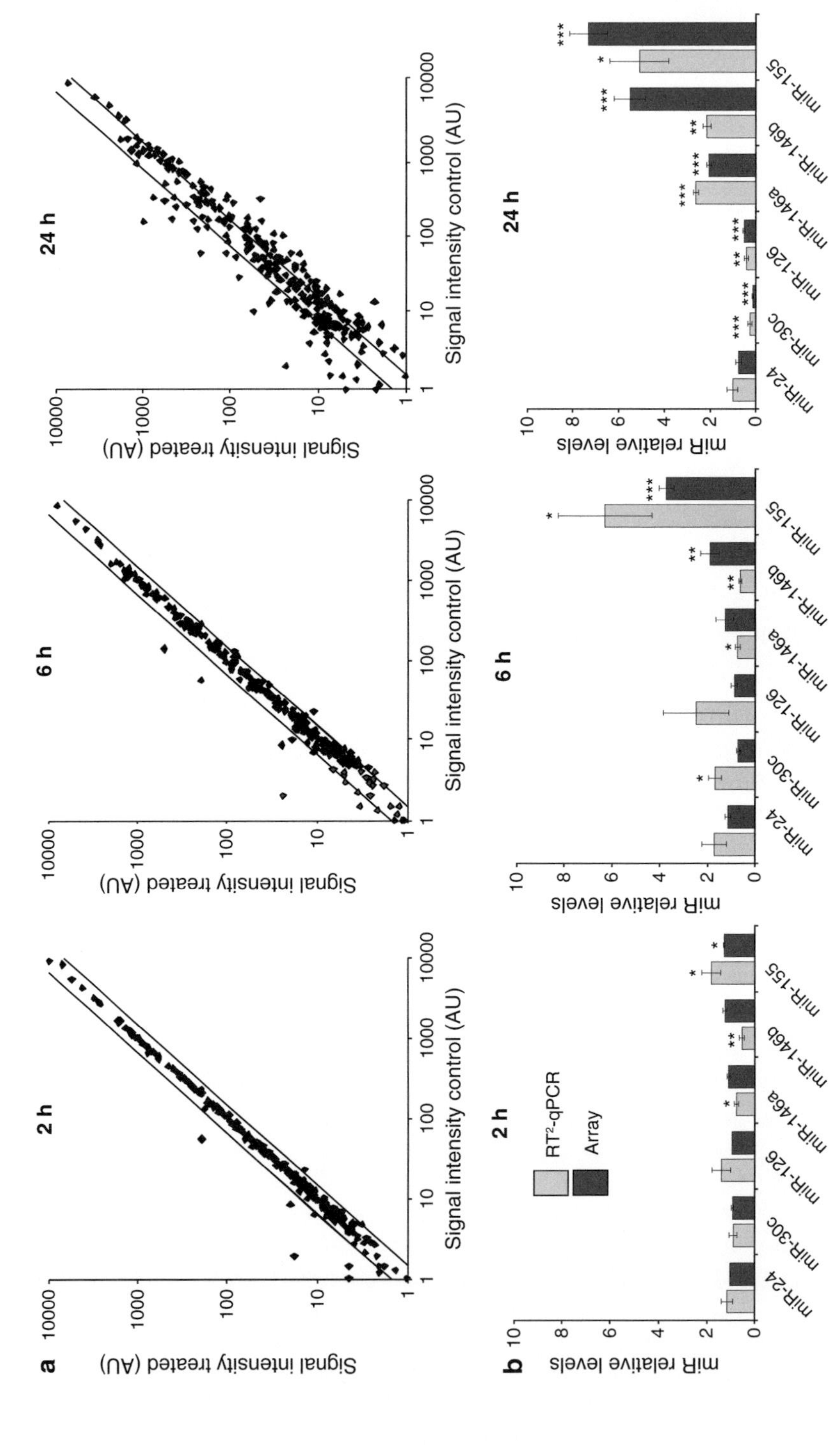

a
b
2 h
6 h
24 h
Signal intensity treated (AU)
Signal intensity control (AU)
1
10
100
1000
10000
miR relative levels
0
2
4
6
8
10
RT²-qPCR
Array
miR-24
miR-30c
miR-126
miR-146a
miR-146b
miR-155

Table 1 Top 6 down- and upregulated miRNAS in hCMEC/D3 stimulated with 100 ng/ml TNFα and IFNγ for 2, 6 and 24 h detected by Agilent v13 miRNA array ($n=3$)

	2 h		6 h		24 h	
Name	Fold change	SEM	Fold change	SEM	Fold change	SEM
Upregulated miRNAs						
miR-155-5p	1.30**	0.02	3.73***	0.01	8.99***	0.85
miR-146b-5p	1.24	0.28	1.90*	0.89	6.92***	0.68
miR-21-3p	3.39***	0.02	3.59***	0.07	4.69***	0.25
miR-572	0.98	0.12	1.08	0.11	3.43***	0.14
miR-575	1.02	0.06	1.00	0.08	3.27***	0.13
miR-663a	1.00	0.16	−1.17	0.20	3.09***	0.23
Down-regulated miRs						
miR-30c-5p	0.93	0.02	0.71*	0.03	0.15***	0.04
miR-324-5p	0.95	0.02	0.92	0.03	0.27***	0.04
miR-361-5p	0.94	0.03	0.92	0.05	0.31***	0.05
miR-31-3p	0.96	0.02	0.9	0.02	0.32***	0.04
miR-27b-3p	0.94	0.04	0.87	0.06	0.36***	0.03
miR-505-3p	1.12	0.02	1.03	0.02	0.36***	*0.06*

$*P<0.05$, $**P<0.01$, $***P<0.001$ compared to unstimulated cells

extravasation of systemic tracers into the CNS. Furthermore, in vitro studies on human brain endothelial cells showed that manipulation of this miRNA led to disorganisation of tight junctions and loss of focal adhesion plaques by targeting molecules involved in cell-cell complexes such as annexin-2 and claudin-1 and focal adhesion components such as DOCK-1 and syntenin-1. Therefore, miR-155 has been proposed to act as a negative regulator for BBB function. More recently, miR-155 inhibition has been shown to promote recovery after experimental stroke in mouse mediated by upregulation of its target protein Rheb [11]. In this study, in vivo loss of miR-155 led to improved preservation of tight junctions in microvessels as

Fig. 3 Microarray profiling of miRNAs in human brain endothelium in response to TNFα/IFNγ. Confluent monolayers of hCMEC/D3 cells were stimulated with TNFα/IFNγ (100 ng/ml) at 2, 6 and 24 h. Total RNA was isolated and microarray analysis was performed to determine miRNAs levels. hCMEC/D3 cells at basal conditions expressed 216 miRNAs. A. The expression levels of miRNAs are represented on a scatter plot, and signal intensities (SI) of individual miRNAs in control cells are plotted on the *X* axis, whereas SI of individual miRNAs in cytokine-treated cells are plotted on the *Y* axis. SI values are presented on a log10 scale ($n=3$). B. Levels of six miRNAs were determined by RT^2-qPCR in RNA samples used in A. Shown are examples of miRNAs that either decreased (miR-30c, miR-126) or increased (miR-146a/miR-146ab, miR-155) and an miRNA that remained unchanged (miR-24) following TNFα/IFNγ treatment. RT^2-qPCR results are shown alongside the miRNA relative levels obtained in the microarray analysis for comparison. Control values were normalised to one, and results are expressed as miRNA levels in TNFα/IFNγ-treated cells relative to those in unstimulated cells. U6B small nuclear RNA was used as an internal standard. $*P<0.05$, $**P<0.01$, $***P<0.001$ compared to unstimulated cells ($n=3$ with duplicate samples)

assessed by electron microscopy techniques and by ZO-1 staining. Furthermore, miR-155-5p has recently found to be involved in monocyte and T cell firm adhesion to human brain endothelium in vitro [16]. In addition, miR-155-5p has been shown to mediate ICAM-1 expression, apoptosis, reactive oxygen species and NO generation in human brain endothelial cells in response to oxidised LDL [51], suggesting that this miRNA has complex functions in brain endothelial pathological states. Altogether, these studies would indicate that inhibition of miR-155 may prove a suitable molecular target for supporting the cerebral microvasculature in neuroinflammation.

3.1.2 microRNA-146a and microRNA-146b

The miR-146 family is composed by two miRNAs, miR-146a and miR-146b, that are encoded on chromosomes 5q34 and 10q24, respectively, and both miR-146 genes have an identical seed region but differ by two nucleotides in their stem-loop secondary structure [80]. Taganov and colleagues investigated miRNAs involved in regulation of innate immune response using a monocytic cell line, THP-1, as a model to investigate its response to microbial components and pro-inflammatory cytokines [80]. Their observations suggested that bacterial mediators and pro-inflammatory cytokines such as TNFα induced rapid expression of miR-146a and miR-146b. Bioinformatic analyses of promoters for miR-146a/miR-146b identified and validated the presence of NF-kB binding sites as the main sequence region to promote miR-146 expression [4].

miR-146a appears to be a negative regulator of inflammation under physiological conditions. Furthermore, several studies suggest miR-146a as a molecular brake on inflammation, oncogenic transformation and myeloid cell proliferation. The anti-inflammatory role of miR-146a and miR-146b has been demonstrated in various cells, e.g. human umbilical vein endothelial cells (HUVECs), monocytes, astrocytes, T cells and, recently, in brain endothelial cells [90]. In fact, although the function of miR-146a in brain endothelium is still unclear, Wu and colleagues demonstrated that miR-146a-5p is upregulated at the neurovascular unit of both MS patients and mice with EAE and in cytokine-activated human BEC. Indeed, using a flow-based assay, it was demonstrated that miR-146a-5p was capable of decreasing leucocyte adhesion to BEC and endothelial expression of VCAM-1 and CCL2 by directly suppressing components of the NF-kB pathway such as TRAF6, IRAK1, RhoA and nuclear factor of activated T cells 5 (NFAT5) [90].

3.1.3 Other Inflammation-Upregulated miRNAs

Interestingly, miR-21-3p was the third highest miRNA upregulated by TNFα and IFNγ in human brain endothelial cells (Table 1). As miR-21-3p has been reported to play significant roles in regulating cellular apoptosis, immuno-inflammatory responses and the expression of angiogenic factors in various diseases, Ge et al. postulated that miR-21-3p could exert an important impact on BBB damage after traumatic brain injury (TBI) as it is highly expressed by brain endothelial cells [28]. In experimental TBI, increasing miR-21-3p levels reduced Evans blue extravasation and loss of occludin and claudin-5 in BECs by activating the angiopoietin-1 and Tie-2 axis, whereas the reverse was observed using antagomirs specific for miR-21-3p. In addition, we have observed upregulation of miR-21-3p in human brain endothelium following stimulation with IL-1β (unpublished results). These observations would suggest that upregulation of miR-21-3p by BECs constitutes a general protective mechanism against inflammation-induced BBB damage.

While this would indicate that miR-21-3p may have generalised anti-inflammatory actions on brain endothelium, other regulated miRNAs may be specific to a particular inflammation-associated pathology. For example, HIV-1 Tat has been shown to induce the expression of miR-101 in human brain endothelium leading to suppression of VE-cadherin and increased barrier permeability [60]. Other miRNAs, although not detected or unchanged in cytokine-activated cultured human brain endothelium, may also play an important role in inflammation at the BBB in other experimental models. For example, expression of miR-26b-5p and miR-28-3p were also significantly upregulated in bEnd.3 cells treated with lupus serum and with the complement protein C5a, respectively, and in cortical brains of a mouse model of lupus [22]. Although the authors performed a bioinformatic analysis of potential miRNA gene targets identifying components involved in inflammation, matrix arrangement, and apoptosis, the functional relevance of these two miRNAs in brain endothelium remains to be confirmed. Another miRNA, miR-144-3p, has been shown to increase in the cerebral microvasculature of aged rats, leading to downregulation of Nrf2, a basic leucine zipper protein that regulates the expression of antioxidant proteins that protect against oxidative damage triggered by injury and inflammation [19]. Interestingly, this effect was partially reversed by caloric restriction, possibly via its significant anti-inflammatory effects. Finally, miR-29b-3p appears to negatively regulate BBB function in a mouse model of hyperhomocysteinemia by suppressing DNMT3b leading to increased levels of MMPs, in particular MMP9, a known regulator of BBB permeability in inflammation [41].

3.2 Cerebrovascular Housekeeping miRNAs

The ten most highly miRNAs expressed in hCMEC/D3 cells under basal conditions were, in this order, miR-21-5p, miR-126-3p, miR-16-5p, let-7a-5p, let-7f-5p, miR-15b-5p and let-7b-5p [67]. Highly abundant small RNAs such as miR-720 have been later defined as fragments of other RNAs, in this case a tRNA, and removed from miRNA databases. However, this miRNA profile may not be specific to cerebral endothelium as other studies have shown that primary endothelial cells from

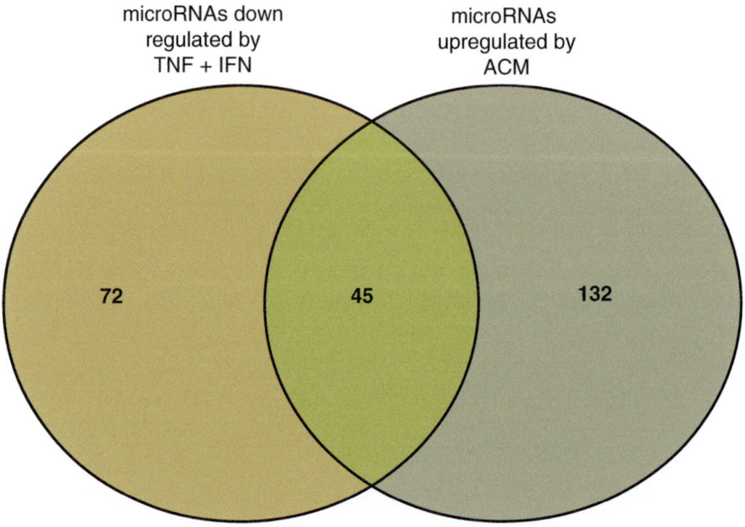

Fig. 4 Summary of miRNAs regulated in hCMEC/D3 cells stimulated with either TNFα and IFNy or astrocyte-conditioned media (ACM). hCMEC/D3 were stimulated with 10 ng/ml cytokines for 24 h or 50 % ACM in supplemented EGM-2 for 48 h. microRNA profiling was analysed by Agilent v13 miRNA array for cytokine-stimulated hCMEC/D3 cells and Exiqon miRNA array for ACM-stimulated ones. Comparison between downregulated miRNAs in cytokine-stimulated cells and upregulated miRNAs in ACM-stimulated cells identified 45 miRNA candidates in common to both treatments [67]

different vascular beds do not differ much in their miRNA profile suggesting a pan-endothelial miRNA signature [58]. Interestingly, the profile of miRNAs in cultured human brain microvascular endothelium appeared most closely related to that of HUVECs, whereas those from other vascular beds appeared to differ in the expression of miR-99b, miR-20b and let-7b. This observation may reflect the well-known fact that brain endothelium, like endothelium of other vascular beds, de-differentiates in culture and preservation of their unique phenotype is dependent on culture conditions. Indeed, the pattern of miRNA expression by hCMEC/D3 cells following incubation with astrocyte-conditioned media, a well-known method of inducing barrier properties in cultured brain endothelium, appears to be the reverse of that for many miRNAs altered by cytokines, a treatment that leads to a decrease in barrier properties [52]. In fact, there was a remarkable overlap in the number of miRNAs decreased by TNFα and IFNγ (barrier-reducing treatment) and those increased by astrocyte-released factors (barrier-inducing treatment) (Fig. 4), including members of the miR-30 and miR-125 families but excluding other endothelial-enriched miRNAs such as miR-126. Nevertheless, it is thus possible that this signature of 45 miRNAs constitutes a BBB-specific maturation phenotype, although this does not preclude the possibility that other endothelial miRNAs decreased by inflammatory stimuli but not increased by astrocyte-conditioned media (e.g. miR-126) may have important roles in inflammation at the BBB.

3.2.1 microRNA-125a

miR-125a is a miRNA located on chromosome 19q.13 in the human genome. It forms part of the miR-125 family together with miR-125b1 and miR-125b2 [71]. The miR-125 family has been reported to be involved in different carcinomas acting as a promoters or repressors of tumorigenesis [77].

In particular, miR-125a-5p is known for being a tumour suppressor miRNA [13, 88]. Furthermore, it also has an anti-inflammatory function in macrophages by inducing the formation of type 2 macrophages [86]. At the neurovascular unit, miR-125a-5p was the first miRNA to be demonstrated to be severely reduced in endothelial cells derived from MS lesions and in cultured human brain endothelium upon inflammatory stimuli and to have a functional effect in BECs. Although its mRNA targets in brain endothelium still remain elusive, miR-125a-5p appeared to directly regulate barrier function in an in vitro BBB model by decreasing transendothelial electrical resistance and inducing disorganisation of tight junctional complexes. In addition, elevating miR-125a levels partially reversed cytokine-induced monocyte migration through brain endothelial cell layers in vitro [67]. These led us to postulate that miR-125a-5p contributed to the maintenance of a quiescent state in brain endothelium although this may not be a unique feature of this miRNA as other miRNAs may have effects (for details, see below).

3.2.2 microRNA-98 and let-7g

Members of the let-7 family, which also include miR-98, formed part of the BBB-specific pattern of miRNA expression [67]. miR-98 and let-7g belong to a highly conserved miRNA family formed by 12 different miRNAs in mammals [63]. The let-7 family of miRNAs has been described to be involved in the pathobiology of cancer, although their role in the endothelium is not clear yet.

Overexpression of miR-98-5p and let-7g-3p in in vitro (BECs) and in vivo (animal model of localised aseptic meningitis) models of inflammation appears to reduce leucocyte adhesion to and migration across endothelium, to decrease expression of pro-inflammatory cytokines and to increase BBB tightness [72]. At least some of the effects mediated by these two miRNAs appear to be related to the direct inhibition of the two inflammatory chemokines CCL2 and CCL5. The significance of this study lies in the fact that this is one of the first attempts to manipulate cerebrovascular miRNAs in vivo suggesting their potential use as a therapeutic tool to prevent BBB dysfunction in neuroinflammation.

3.3 *Cerebrovascular microRNAs in Cell-Cell Communication*

A recent study has profiled the pattern of miRNAs in extracellular microvesicles released by inflammation-activated brain endothelial cells [92]. Here, the authors exposed human brain pericytes to exosomes isolated from supernatants of mouse brain endothelial cells treated with a combination of lipopolysaccharide and cytokines such as TNFα, IFNγ and IL-6. Exosome-exposed pericytes showed highly induced mRNA and protein expression of VEGF-B compared to control pericytes, and this angiogenic factor was assumed to be a downstream target of the miRNAs contained within extracellular microvesicles released following inflammatory stimulation. This study gives the first proof of a functional role of brain endothelial miRNAs in communication with another cell type of the neurovascular unit. However, the profile of miRNAs released by mouse brain endothelial cells in extracellular microvesicles under inflammatory conditions appears very different from the changes observed in intracellular levels of miRNAs of inflammation-activated human brain endothelial cells suggesting that either (a) the miRNA profile in inflammation-activated brain endothelial cells is species and stimulus specific or (b) miRNAs released in exosomes do not reflect the intracellular levels of brain endothelial miRNAs. The latter explanation would imply that individual miRNAs are specifically selected for release into exosomes thereby opening up the possibility of a role for specific miRNAs in paracrine cell-cell communication. Indeed, this appears to be the case in HUVECs exposed to high glucose which appear to selectively transfer miR-503 via microparticles to vascular pericytes thereby reducing their expression of EFNB2 and VEGFA [12].

4 Conclusions and Further Perspectives

The field of BBB and miRNA biology is still in its infancy. Inflammatory stimuli induce changes in the levels of cerebrovascular miRNAs characterised mainly by upregulation of inflammiRs and downregulation of brain endothelial housekeeping miRNAs, apparently contributing to BBB dysfunction (paracellular permeability and leucocyte adhesion). However, we are still far from validating cerebrovascular miRNAs as potential therapeutic or prophylactic targets for BBB dysfunction in inflammatory and/or autoimmune disorders. First, even though the effects of individual miRNAs on cerebrovascular function are slowly being unravelled, we need to know the combinatorial effects of different miRNAs in specific CNS pathologies before we contemplate to develop therapies based on miRNAs. Second, in view of the multiplicity of possible gene targets for each individual miRNA, whether modulating their levels in the cerebrovascular bed may have effects on other endothelial cellular pathways that may contribute to the pathogenesis of the inflammatory disease in question needs to be investigated. Finally, knowledge of side effects due to non-specific delivery of miRNA modulators into other tissues or organs (e.g. liver, kidney) would also be a prerequisite for validating their therapeutic potential unless very specific delivery systems targeted at the cerebrovascular bed are developed as carriers for miRNA modulators (e.g. brain endothelial-specific nanoparticles). Nevertheless, the potential for manipulating this novel class of regulators of gene expression for therapeutic purposes is vast and should be given considerable attention in the near future.

References

1. Abbott NJ, Patabendige AA, Dolman DE, Yusof SR, Begley DJ (2010) Structure and function of the blood-brain barrier. Neurobiol Dis 37:13–25
2. Abdelfattah AM, Park C, Choi MY (2014) Update on non-canonical microRNAs. Biomolecular concepts 5:275–287
3. Almutairi MM, Gong C, Xu YG, Chang Y, Shi H (2016) Factors controlling permeability of the blood-brain barrier. Cell Mol Life Sci 73(1):57–77
4. Baltimore D, Boldin MP, O'Connell RM, Rao DS, Taganov KD (2008) MicroRNAs: new regulators of immune cell development and function. Nat Immunol 9:839–845
5. Bartel DP (2004) MicroRNAs: genomics, biogenesis, mechanism, and function. Cell 116:281–297
6. Bartel DP (2009) MicroRNAs: target recognition and regulatory functions. Cell 136:215–233
7. Bernstein E, Caudy AA, Hammond SM, Hannon GJ (2001) Role for a bidentate ribonuclease in the initiation step of RNA interference. Nature 409:363–366
8. Betel D, Koppal A, Agius P, Sander C, Leslie C (2010) Comprehensive modeling of microRNA targets predicts functional non-conserved and non-canonical sites. Genome Biol 11:R90
9. Bohnsack MT, Czaplinski K, Gorlich D (2004) Exportin 5 is a RanGTP-dependent dsRNA-binding protein that mediates nuclear export of pre-miRNAs. RNA 10:185–191
10. Borchert GM, Lanier W, Davidson BL (2006) RNA polymerase III transcribes human microRNAs. Nat Struct Mol Biol 13:1097–1101
11. Caballero-Garrido E, Pena-Philippides JC, Lordkipanidze T, Bragin D, Yang Y, Erhardt EB, Roitbak T (2015) In Vivo Inhibition of miR-155 Promotes Recovery after Experimental Mouse Stroke. J Neurosci 35:12446–12464
12. Caporali A, Meloni M, Nailor A, Mitić T, Shantikumar S, Riu F, Sala-Newby G, Rose L, Besnier M, Katare R, Voellenkle C, Verkade P, Martelli F, Madeddu P, Emanueli C (2015). p75NTR-dependent activation of NF-κB regulates microRNA-503 transcription and pericyte–endothelial crosstalk in diabetes after limb ischaemia. Nature Communications 6: 8024
13. Carbajal JM, Schaeffer RC Jr (1999) RhoA inactivation enhances endothelial barrier function. Am J Physiol 277:C955–C964
14. Carthew RW, Sontheimer EJ (2009) Origins and Mechanisms of miRNAs and siRNAs. Cell 136:642–655
15. Ceribelli A, Satoh M, Chan EK (2012) MicroRNAs and autoimmunity. Curr Opin Immunol 24:686–691
16. Cerutti C, Soblechero-Martin P, Wu D3, Lopez-Ramirez MA, de Vries H, Sharrack B, Male DK, Romero IA (2016). "MicroRNA-155 contributes to shear-resistant leukocyte adhesion to human brain endothelium in vitro." Fluids Barriers CNS 13(1):8
17. Chatterjee S, Grosshans H (2009) Active turnover modulates mature microRNA activity in Caenorhabditis elegans. Nature 461:546–549
18. Chendrimada TP, Gregory RI, Kumaraswamy E, Norman J, Cooch N, Nishikura K, Shiekhattar R (2005) TRBP recruits the Dicer complex to Ago2 for microRNA processing and gene silencing. Nature 436:740–744
19. Csiszar A, Gautam T, Sosnowska D, Tarantini S, Banki E, Tucsek Z, Toth P, Losonczy G, Koller A, Reglodi D, Giles CB, Wren JD, Sonntag WE, Ungvari Z (2014) Caloric restriction confers persistent anti-oxidative, pro-angiogenic, and anti-inflammatory effects and promotes anti-aging miRNA expression profile in cerebromicrovascular endothelial cells of aged rats. Am J Physiol Heart Circ Physiol 307:H292–H306
20. Danger R, Braza F, Giral M, Soulillou J-P, Brouard S (2014) MicroRNAs, major players in B cells homeostasis and function. Front Immunol 5:18
21. Du C, Liu C, Kang J, Zhao G, Ye Z, Huang S, Li Z, Wu Z, Pei G (2009) MicroRNA miR-326 regulates TH-17 differentiation and is associated with the pathogenesis of multiple sclerosis. Nat Immunol 10:1252–1259

22. Eadon MT, Jacob A, Cunningham PN, Quigg RJ, Garcia JG, Alexander JJ (2014) Transcriptional profiling reveals that C5a alters microRNA in brain endothelial cells. Immunology 143:363–373
23. Eulalio A, Behm-Ansmant I, Schweizer D, Izaurralde E (2007) P-body formation is a consequence, not the cause, of RNA-mediated gene silencing. Mol Cell Biol 27:3970–3981
24. Faraoni I, Antonetti FR, Cardone J, Bonmassar E (2009) miR-155 gene: a typical multifunctional microRNA. Biochim Biophys Acta 1792:497–505
25. Filipowicz W, Bhattacharyya SN, Sonenberg N (2008) Mechanisms of post-transcriptional regulation by microRNAs: are the answers in sight? Nat Rev Genet 9:102–114
26. Friedman JM, Jones PA (2009) MicroRNAs: critical mediators of differentiation, development and disease. Swiss Med Wkly 139:466–472
27. Friedman RC, Farh KK, Burge CB, Bartel DP (2009) Most mammalian mRNAs are conserved targets of microRNAs. Genome Res 19:92–105
28. Ge X, Han Z, Chen F, Wang H, Zhang B, Jiang R, Lei P, Zhang J (2015) MiR-21 alleviates secondary blood-brain barrier damage after traumatic brain injury in rats. Brain Res 1603:150–157
29. Graves P, Zeng Y (2012) Biogenesis of mammalian microRNAs: a global view. Genomics Proteomics Bioinformatics 10:239–245
30. Gregory RI, Yan KP, Amuthan G, Chendrimada T, Doratotaj B, Cooch N, Shiekhattar R (2004) The Microprocessor complex mediates the genesis of microRNAs. Nature 432:235–240
31. Ha M, Kim VN (2014) Regulation of microRNA biogenesis. Nat Rev Mol Cell Biol 15:509–524
32. Han J, Lee Y, Yeom KH, Kim YK, Jin H, Kim VN (2004) The Drosha-DGCR8 complex in primary microRNA processing. Genes Dev 18:3016–3027
33. Harris TA, Yamakuchi M, Ferlito M, Mendell JT, Lowenstein CJ (2008) MicroRNA-126 regulates endothelial expression of vascular cell adhesion molecule 1. Proc Natl Acad Sci U S A 105:1516–1521
34. Huang A, Chen Y, Wang X, Zhao S, Su N, White DW (2004) Functional silencing of hepatic microsomal glucose-6-phosphatase gene expression in vivo by adenovirus-mediated delivery of short hairpin RNA. FEBS Lett 558:69–73
35. Huntzinger E, Izaurralde E (2011). Gene silencing by microRNAs: contributions of translational repression and mRNA decay. Nature Reviews Genetics 12: 99–110
36. Hutvagner G, Simard MJ, Mello CC, Zamore PD (2004) Sequence-specific inhibition of small RNA function. PLoS Biol 2:E98
37. Jeker LT, Bluestone JA (2013) microRNA regulation of T-cell differentiation and function. Immunological Reviews 253:65–81
38. Jin, Yi, Zujian Chen, Xiqiang Liu, Xiaofeng Zhou (2013) Evaluating the MicroRNA targeting sites by Luciferase reporter gene assay. Methods Mol Biol (Clifton, N.J.) 936:117–127
39. John B, Enright AJ, Aravin A, Tuschl T, Sander C, Marks DS (2004) Human MicroRNA targets. PLoS Biol 2:e363
40. Junker A, Krumbholz M, Eisele S, Mohan H, Augstein F, Bittner R, Lassmann H, Wekerle H, Hohlfeld R, Meinl E (2009) MicroRNA profiling of multiple sclerosis lesions identifies modulators of the regulatory protein CD47. Brain 132:3342–3352
41. Kalani A, Kamat PK, Familtseva A, Chaturvedi P, Muradashvili N, Narayanan N, Tyagi SC, Tyagi N (2014) Role of microRNA29b in blood-brain barrier dysfunction during hyperhomocysteinemia: an epigenetic mechanism. J Cereb Blood Flow Metab 34:1212–1222
42. Krek A, Grun D, Poy MN, Wolf R, Rosenberg L, Epstein EJ, MacMenamin P, da Piedade I, Gunsalus KC, Stoffel M, Rajewsky N (2005) Combinatorial microRNA target predictions. Nat Genet 37:495–500
43. Krol J, Loedige I, Filipowicz W (2010) The widespread regulation of microRNA biogenesis, function and decay. Nat Rev Genet 11:597–610
44. Lagos-Quintana M, Rauhut R, Lendeckel W, Tuschl T (2001) Identification of novel genes coding for small expressed RNAs. Science 294:853–858
45. Lau NC, Lim LP, Weinstein EG, Bartel DP (2001) An abundant class of tiny RNAs with probable regulatory roles in Caenorhabditis elegans. Science 294:858–862

46. Lee RC, Ambros V (2001) An extensive class of small RNAs in Caenorhabditis elegans. Science 294:862–864
47. Lee RC, Feinbaum RL, Ambros V (1993) The C. elegans heterochronic gene lin-4 encodes small RNAs with antisense complementarity to lin-14. Cell 75:843–854
48. Lee Y, Kim M, Han J, Yeom KH, Lee S, Baek SH, Kim VN (2004) MicroRNA genes are transcribed by RNA polymerase II. EMBO J 23:4051–4060
49. Lewis BP, Shih IH, Jones-Rhoades MW, Bartel DP, Burge CB (2003) Prediction of mammalian microRNA targets. Cell 115:787–798
50. Lim LP, Lau NC, Weinstein EG, Abdelhakim A, Yekta S, Rhoades MW, Burge CB, Bartel DP (2003) The microRNAs of Caenorhabditis elegans. Genes Dev 17:991–1008
51. Liu Y, Pan Q, Zhao Y, He C, Bi K, Chen Y, Zhao B, Chen YMX (2015). MicroRNA-155 Regulates ROS Production, NO Generation, Apoptosis and Multiple Functions of Human Brain Microvessel Endothelial Cells Under Physiological and Pathological Conditions. J Cell Biochem 116 (12):2870–81
52. Lopez-Ramirez MA, Fischer R, Torres-Badillo CC, Davies HA, Logan K, Pfizenmaier K, Male DK, Sharrack B, Romero IA (2012) Role of caspases in cytokine-induced barrier breakdown in human brain endothelial cells. J Immunol 189:3130–3139
53. Lopez-Ramirez MA, Male DK, Wang C, Sharrack B, Wu D, Romero IA (2013) Cytokine-induced changes in the gene expression profile of a human cerebral microvascular endothelial cell-line, hCMEC/D3. Fluids Barriers CNS 10:27
54. Lopez-Ramirez MA, Wu D, Pryce G, Simpson JE, Reijerkerk A, King-Robson J, Kay O, de Vries HE, Hirst MC, Sharrack B, Baker D, Male DK, Michael GJ, Romero IA (2014) MicroRNA-155 negatively affects blood-brain barrier function during neuroinflammation. FASEB J 28:2551–2565
55. Lund E, Guttinger S, Calado A, Dahlberg JE, Kutay U (2004) Nuclear export of microRNA precursors. Science 303:95–98
56. Ma X, Zhou J, Zhong Y, Jiang L, Mu P, Li Y, Singh N, Nagarkatti M, Nagarkatti P (2014) Expression, regulation and function of microRNAs in multiple sclerosis. Int J Med Sci 11:810–818
57. Martello G, Rosato A, Ferrari F, Manfrin A, Cordenonsi M, Dupont S, Enzo E, Guzzardo V, Rondina M, Spruce T, Parenti AR, Daidone MG, Bicciato S, Piccolo S (2010) A MicroRNA targeting dicer for metastasis control. Cell 141:1195–1207
58. McCall MN, Kent OA, Yu J, Fox-Talbot K, Zaiman AL, Halushka MK (2011) MicroRNA profiling of diverse endothelial cell types. BMC Med Genomics 4:78
59. Miranda KC, Huynh T, Tay Y, Ang YS, Tam WL, Thomson AM, Lim B, Rigoutsos I (2006) A pattern-based method for the identification of MicroRNA binding sites and their corresponding heteroduplexes. Cell 126:1203–1217
60. Mishra R, Singh SK (2013) HIV-1 Tat C modulates expression of miRNA-101 to suppress VE-cadherin in human brain microvascular endothelial cells. J Neurosci 33:5992–6000
61. O'Connell RM, Taganov KD, Boldin MP, Cheng G, Baltimore D (2007) MicroRNA-155 is induced during the macrophage inflammatory response. Proc Natl Acad Sci U S A 104:1604–1609
62. Pasquinelli AE, Reinhart BJ, Slack F, Martindale MQ, Kuroda MI, Maller B, Hayward DC, Ball EE, Degnan B, Muller P, Spring J, Srinivasan A, Fishman M, Finnerty J, Corbo J, Levine M, Leahy P, Davidson E, Ruvkun G (2000) Conservation of the sequence and temporal expression of let-7 heterochronic regulatory RNA. Nature 408:86–89
63. Peter ME (2009) Let-7 and miR-200 microRNAs: guardians against pluripotency and cancer progression. Cell Cycle 8:843–852
64. Pillai RS, Bhattacharyya SN, Filipowicz W (2007) Repression of protein synthesis by miRNAs: how many mechanisms? Trends Cell Biol 17:118–126
65. Popescu BF, Pirko I, Lucchinetti CF (2013) Pathology of multiple sclerosis: where do we stand? Continuum (Minneap Minn) 19:901–921
66. Qiu R, Liu Y, Wu JY, Liu K, Mo W, He R (2009) Misexpression of miR-196a induces eye anomaly in Xenopus laevis. Brain Res Bull 79:26–31
67. Reijerkerk A, Lopez-Ramirez MA, van Het Hof B, Drexhage JA, Kamphuis WW, Kooij G, Vos JB, van der Pouw Kraan TC, van Zonneveld AJ, Horrevoets AJ, Prat A, Romero IA, de

Vries HE (2013) MicroRNAs regulate human brain endothelial cell-barrier function in inflammation: implications for multiple sclerosis. J Neurosci 33:6857–6863
68. Reinhart BJ, Slack FJ, Basson M, Pasquinelli AE, Bettinger JC, Rougvie AE, Horvitz HR, Ruvkun G (2000) The 21-nucleotide let-7 RNA regulates developmental timing in Caenorhabditis elegans. Nature 403:901–906
69. Rink C, Khanna S (2011) MicroRNA in ischemic stroke etiology and pathology. Physiol Genomics 43:521–528
70. Ro S, Song R, Park C, Zheng H, Sanders KM, Yan W (2007) Cloning and expression profiling of small RNAs expressed in the mouse ovary. RNA 13:2366–2380
71. Rodriguez A, Griffiths-Jones S, Ashurst JL, Bradley A (2004) Identification of mammalian microRNA host genes and transcription units. Genome Res 14:1902–1910
72. Rom S, Dykstra H, Zuluaga-Ramirez V, Reichenbach NL, Persidsky Y (2015) miR-98 and let-7g* protect the blood-brain barrier under neuroinflammatory conditions. J Cereb Blood Flow Metab 35:1957–1965
73. Ross J (1995) mRNA stability in mammalian cells. Microbiol Rev 59:423–450
74. Rossi B, Angiari S, Zenaro E, Budui SL, Constantin G (2011) Vascular inflammation in central nervous system diseases: adhesion receptors controlling leukocyte-endothelial interactions. J Leukoc Biol 89:539–556
75. Sassen S, Miska EA, Caldas C (2008) MicroRNA: implications for cancer. Virchows Arch 452:1–10
76. Suarez Y, Sessa WC (2009) MicroRNAs as novel regulators of angiogenesis. Circ Res 104:442–454
77. Sun YM, Lin KY, Chen YQ (2013) Diverse functions of miR-125 family in different cell contexts. J Hematol Oncol 6:6
78. Svoboda P (2015) A toolbox for miRNA analysis. FEBS Letters 589:1694–1701
79. Tabas-Madrid D, Muniategui A, Sanchez-Caballero I, Martinez-Herrera DJ, Sorzano CO, Rubio A, Pascual-Montano A (2014) Improving miRNA-mRNA interaction predictions. BMC Genomics 15 Suppl 10:S2
80. Taganov KD, Boldin MP, Chang KJ, Baltimore D (2006) NF-kappaB-dependent induction of microRNA miR-146, an inhibitor targeted to signaling proteins of innate immune responses. Proc Natl Acad Sci U S A 103:12481–12486
81. Thomson DW, Bracken CP, Goodall GJ (2011) Experimental strategies for microRNA target identification. Nucleic Acids Research 39:6845–6853
82. Tili E, Michaille JJ, Cimino A, Costinean S, Dumitru CD, Adair B, Fabbri M, Alder H, Liu CG, Calin GA, Croce CM (2007) Modulation of miR-155 and miR-125b levels following lipopolysaccharide/TNF-alpha stimulation and their possible roles in regulating the response to endotoxin shock. J Immunol 179:5082–5089
83. Varendi K, Matlik K, Andressoo JO (2015) From microRNA target validation to therapy: lessons learned from studies on BDNF. Cell Mol Life Sci 72:1779–1794
84. Vasudevan S, Tong Y, Steitz JA (2007) Switching from repression to activation: microRNAs can up-regulate translation. Science 318:1931–1934
85. Wani S, Cloonan N (2014) 'Profiling direct mRNA-microRNA interactions using synthetic biotinylated microRNA-duplexes', *bioRxiv*. QIMR Berghofer MRI
86. Watanabe N, Kato T, Fujita A, Ishizaki T, Narumiya S (1999) Cooperation between mDia1 and ROCK in Rho-induced actin reorganization. Nat Cell Biol 1:136–143
87. Wightman B, Ha I, Ruvkun G (1993) Posttranscriptional regulation of the heterochronic gene lin-14 by lin-4 mediates temporal pattern formation in C. elegans. Cell 75:855–862
88. Wojciak-Stothard B, Ridley AJ (2002) Rho GTPases and the regulation of endothelial permeability. Vascul Pharmacol 39:187–199
89. Wu L, Belasco JG (2008) Let me count the ways: mechanisms of gene regulation by miRNAs and siRNAs. Mol Cell 29 (1):1–7
90. Wu D, Cerutti C, Lopez-Ramirez MA, Pryce G, King-Robson J, Simpson JE, van der Pol SM, Hirst MC, de Vries HE, Sharrack B, Baker D, Male DK, Michael GJ, Romero IA (2015) Brain endothelial miR-146a negatively modulates T-cell adhesion through repressing multiple targets to inhibit NF-kappaB activation. J Cereb Blood Flow Metab 35:412–423

91. Xiao C, Rajewsky K (2009) MicroRNA control in the immune system: basic principles. Cell 136:26–36
92. Yamamoto S, Niida S, Azuma E, Yanagibashi T, Muramatsu M, Huang TT, Sagara H, Higaki S, Ikutani M, Nagai Y, Takatsu K, Miyazaki K, Hamashima T, Mori H, Matsuda N, Ishii Y, Sasahara M (2015) Inflammation-induced endothelial cell-derived extracellular vesicles modulate the cellular status of pericytes. Sci Rep 5:8505
93. Yin KJ, Hamblin M, Chen YE (2014) Non-coding RNAs in cerebral endothelial pathophysiology: emerging roles in stroke. Neurochem Int 77:9–16
94. Yu L, Liao Q, Zeng X, Lv Z, Zheng H, Zhao Y, Sun X, Wu Z (2014) MicroRNA expressions associated with eosinophilic meningitis caused by Angiostrongylus cantonensis infection in a mouse model. Eur J Clin Microbiol Infect Dis 33:1457–1465
95. Zeng Y, Yi R, Cullen BR (2005) Recognition and cleavage of primary microRNA precursors by the nuclear processing enzyme Drosha. EMBO J 24:138–148
96. Zhao P, Zhao L, Zhang T, Wang H, Qin C, Yang S, Xia X (2012) Changes in microRNA expression induced by rabies virus infection in mouse brains. Microb Pathog 52:47–54

Blood-Brain Barrier Dysfunction during Central Nervous System Autoimmune Diseases

Jessica L. Williams and Robyn S. Klein

Abstract The blood-brain barrier (BBB) cellular constituents and their molecular interactions critically regulate immune activation within the central nervous system (CNS) such that it is protected from fatal complications of inflammatory processes. While the mechanisms that prevent leukocyte access to the CNS parenchyma were originally ascribed to physical barriers comprised of specialized endothelial cells (ECs) with ensheathing pericytes, astrocyte endfeet, and their basement membranes, it is now established that these walls are, in fact, molecular in nature. Cellular adhesion molecules, chemoattractants, and the receptors that regulate their patterns of expression maintain barrier integrity and function, limiting the entry of leukocytes and their egress into the CNS parenchyma from perivascular spaces. These molecular mechanisms are also critical for neuroinflammatory responses to pathogen invasion within the CNS, promoting immune cell interactions at endothelial barriers that ensure local T cell reactivation, a requirement for their role in pathogen clearance.

J.L. Williams
Department of Medicine, Washington University School of Medicine,
St. Louis, MO 63110, USA

R.S. Klein, MD, PhD (✉)
Department of Medicine, Washington University School of Medicine,
St. Louis, MO 63110, USA

Department of Pathology and Immunology, Washington University School of Medicine,
St. Louis, MO 63110, USA

Department of Anatomy and Neurobiology, Washington University School of Medicine,
St. Louis, MO 63110, USA

Washington University School of Medicine, Departments of Internal Medicine, Neurobiology,
660 S. Euclid Ave, St. Louis, MO 63110, USA
e-mail: rklein@dom.wustl.edu

R. Lyck, G. Enzmann (eds.), *The Blood Brain Barrier and Inflammation*,
Progress in Inflammation Research, DOI 10.1007/978-3-319-45514-3_8

1 Introduction

The blood-brain barrier (BBB) cellular constituents and their molecular interactions critically regulate immune activation within the central nervous system (CNS) such that it is protected from fatal complications of inflammatory processes. While the mechanisms that prevent leukocyte access to the CNS parenchyma were originally ascribed to physical barriers comprised of specialized endothelial cells (ECs) with ensheathing pericytes, astrocyte endfeet, and their basement membranes, it is now established that these walls are, in fact, molecular in nature. Cellular adhesion molecules, chemoattractants, and the receptors that regulate their patterns of expression maintain barrier integrity and function, limiting the entry of leukocytes and their egress into the CNS parenchyma from perivascular spaces. These molecular mechanisms are also critical for neuroinflammatory responses to pathogen invasion within the CNS, promoting immune cell interactions at endothelial barriers that ensure local T cell reactivation, a requirement for their role in pathogen clearance.

The role of the BBB in the neuropathogenesis of autoimmune diseases within the CNS is a controversial topic. While most agree that immune and endothelial cell interactions can contribute to ongoing access of autoreactive mononuclear cells and/or pathologic antibodies to the CNS parenchyma, there is disagreement on whether a primary defect in barrier function underlies these processes (Fig. 1). In neuromyelitis optica (NMO), which is characterized by inflammatory demyelinating lesions in the spinal cord and optic nerve, studies indicate that granulocyte and macrophage infiltration is triggered by the CNS entry of anti-aquaporin-4 gamma globulins (AQP4-IgG), which bind astrocyte endfeet and activate the classical complement cascade [1–3]. While the underlying processes that contribute to the presence of AQP4-IgG within the CNS are unclear, all studies suggest BBB disruption as a component of the process.

Multiple sclerosis (MS) is an autoimmune disease of the CNS in which the CNS infiltration of autoreactive mononuclear cells is associated with areas of demyelination and neurologic disability [4, 5]. In contrast to NMO, MS lesions occur in all regions of the CNS, without detection of AQP4-IgG within the sera, diagnostic criteria that distinguish the two diseases [6–8]. Although controversy exists regarding whether MS is initiated by pathologic events within the CNS versus the immune system, it is clear that stringent regulatory control mechanisms that normally limit immune cell entry at the BBB are dysfunctional. Pathologic abnormalities at the BBB during MS include increased EC expression of cellular adhesion molecules that arrest leukocytes and altered expression of chemoattractants that normally prevent the egress of infiltrating leukocytes from perivascular locations. Studies also indicate critical roles for astrocytes and pericytes in controlling BBB integrity and function during autoimmune inflammation through expression of vascular endothelial growth factor-A, endothelial cell growth factor-1, platelet-derived growth factor-β, transforming growth factor-β (TGF-β), angiopoietins, and Notch as well as formation of gap junctions. This chapter will discuss the role of the BBB during CNS autoimmune diseases, probing the evidence for and against BBB dysfunction as a primary defect in the loss of immune privilege observed.

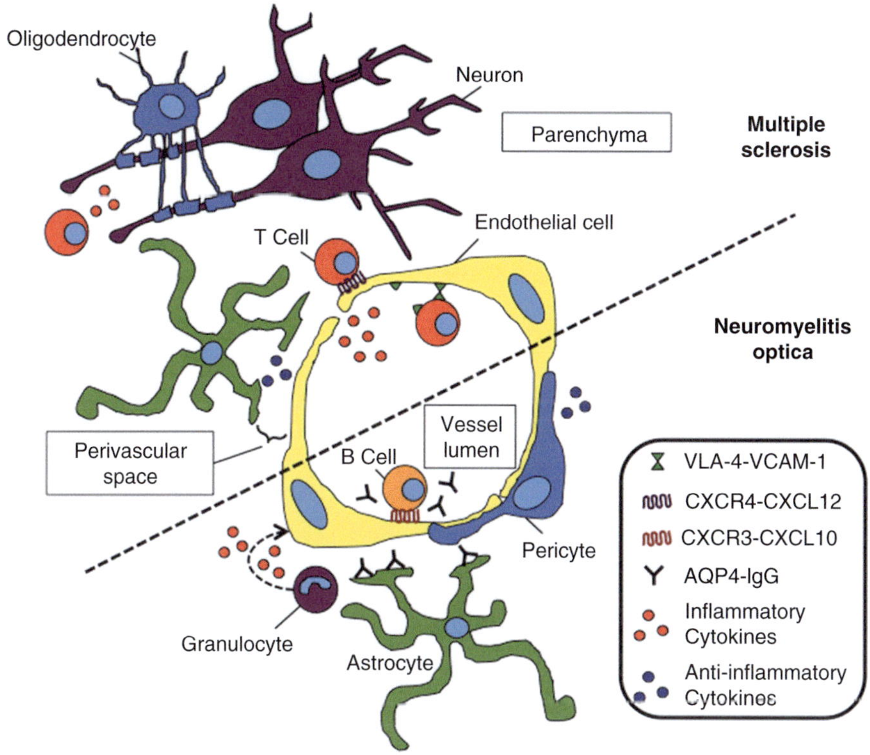

Fig. 1 Mechanisms of BBB breakdown and maintenance during NMO and MS. Activated leukocytes release a number of inflammatory cytokines that induce the rearrangement of multiple functional proteins responsible for tight junction formation. Using adhesion molecules and cytokines, leukocytes are able to crawl along the endothelium of the CNS vasculature and extravasate into the perivascular space. During MS pathogenesis, lymphocytes interact with local APCs presenting cognate antigen and become reactivated. They are then able to invade the CNS parenchyma where they cause damage to oligodendrocytes and neuronal axons. In NMO, AQP4-IgG on astrocyte endfeet and granulocytes perpetuate inflammation allowing lymphocytes to transverse the endothelium using chemokines like CXCL10. Neuroinflammation at the BBB is typically counteracted by anti-inflammatory cytokines secreted by various cell types, including astrocytes. Pericytes also help to maintain the integrity of tight junctions by secreting a number of anti-inflammatory cytokines and trophic factors

2 Cytokines and the Blood-Brain Barrier

Cytokines are small (~5–20 kDa) signaling molecules released by a broad range of cells and act through cell surface receptors to alter cell function. Chemokines, interferons, interleukins, and lymphokines are considered cytokines, while hormones and growth factors are generally excluded from this family of proteins. Although cytokines impact cells in many facets during health and disease, they are particularly important in guiding the immune response and modulate the balance between humoral and cell-based responses following an immune challenge. Cytokines also have very complex interactions with each other in that they can enhance or inhibit the action of other family members. They also have significant impact on constituents of the BBB and mediate autoimmune disease processes. There are several mechanisms by which cytokines are able to exert their effects on the BBB and cells within the CNS parenchyma. Circulating immune cells can become activated and begin secreting cytokines that facilitate their transmigration across the BBB in a very regulated manner by interacting with ECs and then releasing cytokines into the CNS milieu. While still in circulation, blood-borne cytokines can also affect the function of the CNS by crossing the BBB and interacting directly with CNS tissue via saturable transport systems. These types of systems have been described for interleukin (IL)-1α, IL-1β, IL-6, and tumor necrosis factor (TNF)-α [9], which are known to launch the immune response and impact the function of cells that make up the BBB [10]. Of interest, TNF-α and IL-1β differentially regulate the inflammatory phenotype of brain microvascular ECs, with the former eliciting much higher levels of expression of intercellular and vascular cell adhesion molecules (ICAM and VCAM)-1 [11].

Cytokines directly impact the function of the BBB and may contribute to the link between central and peripheral disease. Cytokines released into circulation can limit the integrity of the BBB and increase its permeability without entering the CNS by causing damage and destruction of the tight junctions that serve to limit access of circulating cells and potentially harmful molecules [10]. Exposure of the endothelium to inflammatory mediators like IL-6 and TNF-α has been shown to disorganize the cell–cell junctions and enhances leukocyte endothelial adhesion and migration [12]. In animal studies in which mice develop spontaneous autoimmune disease of the CNS, peripheral upregulation of these T helper-type (Th1) cytokines was observed to precede brain microvascular EC and astrocyte activation [13]. Similar findings are observed in MS lesions with redistribution of junctional proteins.

During NMO pathogenesis, AQP4-IgG within the sera of patients alters polarized expression patterns of aquaporin-4 on astrocytes, which increases the permeability of human endothelial barriers via soluble mediators released from granulocytes [14]. In a large cytokine study, it was found that NMO patients have elevated cytokines in the CSF related to Th17 (IL-6 and IL-8) and Th2 (IL-10 and IL-13) profiles [15–18]. In MS, cytokine levels in the plasma precede a surge of disease activity marked by areas of BBB breakdown and infiltration of immune cells and decrease during remission. In contrast to that seen in NMO patients, the inflammatory cytokines that are characteristic of MS pathogenesis are typically classified as innate, Th1, or Th17 associated. Together they include interferon (IFN)-γ, TNF-α, IL-1β, IL-6, IL-12, IL-17, IL-18, and IL-23; and can be found in patient sera or cerebrospinal fluid (CSF) [12, 15–17, 19]. Levels of TNF-α in the sera of MS patients correlated with active barrier damage as measured by albumin ratios [20], and expression of IFN-γ, TNF-α, and IL-6 by infiltrating cells in the perivascular space of MS lesions is evident [21]. In MS lesions with ongoing immune activity, TNF-α can be detected on ECs and astrocytes [22], indicating these cell types participate in the inflammatory process. Further, IL-17 transcript has been shown to be upregulated in chronic MS brain lesions compared to control patients [23], and numbers of IL-17 mRNA-expressing mononuclear cells were increased in the blood and cerebrospinal fluid of MS patients compared to control individuals [24], suggesting that inflammatory cytokines are prominent in MS pathogenesis and play a critical role in facilitating leukocyte entry beyond the endothelial barrier and into the CNS parenchyma.

Following inflammatory events during the pathogenesis of MS, CNS barrier damage is usually counteracted by anti-inflammatory cytokines like TGF-β1 [25], IL-10, or IFN-β [12]. In fact, IFN-β was one of the first treatments for MS and continues to be a standard-of-care drug [26–29]. IFN-β directly affects the BBB, promoting its efficacy by blocking destruction of endothelial tight junctions and enhancing proteins that facilitate junction integrity, like occludin [30]. Once the endothelial barrier and the basal lamina of the BBB have been compromised, astrocytes, pericytes, and microglia can form a dialogue with the damaged endothelium that may also be important in BBB repair [31]. In both acute and chronic active MS lesions, astrocytes appeared to be the sole source of IL-10, concentrated primarily in association with perivascular endfeet [32]. Interestingly, IL-10 is also expressed by astrocytic endfeet in noninflammatory controls, suggesting a regulatory role for IL-10 at the BBB [22]. Similarly, in MS lesions and in normal controls, TGF-β1 is expressed by CNS ECs; however, this cytokine was found to be highest in acute MS lesions, i.e., in active inflammatory areas [32], again suggesting a regulatory role for TGF-β1 at the BBB. Endothelial cells are also known to secrete the cytokine leukemia inhibitory factor, which induces astrocyte differentiation [33], suggesting there is bidirectional communication between cells that make up the neurovascular unit to maintain barrier integrity during health and disease.

3 Cell Adhesion at the BBB During Autoimmunity

Cellular adhesion molecules (CAMs) and chemokines play essential roles in orchestrating immune and EC interactions at all tissue barriers including the CNS [34]. CAMs, transmembrane proteins located on the cell surfaces of most ECs, mediate binding of cells to the extracellular matrix or other cells. In most tissues, interactions between CAMs allow immune cells to roll and crawl on the luminal face of the vascular endothelium, sampling tissue environments for entry cues. Members of the chemokine superfamily of chemotactic cytokines are among the cues that induce firm adhesion and transendothelial cell migration of immune cells as they exit the blood and enter inflamed tissues [35].

Chemokines comprise a large family of small signaling proteins classified into subfamilies according to positions of conserved cysteine residues at the N-terminus. Chemokines bind members of the Gαi-coupled receptor superfamily. Chemokine signaling induces conformational changes of members of the integrin subfamily that convert low-affinity interactions with their CAM receptors to high-affinity ones that arrest cells at endothelial surfaces. These molecules have a critical role in the maintenance of immune privilege at the BBB and in the recruitment of inflammatory cells across the BBB during acute disease exacerbations in NMO and MS.

The immune privileged status of the CNS has recently been altered to "immune specialized," reflecting increased knowledge of the importance of immune surveillance in preventing opportunistic infections at this site. Thus, it is now clear that lymphocytes routinely survey the CNS and, via interactions with perivascular antigen-presenting cells (APCs), undergo local reactivation in the setting of infection within the CNS that requires cell-mediated immunity for pathogen clearance. In noninfectious states, tight junction adaptor proteins link vascular ECs via adherens junctions, comprised of vascular endothelial cadherin and platelet endothelial cell adhesion molecule (PECAM)-1. This linkage reinforces the barrier against paracellular movement of immune cells. Another feature that distinguishes endothelial cells of the BBB from others is their low level of CAMs, which limits leukocyte interactions that might restrict migration across the endothelium. Thus, most of the lymphocytes that enter the CNS during immune surveillance cross the endothelium in the choroid plexus or meninges, sites where the endothelium does not exhibit the specializations of the BBB and express moderate levels of P-selectin, E-selectin, and intercellular adhesion molecule (ICAM)-1 [36]. Once lymphocytes leave meningeal vessels, they localize along abluminal surfaces, essentially crawling into the CNS along its vasculature. This localization is due to interactions between the chemokine receptor CXCR4, expressed ubiquitously on leukocytes, and its ligand, CXCL12, which is expressed by all CNS ECs and exhibits polarized localization on the parenchymal side of CNS vasculature. CXCL12 is known to provide the localizing cue for CXCR4-bearing leukocytes entering peripheral lymph nodes via high endothelial venules (HEVs). HEVs express CXCL12 on their lumina, capturing leukocytes from the blood. The opposite expression pattern is observed for CXCL12 within the CNS, which serves to limit leukocyte entry, localizing them to perivascular spaces where they may encounter APCs that determine their eventual fate within the CNS.

During autoimmune diseases of the CNS, leukocytes gain inappropriate access to the CNS due to breakdown of the BBB. During inflammation, resident glial cells and ECs display alterations in morphology and expression of CAMs and chemokines that promote leukocyte entry into the CNS parenchyma. Early studies in a murine model for MS, experimental autoimmune encephalomyelitis, demonstrated a critical role for immune cell expression of $\alpha4\beta1$ (VLA-4), an integrin dimer that binds vascular cell adhesion molecule (VCAM)-1 on endothelium, in the CNS entry of autoreactive $CD4^+$ T cells [37]. This finding resulted in the development of a successful drug treatment for MS, natalizumab, a humanized monoclonal antibody against VLA-4 that is administered intravenously at monthly intervals. Few studies have addressed the role of integrin molecules in the pathogenesis of NMO. However, in two reports, use of natalizumab failed to prevent NMO relapse and, in one case, precipitated a severe and catastrophic exacerbation [38, 39]. Further studies and established animal models of NMO are needed to identify the critical adhesion molecules utilized by infiltrating immune cells during the neuropathogenesis of NMO.

Studies of chemokine expression in NMO patients have revealed increased cerebrospinal fluid (CSF) levels of IL-8 and CXCL13, which recruit granulocytes and B cells, respectively [15, 40]. Notably, CXCL13 level in the CSF of NMO patients is higher than that in the CSF of MS patients and correlated with the severity of NMO disease activity as indicated by relapse rate [41, 42]. Given that antibody-independent functions of peripheral B cells derived from NMO patients are markedly impaired [42], it is likely that CXCL13-mediated recruitment of these cells into the CNS plays an important role in NMO pathogenesis. Additionally, there is evidence that other chemokines, like CXCL10, have a role in the pathogenesis of NMO at the BBB as sera from patients in the acute phase of NMO exhibit increased expression of CXCL10 by ECs [43]. CXCL10 is known to recruit multiple types of leukocytes including natural killer cells, T cells, and dendritic cells. The co-recruitment of these immune cells may provide the space and setting for T and B cell reactivation within perivascular spaces.

While numerous studies have identified increased expression of T cell chemokines within the CNS in animal models of MS, few have shown that they directly impact on loss of immune privilege at the BBB. Lymphocytes that enter the CNS during autoimmunity do not appear to pause within perivascular spaces to receive "checkpoint" instructions, including those that influence the activation of autoreactive immune cells. Examining CNS tissue specimens from non-MS and MS patients, we have shown a critical role for receptors that modulate the pattern of expression of CXCL12 at the BBB. Loss of CXCL12 from abluminal surfaces of the BBB and redistribution of CXCL12 to vessel lumina occur specifically within MS lesions compared to other neuroinflammatory diseases. In areas of extensive mononuclear cell infiltration, the expression of CXCL12 on ECs is lost [44], suggesting a role for immune molecules in regulating chemokine signaling at the BBB during the pathogenesis of MS.

The chemokine CCL2 has also been linked to BBB dysfunction during MS. CCL2 is elevated during CNS inflammation and is reduced within the CSF of MS patients during acute relapses [45, 46]. *In vitro* experiments using BBB models indicate that CCL2 induces tight and adherens junction remodeling via redistribution of junctional adhesion molecule-A (JAMA) and association of β-catenin with PECAM-1, the latter of which is then relocated to the plasma membrane [47, 48]. JAMA is a transmembrane protein that normally maintains EC interactions but acts as a leukocyte adhesion molecule during inflammation [49]. JAMA expression is decreased within inflammatory lesions in animal models of MS [50], suggesting that its expression helps maintain BBB integrity and is altered during inflammatory disruptions of the BBB.

4 Immune Cell Trafficking in the CNS

Members of the integrin family are among the most important adhesion molecules to regulate leukocyte attachment and extravasation across the BBB and into inflamed tissue. ICAM-1 is a cytokine-inducible adhesion molecule that is upregulated during CNS inflammatory diseases, like MS [32, 51]. It is expressed by numerous cell types, including ECs and astrocytes, and is the ligand for lymphocyte function-associated antigen 1 (LFA1) and macrophage antigen 1 (MAC1), which are expressed by lymphocytes and monocytes, respectively [52]. Within MS lesions, ICAM-1 is widely expressed on ECs; however, it is also expressed in the surrounding adjacent white matter [32, 53], suggesting soluble mediators, like cytokines, may influence the adhesion molecule's expression on nearby vasculature, dictating locations of new lesion formation. Additionally, during inflammation, TNF-α and IFN-γ induce ECs to express VCAM-1, which is bound by activated T cells that express integrins like VLA-4 to facilitate their entry into the CNS [54]. A detailed review outlining the molecular events orchestrating leukocyte extravasation across the CNS endothelium has been recently published [55].

The initial encounter of T cells with their specific neuroantigen usually occurs in the deep cervical lymph nodes in the context of major histocompatibility complex (MHC) class II [56, 57]. Entrance of T cells beyond the confines of the perivascular space typically requires reactivation by interacting with APCs, like macrophages or dendritic cells, or occasionally pericytes, that express MHC II loaded with cognate antigen [57]. These APCs are rich in the perivascular space, and, in addition, a population of $CCR7^+$ dendritic cells are detected in human cerebrospinal fluid and in inflamed MS lesions [58]. B cells are also able to present antigen and more recently have been described as interacting with T cells in ectopic lymphoid follicles formed around the CNS meningeal endothelium, but not in parenchymal lesions, in secondary progressive MS patients [59]. Of note, large, subpial cortical lesions are often found adjacent to B-cell-containing meningeal follicles, suggesting soluble factors, potentially cytokines, secreted from these structures mediate pathology [60]. Additionally, as opposed to normal microvessels isolated from the human CNS, microvessels from MS patients express markers of activation as well as significant levels of MHC II [61], suggesting ECs of the BBB are also able to present CNS antigens to infiltrating T cells. Together, these studies indicate that restimulation of T cells can occur in multiple locations during the pathogenesis of MS.

Once T cells have extravasated across the BBB and are restimulated, to gain access to the CNS parenchyma, they must pass through a barrier of extracellular matrix, called the glia limitans, which can be broken down by matrix metalloproteinases (MMPs). MMPs are able to perpetuate the inflammatory response at the BBB as they induce the cleavage of TNF-α from a cell-bound to soluble form, are involved in the degradation of the type IV collagen that makes up the glia limitans, and mediate proteolysis of myelin components during MS. In particular, MMP2 and MMP9 are detectable in the CSF of MS patients and are present on endothelial cells, pericytes, and astrocytes in MS lesions [62]. Once infiltrating leukocytes access the CNS parenchyma, beyond the BBB, inflammatory mediators are released, which causes damage to CNS axons, leading to progressive neurodegeneration and persistent disability in MS patients. In an extensive study comparing lesion characteristics in varying MS disease subtypes, it was found that in progressive stages of disease, active demyelination and neurodegeneration are seen only in patients with pronounced inflammation in the CNS [63], confirming that immune cell infiltration beyond the borders of the endothelium and glia limitans can lead to devastating damage in the CNS parenchyma.

5 Conclusion

The pathogenesis of NMO and MS requires the breakdown of the BBB and the entry of leukocytes into the perivascular spaces and surrounding CNS parenchyma. Activation of immune cells in the periphery likely precedes release of inflammatory cytokines and/or autoantibodies, which strongly contribute to the loss of BBB integrity. Inflammatory molecules not only cause rearrangement of tight junction proteins, making the endothelium more permeable to soluble mediators and cells, but also induce the expression of several adhesion molecules and chemokines on cells that make up the neurovascular unit. During NMO pathogenesis, AQP4-IgG binding to astrocyte endfeet induces complement-mediated inflammation, encouraging the recruitment and activation of granulocytes. This perpetuated inflammatory event typically occurs in the optic nerve and spinal cord, results in demyelination, and can lead to loss of sensation, paralysis of the limbs, and blindness. In MS, activated T cells crawl along the CNS vasculature using chemokines and adhesion molecules as molecular highways. Once in the perivascular space, T cells interact with APCs, including macrophages and dendritic cells, to become reactivated and begin producing MMPs. Following degradation of the collagen barrier that makes up the glia limitans, T cells can then invade the CNS parenchyma and cause damage and destruction of the myelin sheath, leading to the motor and sensory impairments commonly seen in MS patients. While the mechanisms underlying the inciting events of NMO and MS may differ, it is evident that the pathology associated with both CNS autoimmune diseases is reliant on the compromise of the BBB.

Acknowledgments This work was supported by the National Multiple Sclerosis Society Postdoctoral Fellowship (J.L. Williams) and the National Institutes of Health/National Institute of Neurological Disorders and Stroke Grant P01 NS059560 (R.S. Klein).

References

1. Hinson SR, McKeon A, Lennon VA (2010) Neurological autoimmunity targeting aquaporin-4. Neuroscience 168:1009–1018
2. Marignier R, Giraudon P, Vukusic S, Confavreux C, Honnorat J (2010) Anti-aquaporin-4 antibodies in Devic's neuromyelitis optica: therapeutic implications. Ther Adv Neurol Disord 3:311–321
3. Graber DJ, Levy M, Kerr D, Wade WF (2008) Neuromyelitis optica pathogenesis and aquaporin 4. J Neuroinflammation 5:22
4. Frohman EM, Racke MK, Raine CS (2006) Multiple sclerosis–the plaque and its pathogenesis. N Engl J Med 354:942–955
5. Wingerchuk DM, Lucchinetti CF, Noseworthy JH (2001) Multiple sclerosis: current pathophysiological concepts. Lab Invest 81:263–281
6. Lennon VA, Wingerchuk DM, Kryzer TJ, Pittock SJ, Lucchinetti CF, Fujihara K et al (2004) A serum autoantibody marker of neuromyelitis optica: distinction from multiple sclerosis. Lancet 364:2106–2112
7. Lennon VA, Kryzer TJ, Pittock SJ, Verkman AS, Hinson SR (2005) IgG marker of optic-spinal multiple sclerosis binds to the aquaporin-4 water channel. J Exp Med 202:473–477

8. Roemer SF, Parisi JE, Lennon VA, Benarroch EE, Lassmann H, Bruck W et al (2007) Pattern-specific loss of aquaporin-4 immunoreactivity distinguishes neuromyelitis optica from multiple sclerosis. Brain 130:1194–1205
9. Banks WA, Kastin AJ, Broadwell RD (1995) Passage of cytokines across the blood-brain barrier. Neuroimmunomodulation 2:241–248
10. Dantzer R, Kelley KW (2007) Twenty years of research on cytokine-induced sickness behavior. Brain Behav Immun 21:153–160
11. O'Carroll SJ, Kho DT, Wiltshire R, Nelson V, Rotimi O, Johnson R et al (2015) Pro-inflammatory TNFalpha and IL-1beta differentially regulate the inflammatory phenotype of brain microvascular endothelial cells. J Neuroinflammation 12:131
12. Minagar A, Alexander JS (2003) Blood-brain barrier disruption in multiple sclerosis. Mult Scler 9:540–549
13. Alvarez JI, Saint-Laurent O, Godschalk A, Terouz S, Briels C, Larouche S et al (2015) Focal disturbances in the blood-brain barrier are associated with formation of neuroinflammatory lesions. Neurobiol Dis 74:14–24
14. Vincent T, Saikali P, Cayrol R, Roth AD, Bar-Or A, Prat A et al (2008) Functional consequences of neuromyelitis optica-IgG astrocyte interactions on blood-brain barrier permeability and granulocyte recruitment. J Immunol 181:5730–5737
15. Uzawa A, Mori M, Arai K, Sato Y, Hayakawa S, Masuda S et al (2010) Cytokine and chemokine profiles in neuromyelitis optica: significance of interleukin-6. Mult Scler 16:1443–1452
16. Matsushita T, Tateishi T, Isobe N, Yonekawa T, Yamasaki R, Matsuse D et al (2013) Characteristic cerebrospinal fluid cytokine/chemokine profiles in neuromyelitis optica, relapsing remitting or primary progressive multiple sclerosis. PLoS One 8:e61835
17. Kothur K, Wienholt L, Brilot F, Dale RC (2016) CSF cytokines/chemokines as biomarkers in neuroinflammatory CNS disorders: a systematic review. Cytokine 77:227–237
18. Uzawa A, Mori M, Ito M, Uchida T, Hayakawa S, Masuda S et al (2009) Markedly increased CSF interleukin-6 levels in neuromyelitis optica, but not in multiple sclerosis. J Neurol 256:2082–2084
19. Wen SR, Liu GJ, Feng RN, Gong FC, Zhong H, Duan SR et al (2012) Increased levels of IL-23 and osteopontin in serum and cerebrospinal fluid of multiple sclerosis patients. J Neuroimmunol 244:94–96
20. Sharief MK, Thompson EJ (1992) In vivo relationship of tumor necrosis factor-alpha to blood-brain barrier damage in patients with active multiple sclerosis. J Neuroimmunol 38:27–33
21. Woodroofe MN, Cuzner ML (1993) Cytokine mRNA expression in inflammatory multiple sclerosis lesions: detection by non-radioactive in situ hybridization. Cytokine 5:583–588
22. Cannella B, Raine CS (1995) The adhesion molecule and cytokine profile of multiple sclerosis lesions. Ann Neurol 37:424–435
23. Lock C, Hermans G, Pedotti R, Brendolan A, Schadt E, Garren H et al (2002) Gene-microarray analysis of multiple sclerosis lesions yields new targets validated in autoimmune encephalomyelitis. Nat Med 8:500–508
24. Matusevicius D, Kivisakk P, He B, Kostulas N, Ozenci V, Fredrikson S et al (1999) Interleukin-17 mRNA expression in blood and CSF mononuclear cells is augmented in multiple sclerosis. Mult Scler 5:101–104
25. Chen Y, Kuchroo VK, Inobe J, Hafler DA, Weiner HL (1994) Regulatory T cell clones induced by oral tolerance: suppression of autoimmune encephalomyelitis. Science 265:1237–1240
26. Interferon beta-1b is effective in relapsing-remitting multiple sclerosis. I. Clinical results of a multicenter, randomized, double-blind, placebo-controlled trial. The IFNB Multiple Sclerosis Study Group (1993). Neurology 43:655–61.
27. Paty DW, Li DK (1993) Interferon beta-1b is effective in relapsing-remitting multiple sclerosis. II. MRI analysis results of a multicenter, randomized, double-blind, placebo-controlled trial. UBC MS/MRI Study Group and the IFNB Multiple Sclerosis Study Group. Neurology 43:662–667
28. Jacobs LD, Cookfair DL, Rudick RA, Herndon RM, Richert JR, Salazar AM et al (1996) Intramuscular interferon beta-1a for disease progression in relapsing multiple sclerosis. The Multiple Sclerosis Collaborative Research Group (MSCRG). Ann Neurol 39:285–294

29. Jacobs LD, Beck RW, Simon JH, Kinkel RP, Brownscheidle CM, Murray TJ et al (2000) Intramuscular interferon beta-1a therapy initiated during a first demyelinating event in multiple sclerosis. CHAMPS Study Group. N Engl J Med 343:898–904
30. Minagar A, Long A, Ma T, Jackson TH, Kelley RE, Ostanin DV et al (2003) Interferon (IFN)-beta 1a and IFN-beta 1b block IFN-gamma-induced disintegration of endothelial junction integrity and barrier. Endothelium J Endothelial Cell Res 10:299–307
31. Abbott NJ, Ronnback L, Hansson E (2006) Astrocyte-endothelial interactions at the blood-brain barrier. Nat Rev Neurosci 7:41–53
32. Brosnan CF, Cannella B, Battistini L, Raine CS (1995) Cytokine localization in multiple sclerosis lesions: correlation with adhesion molecule expression and reactive nitrogen species. Neurology 45:S16–S21
33. Abbott NJ (2002) Astrocyte-endothelial interactions and blood-brain barrier permeability. J Anat 200:629–638
34. Engelhardt B (2008) Immune cell entry into the central nervous system: involvement of adhesion molecules and chemokines. J Neurol Sci 274:23–26
35. Williams JL, Holman DW, Klein RS (2014) Chemokines in the balance: maintenance of homeostasis and protection at CNS barriers. Front Cell Neurosci 8:154
36. Kivisakk P, Mahad DJ, Callahan MK, Trebst C, Tucky B, Wei T et al (2003) Human cerebrospinal fluid central memory CD4+ T cells: evidence for trafficking through choroid plexus and meninges via P-selectin. Proc Natl Acad Sci U S A 100:8389–8394
37. Yednock TA, Cannon C, Fritz LC, Sanchez-Madrid F, Steinman L, Karin N (1992) Prevention of experimental autoimmune encephalomyelitis by antibodies against alpha 4 beta 1 integrin. Nature 356:63–66
38. Kleiter I, Hellwig K, Berthele A, Kumpfel T, Linker RA, Hartung HP et al (2012) Failure of natalizumab to prevent relapses in neuromyelitis optica. Arch Neurol 69:239–245
39. Kitley J, Evangelou N, Kuker W, Jacob A, Leite MI, Palace J (2014) Catastrophic brain relapse in seronegative NMO after a single dose of natalizumab. J Neurol Sci 339:223–225
40. Zhong X, Wang H, Dai Y, Wu A, Bao J, Xu W et al (2011) Cerebrospinal fluid levels of CXCL13 are elevated in neuromyelitis optica. J Neuroimmunol 240–241:104–108
41. Alvarez E, Piccio L, Mikesell RJ, Klawiter EC, Parks BJ, Naismith RT et al (2013) CXCL13 is a biomarker of inflammation in multiple sclerosis, neuromyelitis optica, and other neurological conditions. Mult Scler 19:1204–1208
42. Quan C, Yu H, Qiao J, Xiao B, Zhao G, Wu Z et al (2013) Impaired regulatory function and enhanced intrathecal activation of B cells in neuromyelitis optica: distinct from multiple sclerosis. Mult Scler 19:289–298
43. Shimizu F, Nishihara H, Sano Y, Takeshita Y, Takahashi S, Maeda T et al (2015) Markedly increased IP-10 production by blood-brain barrier in neuromyelitis optica. PLoS One 10:e0122000
44. McCandless EE, Piccio L, Woerner BM, Schmidt RE, Rubin JB, Cross AH et al (2008) Pathological expression of CXCL12 at the blood-brain barrier correlates with severity of multiple sclerosis. Am J Pathol 172:799–808
45. Scarpini E, Galimberti D, Baron P, Clerici R, Ronzoni M, Conti G et al (2002) IP-10 and MCP-1 levels in CSF and serum from multiple sclerosis patients with different clinical subtypes of the disease. J Neurol Sci 195:41–46
46. Malmestrom C, Andersson BA, Haghighi S, Lycke J (2006) IL-6 and CCL2 levels in CSF are associated with the clinical course of MS: implications for their possible immunopathogenic roles. J Neuroimmunol 175:176–182
47. Stamatovic SM, Sladojevic N, Keep RF, Andjelkovic AV (2012) Relocalization of junctional adhesion molecule A during inflammatory stimulation of brain endothelial cells. Mol Cell Biol 32:3414–3427
48. Roberts TK, Eugenin EA, Lopez L, Romero IA, Weksler BB, Couraud PO et al (2012) CCL2 disrupts the adherens junction: implications for neuroinflammation. Lab Invest 92:1213–1233
49. Nourshargh S, Krombach F, Dejana E (2006) The role of JAM-A and PECAM-1 in modulating leukocyte infiltration in inflamed and ischemic tissues. J Leukoc Biol 80:714–718

50. Yeung D, Manias JL, Stewart DJ, Nag S (2008) Decreased junctional adhesion molecule-A expression during blood-brain barrier breakdown. Acta Neuropathol 115:635–642
51. Sobel RA, Mitchell ME, Fondren G (1990) Intercellular adhesion molecule-1 (ICAM-1) in cellular immune reactions in the human central nervous system. Am J Pathol 136:1309–1316
52. Carlos TM, Harlan JM (1994) Leukocyte-endothelial adhesion molecules. Blood 84:2068–2101
53. Bo L, Peterson JW, Mork S, Hoffman PA, Gallatin WM, Ransohoff RM et al (1996) Distribution of immunoglobulin superfamily members ICAM-1, −2, −3, and the beta 2 integrin LFA-1 in multiple sclerosis lesions. J Neuropathol Exp Neurol 55:1060–1072
54. Steinman L (2001) Multiple sclerosis: a two-stage disease. Nat Immunol 2:762–764
55. Vestweber D (2015) How leukocytes cross the vascular endothelium. Nat Rev Immunol 15:692–704
56. Weller RO, Engelhardt B, Phillips MJ (1996) Lymphocyte targeting of the central nervous system: a review of afferent and efferent CNS-immune pathways. Brain Pathol 6:275–288
57. Engelhardt B, Ransohoff RM (2005) The ins and outs of T-lymphocyte trafficking to the CNS: anatomical sites and molecular mechanisms. Trends Immunol 26:485–495
58. Kivisakk P, Mahad DJ, Callahan MK, Sikora K, Trebst C, Tucky B et al (2004) Expression of CCR7 in multiple sclerosis: implications for CNS immunity. Ann Neurol 55:627–638
59. Serafini B, Rosicarelli B, Magliozzi R, Stigliano E, Aloisi F (2004) Detection of ectopic B-cell follicles with germinal centers in the meninges of patients with secondary progressive multiple sclerosis. Brain Pathol 14:164–174
60. Magliozzi R, Howell O, Vora A, Serafini B, Nicholas R, Puopolo M et al (2007) Meningeal B-cell follicles in secondary progressive multiple sclerosis associate with early onset of disease and severe cortical pathology. Brain 130:1089–1104
61. Washington R, Burton J, Todd RF 3rd, Newman W, Dragovic L, Dore-Duffy P (1994) Expression of immunologically relevant endothelial cell activation antigens on isolated central nervous system microvessels from patients with multiple sclerosis. Ann Neurol 35:89–97
62. Conlon P, Oksenberg JR, Zhang J, Steinman L (1999) The immunobiology of multiple sclerosis: an autoimmune disease of the central nervous system. Neurobiol Dis 6:149–166
63. Frischer JM, Bramow S, Dal-Bianco A, Lucchinetti CF, Rauschka H, Schmidbauer M et al (2009) The relation between inflammation and neurodegeneration in multiple sclerosis brains. Brain 132:1175–1189

Pathways Across the Blood-Brain Barrier

Michael Abadier and Ruth Lyck

Abstract The parenchymal blood-brain barrier (BBB) is formed by highly specialized vascular endothelial cells of the central nervous system (CNS). As part of the neurovascular unit (NVU), the BBB builds up a tight barrier between the changing milieu of the bloodstream and the vulnerable CNS. Yet, during inflammatory diseases of the CNS, immune cells are recruited into the CNS and thus migrate across the inflamed BBB. In particular, effector T (T_{eff}) cells critically contribute to autoimmune neuroinflammation such as multiple sclerosis (MS). Extravasation of T_{eff} cells across the inflamed BBB is a well-coordinated multistep process tightly regulated through cell adhesion molecules, chemotactic factors, and their receptors. An initial contact between the circulating T_{eff} cell and the inflamed endothelial cells of the BBB mediates slowing down of T_{eff} cells. Then, integrins on the T_{eff} cell surface acquire an activated conformation. This in turn is prerequisite for shear-resistant arrest that transforms into firm adhesion, crawling, and finally diapedesis. Following diapedesis, T_{eff} cells accumulate in the perivascular space between the two basement membranes of the NVU. Only after reactivation with their cognate antigen by antigen-presenting cells (APCs), T_{eff} cells can breach the parenchymal basement membrane and infiltrate the CNS parenchyma. Interfering with pathological T_{eff} cell recruitment into the CNS has been successfully translated into the clinic for the treatment of MS patients through natalizumab, which blocks extravasation of immune cells across the BBB. This review introduces the molecular players and discusses the cellular pathway of T_{eff} cell extravasation across the inflamed BBB.

M. Abadier
Division of Inflammation Biology, La Jolla Institute for Allergy and Immunology, La Jolla, CA 92037, USA

R. Lyck (✉)
Theodor Kocher Institute, University of Bern, Freiestrasse 1, 3012 Bern, Switzerland
e-mail: Ruth.lyck@tki.unibe.ch

R. Lyck, G. Enzmann (eds.), *The Blood Brain Barrier and Inflammation*, Progress in Inflammation Research, DOI 10.1007/978-3-319-45514-3_9

1 Introduction

The CNS was traditionally regarded as an immune-privileged site where immunosurveillance is absent. It was believed that CNS homeostasis would not endure patrolling of immune cells in search for their specific antigens. Yet, the concept of CNS immunological ignorance was challenged by evidences from immune-mediated CNS diseases such as MS or its animal model experimental autoimmune encephalomyelitis (EAE), in which autoaggressive T_{eff} cells migrate into the CNS parenchyma suggesting that inflammatory neuro-antigen-specific T_{eff} cells can successfully cross the CNS barriers [95]. The first direct evidence that T_{eff} cells can cross the BBB came from ultrastructural analysis, in which radioactively labeled encephalitogenic $CD4^+$ T cell blasts were observed beyond the BBB 6 h after injection into healthy Lewis rats [67]. Hence, although low in number, activated rather than resting $CD4^+$ T lymphocytes were capable of breaching the BBB.

Modern in vitro and in vivo imaging tools, but also early classical electron microscopy, have been fundamental in resolving the molecular and cellular mechanism of immune cell extravasation across the BBB. In this review we will guide you through a journey of immune cell extravasation across the inflamed BBB. Thereby, we will set a particular focus on $CD4^+$ T_{eff} cells that have been subject of our own research because of their importance in the pathogenesis of MS. Notably, the extravasation of $CD4^+$ or $CD8^+$ T cells across the BBB is mediated by discrete molecular and cellular mechanism [132]. First, we will briefly reiterate barriers of the CNS followed by an introduction of cytokine-induced inflammatory changes of the BBB and the molecules involved in T_{eff} cell extravasation. Second, we will discuss the individual steps leading to the successful extravasation of T_{eff} cells. Lastly, we will describe signaling events that proof the active involvement of BBB endothelial cells to T_{eff} cell extravasation.

2 Barriers of the CNS

Barriers between the blood and CNS protect the tightly regulated milieu within the CNS from the more changeable conditions in the bloodstream [3, 54, 133]. The brain barriers include the endothelial BBB of the CNS parenchymal microvessels, the blood-leptomeningeal barrier (BLMB), and the blood-cerebrospinal fluid barrier (BCSFB) established by choroid plexus epithelial cells [53]. At the BBB, endothelial cells establish a physical, transport, and metabolic barrier [3, 54]. Here, the physical strength of endothelial cell-cell junctions is established by adherens junctions (AJs) and uniquely complex tight junctions (TJs) that restrict paracellular diffusion [156]. In fact, the BBB limits and selectively regulates the exchange of substances between the blood and the CNS. However, during inflammation immune cells, e.g., antigen-specific $CD4^+$ T_{eff} cells, effectively breach the BBB and critically contribute to neuroinflammatory disease progression, including MS [100], stroke

[46, 127, 171], Alzheimer's disease [26, 93, 105, 115, 158, 159], and Parkinson's disease [116].

The BBB endothelial cells are part of the NVU, which further harbors the endothelial basement membrane (BM), pericytes embedded into the endothelial BM, and a second parenchymal BM formed by astrocytes and covered by astrocytic end feet and neurons [2, 4, 133]. The localization of the BBB is restricted to CNS microvessels, i.e., the capillaries, precapillary arterioles, and postcapillary venules. One unique feature of the BBB microvessels is the presence of two distinct BMs, which differ in their cellular origin and molecular composition [54]. The endothelial BM is produced by the BBB endothelial cells and contains the extracellular matrix proteins laminin $\alpha 4$ and $\alpha 5$, while the parenchymal BM is produced by the astrocytic end feet and contains laminin $\alpha 1$ and $\alpha 2$ isoforms [14, 54]. The parenchymal BM together with the astrocytic end feet forms the glia limitans, which covers the entire abluminal surface of the brain and spinal cord vessels [53]. At the level of the capillaries, both BMs fuse to appear as a single BM, while at the postcapillary level, the BMs can be identified as separate layers. During neuroinflammation the two BMs delimit the so-called perivascular cuff, filled with immune cells after their migration across the inflamed BBB. However, infiltration of T_{eff} cells into the CNS parenchyma requires breaching of the glia limitans and only occurs upon restimulation of T_{eff} with their cognate antigen by APCs residing in the perivascular space between both BMs [53, 54].

3 Cytokines Involved in CNS Inflammation

One hallmark of inflammatory diseases of the CNS is the inflamed phenotype of the BBB. Inducers of inflammation are cytokines, among them tumor necrosis factor (TNF)-α, interleukin (IL)-1β, and interferon (IFN)-γ. The source of cytokines within the CNS includes microglia, and the cellular components of the NVU plus infiltrating immune cells during acute and chronic neuroinflammation [17]. Cytokine-induced inflammation alters the BBB characteristics associated with compromised barrier properties, shedding of the glycocalyx, and increased expression and cell surface presentation of cell adhesion molecules and chemoattractant factors. These changes altogether have direct effect on the recruitment of T_{eff} cells across the inflamed BBB into the CNS parenchyma.

TNF-α is most prominently expressed by macrophages but also by many other cell types such as neutrophils, eosinophils, T and B cells, and neurons [94]. TNF-α exists as a membrane bound and as a secreted form [6, 113] and mediates its biological activities through two different receptors, TNFR1 and TNFR2 [5, 113]. While TNFR1 is widely expressed within all cell types, the expression of TNFR2 within the brain is limited to immune cells, endothelial cells, microglia, and neurons. The involvement of TNF-α in MS and EAE pathogenesis has been reported in many different clinical and preclinical studies, respectively. In MS patients TNF-α is significantly elevated in serum [65, 109, 161, 172] and cerebrospinal fluid (CSF) [138, 172] during different phases of disease progression. Postmortem in situ analysis demonstrated elevated TNF-α at sites of active MS lesions, where TNF-α is associated with astrocytes and macrophages [69, 113]. In the actively induced EAE mouse model of MS, TNFR1- and TNFR2-deficient C57Bl/6 mice develop attenuated EAE with less inflammatory infiltrates compared to control animals [135]. Neutralizing TNF-α with function-blocking antibodies was shown protective in EAE [137].

IL-1β is mainly produced by monocytes and macrophages and found systemically increased in the circulation during inflammation [142]. IL-1β binds to its receptors IL-1RI or IL-1RII. While IL-1RI is ubiquitously expressed, IL-1RII is mainly expressed by macrophages and B cells. The role of IL-1β during inflammation has been studied in the context of MS pathogenesis. IL-1β was shown to be elevated in serum samples of MS patients [109]. Among the NVU cells, astrocytes are considered prominent producers of IL-1β [4]. Notably, IL-1β synthesis by astrocytes can be induced in a paracrine fashion by TNF-α in a co-culture setup in vitro [48]. Moreover, IL-1β has a direct effect on the organization of TJs and increases BBB permeability in vitro [1, 73]. It has been reported that C57Bl/6 mice lacking IL-1RI are resistant to MOG-induced active EAE suggesting a crucial role for IL-1β to promote the activity of encephalitogenic T cells [150]. Another study showed that mice deficient for IL-1α and IL-1β exhibited significant resistance to EAE, probably due to a failure in inducing Th17 response, whereas single knockout animals did not [110]. On the other hand, serum level of the naturally occurring IL-1R antagonist IL-1Ra was shown to be elevated during MS exacerbation or in response to IFN-β treatment of MS patients [119].

Several inflammatory cell subsets including T and B lymphocytes, NK, and antigen-presenting cells secrete IFN-γ, which binds to its ubiquitously expressed IFN-γ receptor. Co-stimulation of TNF-α with IFN-γ increases peripheral vascular permeability by disrupting junctional distribution of some molecules such as junctional adhesion molecules (JAMs) and vascular endothelial (VE)-cadherin [123, 175]. IFN-γ is a signature of $CD4^+$ Th1 cells that are believed to be the driving force for autoimmunity in MS and EAE. Indeed, IFN-γ and other Th1 cytokines such as IL-12 and IL-2 are highly upregulated in serum of MS patients [160]. Also, IFN-γ is detected in chronic active lesions in MS patients. While IFN-γ plays a defined role in the progression of MS, type I IFNs (IFN-α and IFN-β) exert immunosuppressive effects. Noteworthy, the role of IFN-γ in MS remains vague due to its opposing function as proinflammatory or protective cytokine [11, 122].

4 Molecules Involved in Immune Cell Extravasation During Inflammation

Immune cell extravasation is tightly regulated in terms of immune cell subsets, target organ, and inflammatory condition. Accordingly, the molecular key players involved in this process need to be tightly regulated with respect to availability and activation status. Selectins, G-protein-coupled receptors (GPCRs), chemokines, integrins, and immunoglobulin-like (Ig-like) cell adhesion molecule (CAM) protein families take center stage in the cross talk and adhesive interactions between immune cells and endothelial cells. Additional molecules of different character such as VAP-1 and Ninjurin-1 were described to take part in extravasation of some leukocyte subsets across the BBB [74, 84, 124]. In this chapter, we will introduce the molecules that fulfill critical roles in immune cell extravasation across an inflamed endothelial layer and in particular across the inflamed BBB.

4.1 Selectins and Selectin Ligands

Selectins belong to the family of type I transmembrane glycoproteins, including L-, E-, and P-selectin, reviewed in [91, 114]. The basic structure of selectins consists of a long extracellular N-terminus, a transmembrane domain, and a short cytoplasmic tail. The N-terminal domain bears a Ca^{2+}-dependent (C-type) lectin domain fused to an epidermal growth factor (EGF)-like domain and two to nine consensus repeats with homology to the complementary regulatory (CR) proteins. The lectin domain binds sugar moieties and is more conserved among the three selectins. L-selectin is expressed by granulocytes, monocytes, and most lymphocytes, whereas E-/P--selectins are expressed by inflamed endothelial cells. At the level of meningeal microvascular endothelial cells, P-selectin is stored in Weibel-Palade bodies and can be rapidly translocated to the endothelial surface upon inflammatory stimulus. In contrast, parenchymal endothelial cells of the BBB do not contain an intracellular reservoir of P-selectin and require de novo synthesis upon inflammation [49, 51, 53].

Although several ligand candidates for selectins exist, P-selectin glycoprotein ligand (PSGL)-1 is the most studied selectin ligand. PSGL-1 exists as homodimer through a stabilized interaction of the transmembrane and cytoplasmic domains by juxtamembrane disulfide bonds. The transmembrane domain is composed of an extracellular N-terminal end bearing tyrosine sulfates, followed by a long glycoprotein backbone with many O-linked carbohydrates. Importantly, the critical O-glycans are located at threonine 57 near the N-terminal end [98]. The tetrasaccharide sialyl Lewis X (sLe^x) motif located within the O-linked carbohydrates is the binding recognition site of selectins. PSGL-1-to-P-selectin binding is enhanced by sulfated tyrosines, fucose, sialic acid, and galactose on a single core 2 O-glycan near N-terminus [98]. Notably, naïve $CD4^+$ T cells express nonfunctional PSGL-1 and fail to roll on P-selectin [92]. Upon activation, PSGL-1 on $CD4^+$ Th1 cells but not on Th2 cells becomes functional and supports rolling on inflamed endothelial cells [12, 24, 68, 176]

4.2 Chemokines and Chemokine Receptors

Chemokines are chemotactic cytokines that constitute a family of small proteins [15, 154]. The chemokines are produced and secreted by various cells, including stromal cells, fibroblasts, adipocytes, neural cells, and various leukocytes such as dendritic cells, monocytes, macrophages, and T helper and effector cells, and fulfill a multitude of essential roles in development and homeostasis of the body [37]. In inflammation, chemokines are key players as they orchestrate leukocyte trafficking by controlling adhesion and chemotaxis. To date, more than 50 chemokines and nearly 20 chemokine receptors have been identified [64]. Based on the number and position of highly conserved cysteines in the proteins, chemokines are subdivided into CXC-, CC-, CX_3C- and C-chemokines, where C represents the cysteine and X any other amino acid(s) in between. With the exception of C-chemokines that possess only two conserved cysteines, all others have four conserved cysteines forming disulfide bridges, which determine the tertiary structure of the protein. For immune cell trafficking, chemokines can be presented on the luminal face of endothelial cells, reside in vesicles of endothelial cells, or be deposited on the abluminal side of endothelial cells. For luminal presentation, chemokines bind to glycosaminoglycans (GAGs) that are part of the carbohydrate-rich endothelial glycocalyx [131]. A network of proteoglycans and glycoproteins forms the backbone and connect the glycocalyx to the endothelium. Chemokine presentation by the vascular cell surface via binding to GAG of the glycocalyx is a limiting step for chemokine-triggered leukocyte recruitment [128, 149].

Chemokines bind to chemokine receptors, which are transmembrane proteins of the rhodopsin subfamily of GPCRs [82]. Chemokine receptors either belong to the conventional GPCRs or to the atypical chemokine receptors (ACKRs) [163]. The conventional GPCR chemokine receptors include in humans at least 19 members and are classified according to the chemokines recognized into CXC-, CC-, CX_3C-, and C-chemokine receptors. The ACKRs include at least four members [13]. They were assigned "atypical" due to alternative signaling pathways compared to the conventional chemokine receptors. In general, heterotrimeric G proteins can be activated by GPCRs upon ligand engagement and thereby act as molecular switches that turn on intracellular signaling transduction cascades [121]. Pertussis toxin, a well-known compound that sequesters the Gαi subunit and therefore often used to interfere with conventional GPCR signaling in the analysis of immune cell trafficking, might not interfere with signaling induced through ACKRs [170].

Several chemokines are involved in T_{eff} cell extravasation across the BBB [70]. The lymphoid chemokines, CCL19 and CCL21, are upregulated in BBB endothelial cells during EAE [8, 41, 85]. Apparently, CCL21 is restricted to inflamed blood vessels, whereas CCL19 is expressed by leukocytes, astrocytes, and microglia [41]. Additionally, CCR7-positive T_{eff} cells, receptor for CCL19 and CCL21, are enriched within the inflammatory cuffs in the brain and spinal cord of EAE mice [8]. Furthermore, CXCL12 expression levels are elevated in the CSF of MS patients [80]. In mice afflicted with EAE, CXCL12 is elevated in the spinal cord at the peak of disease [112]. Importantly, the polarized abluminal expression of CXCL12 in the healthy CNS is altered during neuroinflammation. Hence, T_{eff} cells are no longer entrapped in the perivascular space and infiltrate the CNS during EAE or MS [111, 112]. Similar to homeostatic chemokines, several inflammatory chemokines play a role in the BBB endothelial cells during inflammation such as CCL2, CCL4, and

CCL5. Indeed, these chemokines diffuse toward the luminal surface of BBB endothelial cells during inflammation and contribute to T cell recruitment [129]. Furthermore, mice lacking CCR2 [57, 76] or CCL2 [71] were resistant to EAE development upon active immunization, yet adoptively transferred wild-type MOG_{35-55}-specific T_{eff} cells into CCR2 [57] or CCL2 [71] knockout mice, respectively, failed to induce disease. Latter indicates that the expression of CCR2 or CCL2 in host-derived cells might be central for EAE induction [57, 71]. Additionally, a selective expression for CCR1 was detected at the peak of EAE but lost during subsequent relapsing phase of EAE [58]. In line, mice treated with anti-CCL3, ligand for CCR1, showed reduced accumulation of $CD4^{+}$ T_{eff} cells in the CNS [58]. Taken together, chemokines within the NVU and their chemokine receptors on the immune cells essentially contribute to both homeostatic immune cell trafficking and pathological immune cell recruitment during neuroinflammation.

4.3 Integrins

Integrins are cell adhesion receptors that play pivotal roles in multiple cellular processes. They exist as heterodimeric glycoproteins formed by non-covalently linked α and β subunits [99]. In total 18 α-integrin and 8 β-integrin subunits form 24 αβ integrin heterodimers [29, 99]. Each integrin subunit consists of a large extracellular part, a single transmembrane domain, and a short cytoplasmic tail [55]. Some integrins carry a unique ligand-binding site of ~200 amino acids in their α-subunit, the so-called inserted (I) domain, and engage their ligands through this α-integrin I domain. Integrins can adopt three different conformations: bent with closed headpiece, extended with closed headpiece, and extended with open headpiece. These conformations allow binding of the ligand with low, intermediate, or high affinity, respectively [143]. They are named "integrin" because they integrate the intracellular and extracellular environments together. The intracellular domains of the integrin heterodimer bind to cytoskeletal adapter proteins, whereas the extracellular domains form the binding site to proteins of the extracellular matrix (ECM) or to cell adhesion molecules on the surface of other cells. Key integrin molecules involved in immune cell trafficking are lymphocyte function-associated antigen (LFA)-1 (αLβ2, CD11a/CD18), macrophage-1 antigen (Mac)-1 (αMβ2, CD11b/CD18), very late antigen (VLA)-4 (α4β1, CD49d/CD29), and α4β7.

Signaling of integrins across the plasma membrane is bidirectional. Outside-in signaling is induced through integrin binding to its counterpart and transduces signaling into the cell. Inside-out signaling is initiated elsewhere in the cell and results into a change of integrin conformation [139]. Intracellular integrin adaptor molecules are essential for both kinds of signaling pathways [77, 83, 139]. In immune cell trafficking, the adaptor proteins Talin-1 and Kindlin-3 binding to the integrin β-subunit play important roles in the induction and stabilization of integrin high-affinity conformation [43, 60, 90, 118, 151].

4.4 Immunoglobulin-Like Cell Adhesion Molecules

Immunoglobulin-like cell adhesion molecules (Ig-like CAMs) are part of the immunoglobulin superfamily (IgSF) that consists of various groups of cell surface or secreted molecules and fulfill various functions in innate and adaptive immunity [18, 30]. A signature structure of IgSF members is the presence of one or more Ig-like domains. Ig-like CAMs are type I transmembrane glycoproteins characterized by an N-terminal extracellular domain, a single spanning transmembrane domain, and a short C-terminal cytoplasmic tail. They mediate cell-cell adhesion and outside-in signal transduction [10]. Luminal expression of intercellular adhesion molecule (ICAM)-1, ICAM-2, and vascular cell adhesion molecule (VCAM)-1 on the inflamed endothelium is central in T_{eff} cell extravasation across the BBB under neuroinflammatory conditions [62].

The extracellular domains of ICAM-1 (CD54) and ICAM-2 (CD102) contain five and two Ig-like domains, respectively [101]. The short cytoplasmic domains of ICAM-1 and ICAM-2 harbor 28 and 26 amino acids, respectively. LFA-1 is the T_{eff} cell integrin binding to the first extracellular Ig-domains of ICAM-1 and ICAM-2 [47, 103, 167]. Additionally, ICAM-1 is a ligand for the integrin Mac-1. Some reports ascribe a role to Mac-1 in binding ICAM-2, while others fail to identify interaction between Mac-1 and ICAM-2 [101]. ICAM-1 is uniformly localized to the surface of endothelial cells, while ICAM-2 is localized to both the junctions and the surface of endothelial cells [72, 155]. Of note, within the CNS ICAM-1 is also expressed on non-endothelial cells such as leukocytes, pericytes, epithelial cells, and glial cells [130]. On in vitro cultured BBB endothelial cells of the mouse, ICAM-2 is constitutively expressed, whereas the level of ICAM-1 is low under homeostatic conditions but strongly upregulated upon cytokine stimulation [1, 147]. In MS and or EAE lesions, high levels of ICAM-1 were detected in inflamed CNS microvessels, microglia, and astrocytes [89, 126, 146] and accompanied by LFA-1-positive infiltrating mononuclear cells [144].

VCAM-1 (CD106) is made of six or seven extracellular Ig-like domains, a transmembrane domain, and a short cytoplasmic tail of 19 amino acids [42]. VCAM-1 is the endothelial ligand for α4-integrins on the leukocytes, which binds to the first and the fourth Ig-like domains of VCAM-1 [81]. Only few brain capillaries in vivo or scattered BBB endothelial cells in vitro express VCAM-1 under unstimulated conditions. Upon inflammation VCAM-1 is strongly upregulated in vivo and in vitro on BBB endothelial cells [45, 146, 147]. VCAM-1 interaction with VLA-4 or α4β7 plays a crucial role in lymphocyte extravasation across the inflamed BBB in vivo and cytokine-stimulated human umbilical vein endothelial cells (HUVECs) in vitro [32, 44, 88].

Some additional Ig-like CAMs were identified contributing to extravasation of encephalitogenic immune cell subsets across the inflamed BBB. Namely, these are ALCAM (CD166) [35], junctional adhesion molecule-like (JAM-L) [9], and melanoma cell adhesion molecule (MCAM, CD146) [59, 86, 87]. The extracellular domains of ALCAM and MCAM are composed of five Ig-like domains, while the extracellular domain of JAM-L is composed of two Ig-like domains. In particular, endothelial ALCAM on the inflamed BBB was suggested as a BBB-specific and immune cell subset selective trafficking cue [35]. ALCAM can undergo homophilic interactions or act as a ligand for CD6, a molecule present on T_{eff} cells and involved in the formation of the immune synapse between lymphocytes and antigen-presenting cells [25, 153, 179]. Since information on the precise step of immune cell extravasation across the inflamed BBB supported by ALCAM, MCAM, or JAM-L is scarce, their molecular function will not be discussed at present.

5 Trafficking Laws: How T_{eff} Cells Breach the Inflamed BBB?

Immune cells use the blood vasculature as a freeway to continuously circulate through the body. During neuroinflammation T_{eff} cells migrate into the CNS. To this end, they have to cross one of the barriers of the CNS. Extravasation across the inflamed BBB occurs at the level of postcapillary venules, where shear flow is minimal. Importantly, extravasation across the BBB follows sequential events of cell-cell adhesion and cell-to-cell signaling between T_{eff} lymphocytes and the inflamed endothelial cells of the BBB. The individual steps are named according to the behavior of the immune cell. The initial contact formation is mediated through capture and rolling. In the next step, integrin molecules on the surface of the immune cell become activated. This allows the immune cell to fully withstand shear exerted by the blood flow for sustained firm adhesion. Importantly, the event of integrin activation should not be mixed up with the stimulation of T_{eff} cells by APCs with their cognate antigen, as both processes have been referred to as T_{eff} cell activation but differ fundamentally. Finally, T_{eff} cell diapedesis is initiated by crossing the endothelial monolayer of the BBB and its endothelial basement membrane. At this point the T_{eff} cells are entrapped in the perivascular space. Breaching of the parenchymal basement membrane only occurs after successful restimulation by APCs. In the previous section, we introduced the molecules involved in the process of leukocyte trafficking. Now, we will delineate these individual steps of extravasation and describe the role of each molecule with a specific focus on T_{eff} cell extravasation across the inflamed BBB.

5.1 *Capture and Rolling*

The initial transient contact between circulating T_{eff} cells and the inflamed BBB occurs through capture and rolling. Capture can be described as an abrupt stop of the T_{eff} cell and is mediated through VLA-4 binding to the high level of VCAM-1 on the inflamed CNS microvessels [22, 88, 146]. Capture is a process unique to the

BBB and has also been observed under noninflamed conditions of the BBB [164]. Rolling of T_{eff} cells along the luminal face of the inflamed BBB occurs at a speed of 6–7 μm/s in inflamed leptomeningeal vessels [126] and 12–18 μm/s in inflamed spinal cord vessels [21, 134]. Latter is remarkably reduced compared to the blood flow in inflamed CNS microvessels equaling 1000 μm/s [40, 134]. At the molecular level, capture is mediated through VLA-4 binding to VCAM-1 on the BBB [40, 164]. Rolling is specifically mediated through binding of leukocyte PSGL-1 to endothelial E- and P-selectin [78, 79, 134]. Remarkably, absence of E- and P-selectin and PSGL-1 in a triple knockout mouse model does not ameliorate EAE pathogenesis even though T_{eff} cell rolling is completely absent [78, 79, 134]. Despite their relative low number, captured T_{eff} cells rather seem sufficient to induce disease.

5.2 Integrin Activation

After initial transient contact, T_{eff} cells form stronger interactions with the luminal face of the inflamed BBB leading to full shear-resistant arrest. In vivo live cell imaging showed complete abrogation of this step when the T_{eff} cells were treated with pertussis toxin, which blocks GPCR signaling [164]. In contrast, recent in vitro studies showed pertussis toxin-independent shear-resistant arrest and sustained adhesion of freshly stimulated T_{eff} cells on inflamed endothelial cells [132, 141, 149]. Moreover, stimulated T_{eff} cells were found to arrest on recombinant ICAM-1 in the absence of chemokines [147]. Thus, the role of chemotactic factors inducing a signaling cascade that results in a fast activation of preformed integrins on the T_{eff} cell warrants further investigation. Regulation of LFA-1 and VLA-4 integrin conformation in T_{eff} cells is achieved by the cytoplasmic adaptor proteins Talin-1 and Kindlin-3 [83]. In particular, Kindlin-3-regulated LFA-1 and VLA-4 activation is critically involved in the extravasation of T_{eff} cells across the inflamed BBB as proven by the lack of EAE development after adoptive transfer of Kindlin-3 mutant T_{eff} cells [117]. Thus, activated integrins and in particular the tight regulation of activation by intracellular adaptors in T_{eff} cells take center stage in the interplay with the inflamed BBB.

5.3 Firm Adhesion

Successful shear-resistant arrest of the T_{eff} cells transforms into firm adhesion. The critical role of LFA-1 and VLA-4 and their endothelial ligands ICAM-1, ICAM-2, and VCAM-1 in T_{eff} cell firm adhesion to the inflamed BBB has been confirmed by imaging experiments in vitro and in vivo [20, 126, 147]. A detailed in vitro analysis revealed that shear-resistant arrest of the T_{eff} cells is mediated through either ICAM-1 or VCAM-1, whereas their sustained adhesion relies on endothelial ICAM-1 and ICAM-2 [147]. Genetic or functional ablation of VCAM-1/VLA-4 or LFA-1/

ICAM-1 and ICAM-2 demonstrated amelioration or even abrogation of EAE pathogenesis [1, 27, 28, 50, 130, 178]. In conclusion, the high level of ICAM-1 and VCAM-1 on the inflamed BBB must be attributed important roles in T_{eff} cell recruitment into the CNS.

While adherent to the luminal face of the inflamed BBB, T_{eff} cells behave highly dynamic: Within seconds after arrest, T_{eff} cells spread and polarize and then crawl with preferential direction against the blood flow [20, 147]. Crawling of T_{eff} cells requires endothelial ICAM-1 or ICAM-2 [147, 165]. Lymphocytes display crawling distances well above 100 μm in vivo during onset of EAE in leptomeningeal microvessels and in vitro on TNF-α-stimulated BBB endothelial cells [1, 20, 147]. The exceptionally long crawling distances covered by T_{eff} cells is a remarkable characteristic of extravasation across barrier-forming CNS microvascular endothelial cells when compared to non-barrier-forming endothelial cells [148]. Importantly, variants of the BBB inflammatory conditions with an increased level of ICAM-1 coincided with a reduced crawling speed and distance of T_{eff} cells and earlier initiation of diapedesis [1]. Thus, the adhesion molecule profile of the inflamed BBB tightly controls T_{eff} cell shear-resistant arrest, firm adhesion, and the dynamic interaction behavior of adherent T_{eff} cells.

5.4 *Diapedesis*

Diapedesis of T_{eff} cells, but also of other leukocyte subsets, across an endothelial layer occurs either through the endothelial cell-cell junction, via the paracellular pathway, or through a pore formed across the endothelial cell body itself, via the transcellular pathway [1, 31, 33, 100, 108, 120]. It is suggested that paracellular diapedesis occurs through a zipper-like replacement of homophilic interactions of junctional proteins. In contrast, transcellular diapedesis of T_{eff} cells presumably leaves the endothelial junction integrity intact but might still involve junctional molecules recruited to the transcellular pore [31]. With respect to the inflamed BBB, electron microscopic studies strongly supported predominant transcellular diapedesis of mononuclear immune cell subsets (monocytes, T cells) in situ across the inflamed BBB [96, 97, 174] or inflamed blood-retina barrier (BRB) [16, 63].

In vitro live cell imaging revealed formation of invasive protrusions by T_{eff} cells during crawling on cytokine-stimulated BBB endothelium (Fig. 1) [1] and non-BBB endothelial cells [33, 140]. Invasive protrusions contact the apical face of the endothelial cells, form a pore across the endothelial cell proper, extend toward the endothelial cell abluminal face, and, presumably, precede both trans- and paracellular diapedesis. It is hypothesized that T_{eff} cells probe the endothelial cell surface to detect permissive sites for diapedesis [31, 100, 120, 145]. Importantly, a mild but significant loss of barrier properties in IL-1β-stimulated compared to TNF-α-stimulated BBB endothelial cells did not simply translate into increased paracellular diapedesis. Rather, increased levels of ICAM-1 on the surface of the

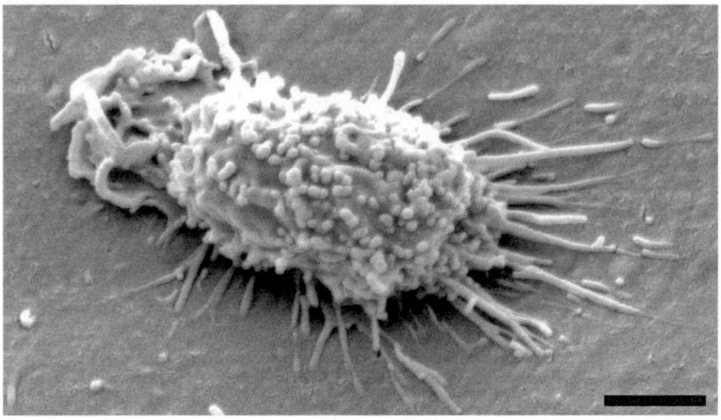

Fig. 1 Ultrastructural analysis of a CD4$^+$ T_{eff} cell adherent and crawling on IL-1β-stimulated primary mouse brain microvascular endothelial cells by scanning electron microscopy. Note the filopodia-like protrusions that form contact between the crawling CD4$^+$ T_{eff} cell and endothelial cell surface. Scale bar equals 2 μm

IL-1β-stimulated BBB endothelial cells correlated with shorter T_{eff} cell crawling distances, increased numbers of diapedesis events, and preferential diapedesis via the transcellular pathway [1]. On the other hand, lack of endothelial ICAM-1 and ICAM-2 on cytokine-stimulated BBB endothelial cells derived from ICAM-1^{null}/ICAM-2$^{-/-}$mice revealed a reduced number and length of invasive protrusions formed by the T_{eff} cell [1]. Unexpectedly, in the rare event of T_{eff} cell diapedesis in the absence of ICAM-1 and ICAM-2 and therefore in the absence of T cell polarization and crawling, transmigration occurred mainly via the transcellular pathway [1]. Taken together, endothelial ICAM-1 together with ICAM-2 fulfills a critical role in the regulation of the rate and the pathway of T_{eff} cell diapedesis across the inflamed BBB.

6 Endothelial Signaling Pathways in Immune Cell Extravasation Across the BBB

We have seen in the previous section that inflamed endothelium of the BBB plays an active role in orchestrating immune cell recruitment through the expression of a variety of CAMs and presentation of chemokines [1, 36, 66]. The active role of endothelial cells during diapedesis has many faces including junctional disassembly, cytoskeletal reorganization, formation of docking structures, and transcellular pores and likewise important the fast closure of the pore after diapedesis. Furthermore, several endothelial CAM mediating outside-in signaling events contribute to leukocyte extravasation. Many findings presented here have been acquired by in vitro imaging or biochemical analysis and therefore are most often based on in vitro cultured endothelial cells. Due to its outstanding importance for extravasation, aspects of ICAM-1-mediated downstream signaling are highlighted.

6.1 ICAM-1 Downstream Signaling Is a Key Event in Leukocyte Diapedesis

Upon leukocyte adhesion, the interaction between ICAM-1 and LFA-1 leads to ICAM-1 clustering on the surface of human umbilical vein endothelial cells (HUVECs) forming docking structures, also referred to as transmigratory cup [19, 34]. Analysis of transfected mouse brain endothelioma cell lines revealed that the cytoplasmic tail of endothelial ICAM-1 and downstream Rho-GTPase signaling are involved in the diapedesis but not adhesion of T_{eff} cells [102]. Experimental clustering of endothelial ICAM-1 induces downstream signaling events, such as the activation of small Rho-GTPases, activation of endothelial nitric oxide synthase, phosphorylation of VE-cadherin, phosphorylation of phosphoinositide-specific phospholipase C (PLC)-γ1, production of phosphatidylinositol phosphate, and increased intracellular calcium level [62, 106]. Increased intracellular calcium in turn activates myosin light-chain kinase and actin cytoskeletal cell contractility and affects endothelial barrier properties [157, 177]. Thus, ICAM-1 induced signaling harbors many important features for successful diapedesis of various leukocyte subsets across vascular endothelial cells.

6.2 VE-Cadherin Phosphorylation Tightly Regulates Immune Cell Extravasation

The adherens junction protein VE-cadherin is considered essential for maintaining endothelial cell-to-cell integrity [168]. Antibodies against VE-cadherin disrupt endothelial junctions and lead to enhanced leukocyte recruitment into inflamed mouse peritoneum in vivo [61]. Replacement of endogenous VE-cadherin by VE-cadherin/α-catenin, a non-dissociable fusion construct, led to stabilized junctions that resisted permeability increase during inflammation and reduced leukocyte recruitment in vivo [136]. Distinct phosphorylation pattern of tyrosine residues in the VE-cadherin cytoplasmic tail has been identified for the regulation of permeability changes of the endothelial layer or leukocyte diapedesis [7, 169, 173]. Some conflicting data on the distinct VE-cadherin phosphorylation pattern affecting leukocyte diapedesis might be caused by variations in the specificity of antibody epitopes [168]. Again, endothelial ICAM-1 engagement might be involved in the tyrosine phosphorylation of VE-cadherin through the non-receptor tyrosine kinase Src [166]. However, this important link between endothelial ICAM-1 and VE-cadherin phosphorylation awaits further proof particular in the context of inflamed BBB endothelial cells. Taken together, the dynamic spatiotemporal association between VE-cadherin and its intracellular adaptor molecules depends on distinct VE-cadherin phosphorylation patterns and is critical for changes of vascular permeability during inflammation and, in consequence, is a prime candidate for the regulation of paracellular diapedesis.

6.3 Pore Formation

T_{eff} cells probe the endothelial monolayer by sending long invasive protrusions to sense a site permissive for diapedesis [1, 33, 141]. As outlined before, T_{eff} cell diapedesis across a tight in vitro model of the BBB is via both routes, the paracellular and the transcellular pathway. Transcellular pores for T_{eff} cell diapedesis range from 3 to 6 μm and are referred to as micro-wounds [107, 108]. Mechanistically, endothelial cells form ventral lamellipodia enriched in integrins α5 and β3 that initiate at the abluminal side and then grow across the pores and close them to reestablish endothelial cell integrity. F-actin accumulates around the pore and contracts or forms at the one side of the pore and then closes the pore by ventral movement. Thus, actin remodeling in the ventral lamellipodium is the key for healing of diapedesis-induced micro-wounds, which is a fast process that takes only 5–10 min in vitro. Remarkably, only one third of inflammatory cuffs in the brain of EAE mice show a compromised barrier through leakage of a small molecular tracer in vivo [125]. The fast closure of the pore after T_{eff} cell diapedesis might represent one important mechanism to restore BBB integrity during inflammatory T_{eff} cell recruitment into the CNS.

7 Conclusion

In recent years various aspects of T_{eff} cell migration across the inflamed BBB have been resolved. Despite similarities to the multistep cascade of immune cell extravasation in the periphery, extravasation of T_{eff} cells across the inflamed BBB displays striking peculiarities. Multiple steps follow each other in a tightly regulated and concerted action to guide the extravasation of immune cells from the circulation into the CNS parenchyma in a highly dynamic process. Importantly, the inflamed BBB actively contributes to this process and ensures maintenance of barrier integrity to its best.

Pharmaceutical targeting of immune cell extravasation to prevent or ameliorate neuroinflammation has been matter of intense research since the seminal finding of the critical role of VCAM-1/α4-integrin as trafficking cue in the development of EAE in the 1990s [23, 178]. Many important studies in the field finally translated into a humanized monoclonal blocking antibody named natalizumab and marketed as Tysabri®, which is applied to reduce relapse rates of MS patients in the clinic today [52]. In contrast, all clinical trials targeting ICAM-1 and ICAM-2/LFA-1 interaction with function-blocking antibodies, namely, Hu23f2G, enlimomab, and efalizumab, aiming to reduce neutrophil infiltration in ischemic stroke and recruitment of encephalitogenic immune cell subsets, respectively, remained controversial [39]. Solely, the humanized antibody efalizumab, targeting the αL-integrin, was approved for treatment of psoriasis patients [152].

A different strategy to block infiltration of immune cells into the CNS is achieved through dampening of the inflammatory condition of the BBB with statins, which are inhibitors of the 3-hydroxy-3-methyl-glutaryl-coenzyme (HMG-Co) A reductase. HMG-CoA is involved in cholesterol synthesis pathway but presumably exerts its immunomodulatory and anti-inflammatory effect rather by interfering with prenylation and thus blockade of Rho-GTPase signaling [38]. In fact, a beneficial effect of statins for a panel of neurological diseases with inflammatory component is under investigation and might result in new therapeutic uses [104]. A third variant to target trafficking of encephalitogenic immune cells was discovered through targeting the Sphingosine-1-phosphate (S1P) receptor with its functional antagonist FTY720 or fingolimod. FTY720 was found to significantly decrease the rate of relapses in relapsing-remitting MS patients. Originally, its effect was explained by a blockade of lymphocyte egress from secondary lymphoid tissues. However, recent studies suggested additional immunomodulatory effects in the treatment of inflammatory diseases of the CNS and other organs [162]. In fact, preclinical investigations in the mouse model revealed protection of the BBB barrier properties and dampening of VCAM-1 and ICAM-1 expression by FTY720 [75].

Unfortunately, in rare cases treatment with natalizumab or FTY720 correlated with severe side effects, i.e., the development of progressive multifocal leukoencephalopathy (PML) [56]. Thus, continued research is required to further elucidate the elaborate process of immune cell migration across the inflamed BBB. Only distinguished knowledge will allow the design of therapeutics that take advantage of unique properties of immune cell extravasation across the inflamed BBB without harmful side effects elsewhere or suppression of homeostatic immune surveillance.

Acknowledgment We thankfully acknowledge financial support by the Swiss Multiple Sclerosis Society, Zürich, Switzerland, and the Foundation for Clinical and Experimental Cancer Research, Bern, Switzerland, to RL. MA was supported by Swiss National Science Foundation Early Postdoc Mobility Fellowship. We are grateful to Gaby Enzmann for critical reading of the manuscript.

References

1. Abadier M, Haghayegh Jahromi N, Cardoso Alves L, Boscacci R, Vestweber D, Barnum S, Deutsch U, Engelhardt B, Lyck R (2015) Cell surface levels of endothelial ICAM-1 influence the transcellular or paracellular T-cell diapedesis across the blood–brain barrier. Eur J Immunol 45(4):1043–1058. doi:10.1002/eji.201445125
2. Abbott NJ (2013) Blood–brain barrier structure and function and the challenges for CNS drug delivery. J Inherit Metab Dis 36(3):437–449. doi:10.1007/s10545-013-9608-0
3. Abbott NJ, Patabendige AA, Dolman DE, Yusof SR, Begley DJ (2010) Structure and function of the blood–brain barrier. Neurobiol Dis 37(1):13–25
4. Abbott NJ, Ronnback L, Hansson E (2006) Astrocyte-endothelial interactions at the blood–brain barrier. Nat Rev 7(1):41–53
5. Aggarwal BB (2003) Signalling pathways of the TNF superfamily: a double-edged sword. Nat Rev Immunol 3(9):745–756. doi:10.1038/nri1184

6. Aggarwal BB, Gupta SC, Kim JH (2012) Historical perspectives on tumor necrosis factor and its superfamily: 25 years later, a golden journey. Blood 119(3):651–665. doi:10.1182/blood-2011-04-325225
7. Allingham MJ, van Buul JD, Burridge K (2007) ICAM-1-mediated, Src- and Pyk2-dependent vascular endothelial cadherin tyrosine phosphorylation is required for leukocyte transendothelial migration. J Immunol 179(6):4053–4064
8. Alt C, Laschinger M, Engelhardt B (2002) Functional expression of the lymphoid chemokines CCL19 (ELC) and CCL 21 (SLC) at the blood–brain barrier suggests their involvement in G-protein-dependent lymphocyte recruitment into the central nervous system during experimental autoimmune encephalomyelitis. Eur J Immunol 32(8):2133–2144
9. Alvarez JI, Kebir H, Cheslow L, Chabarati M, Larochelle C, Prat A (2015) JAML mediates monocyte and CD8 T cell migration across the brain endothelium. Ann Clin Transl Neurol 2(11):1032–1037. doi:10.1002/acn3.255
10. Aplin AE, Howe A, Alahari SK, Juliano RL (1998) Signal transduction and signal modulation by cell adhesion receptors: the role of integrins, cadherins, immunoglobulin-cell adhesion molecules, and selectins. Pharmacol Rev 50(2):197–263
11. Arellano G, Ottum PA, Reyes LI, Burgos PI, Naves R (2015) Stage-specific role of interferon-gamma in experimental autoimmune encephalomyelitis and multiple sclerosis. Front Immunol 6:492. doi:10.3389/fimmu.2015.00492
12. Austrup F, Vestweber D, Borges E, Löhning M, Bräuer R, Herz U, Renz H, Hallman R, Scheffold A, Radbruch A, Hamann A (1997) P- and E-selectin mediate recruitment of T-helper-1 but not T-helper-2 cells into inflamed tissues. Nature 385:81–83
13. Bachelerie F, Graham GJ, Locati M, Mantovani A, Murphy PM, Nibbs R, Rot A, Sozzani S, Thelen M (2015) An atypical addition to the chemokine receptor nomenclature: IUPHAR Review 15. Br J Pharmacol 172(16):3945–3949. doi:10.1111/bph.13182
14. Baeten KM, Akassoglou K (2011) Extracellular matrix and matrix receptors in blood–brain barrier formation and stroke. Dev Neurobiol 71(11):1018–1039. doi:10.1002/dneu.20954
15. Baggiolini M (2001) Chemokines in pathology and medicine. J Intern Med 250(2):91–104
16. Bamforth SD, Lightman SL, Greenwood J (1997) Ultrastructural analysis of interleukin-1 beta-induced leukocyte recruitment to the rat retina. Invest Ophthalmol Vis Sci 38(1):25–35
17. Banks WA (2015) The blood–brain barrier in neuroimmunology: tales of separation and assimilation. Brain Behav Immun 44:1–8. doi:10.1016/j.bbi.2014.08.007
18. Barclay AN (2003) Membrane proteins with immunoglobulin-like domains–a master superfamily of interaction molecules. Semin Immunol 15(4):215–223
19. Barreiro O, Yanez-Mo M, Serrador JM, Montoya MC, Vicente-Manzanares M, Tejedor R, Furthmayr H, Sanchez-Madrid F (2002) Dynamic interaction of VCAM-1 and ICAM-1 with moesin and ezrin in a novel endothelial docking structure for adherent leukocytes. J Cell Biol 157(7):1233–1245
20. Bartholomaus I, Kawakami N, Odoardi F, Schlager C, Miljkovic D, Ellwart JW, Klinkert WE, Flugel-Koch C, Issekutz TB, Wekerle H, Flugel A (2009) Effector T cell interactions with meningeal vascular structures in nascent autoimmune CNS lesions. Nature 462:94–98
21. Battistini L, Piccio L, Rossi B, Bach S, Galgani S, Gasperini C, Ottoboni L, Ciabini D, Caramia MD, Bernardi G, Laudanna C, Scarpini E, McEver RP, Butcher EC, Borsellino G, Constantin G (2003) CD8+ T cells from patients with acute multiple sclerosis display selective increase of adhesiveness in brain venules: a critical role for P-selectin glycoprotein ligand-1. Blood 101(12):4775–4782. doi:10.1182/blood-2002-10-3309
22. Bauer M, Brakebusch C, Coisne C, Sixt M, Wekerle H, Engelhardt B, Fassler R (2009) Beta1 integrins differentially control extravasation of inflammatory cell subsets into the CNS during autoimmunity. Proc Natl Acad Sci U S A 106(6):1920–1925
23. Bittner S, Wiendl H (2016) Neuroimmunotherapies targeting T cells: from pathophysiology to therapeutic applications. Neurotherapeutics 13(1):4–19. doi:10.1007/s13311-015-0405-3
24. Borges E, Tietz W, Steegmaier M, Moll T, Hallmann R, Hamann A, Vestweber D (1997) P-selectin glycoprotein ligand-1 (PSGL-1) on T helper 1 but not on T helper 2 cells binds to P-selectin and supports migration into inflamed skin. J Exp Med 185:573–578

25. Bowen MA, Patel DD, Li X, Modrell B, Malacko AR, Wang WC, Marquardt H, Neubauer M, Pesando JM, Francke U et al (1995) Cloning, mapping, and characterization of activated leukocyte-cell adhesion molecule (ALCAM), a CD6 ligand. J Exp Med 181(6):2213–2220
26. Buckwalter MS, Coleman BS, Buttini M, Barbour R, Schenk D, Games D, Seubert P, Wyss-Coray T (2006) Increased T cell recruitment to the CNS after amyloid beta 1–42 immunization in Alzheimer's mice overproducing transforming growth factor-beta 1. J Neurosci 26(44):11437–11441. doi:10.1523/JNEUROSCI.2436-06.2006
27. Bullard DC, Hu X, Crawford D, McDonald K, Ramos TN, Barnum SR (2014) Expression of a single ICAM-1 isoform on T cells is sufficient for development of experimental autoimmune encephalomyelitis. Eur J Immunol 44(4):1194–1199. doi:10.1002/eji.201344023
28. Bullard DC, Hu X, Schoeb TR, Collins RG, Beaudet AL, Barnum SR (2007) Intercellular adhesion molecule-1 expression is required on multiple cell types for the development of experimental autoimmune encephalomyelitis. J Immunol 178(2):851–857
29. Campbell ID, Humphries MJ (2011) Integrin structure, activation, and interactions. Cold Spring Harb Perspect Biol 3(3). doi:10.1101/cshperspect.a004994
30. Carlos TM, Harlan JM (1994) Leukocyte-endothelial adhesion molecules. Blood 7:2068–2101
31. Carman CV (2009) Mechanisms for transcellular diapedesis: probing and pathfinding by 'invadosome-like protrusions'. J Cell Sci 122(Pt 17):3025–3035
32. Carman CV, Jun CD, Salas A, Springer TA (2003) Endothelial cells proactively form microvilli-like membrane projections upon intercellular adhesion molecule 1 engagement of leukocyte LFA-1. J Immunol 171(11):6135–6144
33. Carman CV, Sage PT, Sciuto TE, de la Fuente MA, Geha RS, Ochs HD, Dvorak HF, Dvorak AM, Springer TA (2007) Transcellular diapedesis is initiated by invasive podosomes. Immunity 26(6):784–797
34. Carman CV, Springer TA (2004) A transmigratory cup in leukocyte diapedesis both through individual vascular endothelial cells and between them. J Cell Biol 167(2):377–388
35. Cayrol R, Wosik K, Berard JL, Dodelet-Devillers A, Ifergan I, Kebir H, Haqqani AS, Kreymborg K, Krug S, Moumdjian R, Bouthillier A, Becher B, Arbour N, David S, Stanimirovic D, Prat A (2008) Activated leukocyte cell adhesion molecule promotes leukocyte trafficking into the central nervous system. Nat Immunol 9(2):137–145
36. Chai Q, He WQ, Zhou M, Lu H, Fu ZF (2014) Enhancement of blood–brain barrier permeability and reduction of tight junction protein expression are modulated by chemokines/cytokines induced by rabies virus infection. J Virol 88(9):4698–4710. doi:10.1128/JVI.03149-13
37. Charo IF, Ransohoff RM (2006) The many roles of chemokines and chemokine receptors in inflammation. N Engl J Med 354(6):610–621. doi:10.1056/NEJMra052723
38. Ciurleo R, Bramanti P, Marino S (2014) Role of statins in the treatment of multiple sclerosis. Pharmacol Res 87:133–143. doi:10.1016/j.phrs.2014.03.004
39. Coisne C, Lyck R, Engelhardt B (2007) Therapeutic targeting of leukocyte trafficking across the blood–brain barrier. Inflamm Allergy Drug Targets 6(4):210–222
40. Coisne C, Mao W, Engelhardt B (2009) Cutting edge: natalizumab blocks adhesion but not initial contact of human T cells to the blood–brain barrier in vivo in an animal model of multiple sclerosis. J Immunol 182(10):5909–5913
41. Columba-Cabezas S, Serafini B, Ambrosini E, Aloisi F (2003) Lymphoid chemokines CCL19 and CCL21 are expressed in the central nervous system during experimental autoimmune encephalomyelitis: implications for the maintenance of chronic neuroinflammation. Brain Pathol 13(1):38–51
42. Cook-Mills JM, Marchese ME, Abdala-Valencia H (2011) Vascular cell adhesion molecule-1 expression and signaling during disease: regulation by reactive oxygen species and antioxidants. Antioxidants & redox signaling 15(6):1607–1638. doi:10.1089/ars.2010.3522
43. Critchley DR, Gingras AR (2008) Talin at a glance. J Cell Sci 121(Pt 9):1345–1347. doi:10.1242/jcs.018085
44. Crook MF, Southgate KM, Newby AC (2002) Both ICAM-1- and VCAM-1-integrin interactions are important in mediating monocyte adhesion to human saphenous vein. J Vasc Res 39(3):221–229. doi:63687

45. Deem TL, Cook-Mills JM (2004) Vascular cell adhesion molecule 1 (VCAM-1) activation of endothelial cell matrix metalloproteinases: role of reactive oxygen species. Blood 104(8):2385–2393. doi:10.1182/blood-2004-02-0665
46. del Zoppo GJ, Mabuchi T (2003) Cerebral microvessel responses to focal ischemia. J Cereb Blood Flow Metab 23(8):879–894. doi:10.1097/01.WCB.0000078322.96027.78
47. Diamond MS, Staunton DE, Marlin SD, Springer TA (1991) Binding of the integrin Mac-1 (CD11b/CD18) to the third immunoglobulin-like domain of ICAM-1 (CD54) and its regulation by glycosylation. Cell 65(6):961–971
48. Didier N, Romero IA, Creminon C, Wijkhuisen A, Grassi J, Mabondzo A (2003) Secretion of interleukin-1beta by astrocytes mediates endothelin-1 and tumour necrosis factor-alpha effects on human brain microvascular endothelial cell permeability. J Neurochem 86(1):246–254
49. Doring A, Wild M, Vestweber D, Deutsch U, Engelhardt B (2007) E- and P-selectin are not required for the development of experimental autoimmune encephalomyelitis in C57BL/6 and SJL mice. J Immunol 179(12):8470–8479
50. Engelhardt B (1998) The role of alpha 4-integrin in T lymphocyte migration into the inflamed and noninflamed central nervous system. Curr Top Microbiol Immunol 231:51–64
51. Engelhardt B (2010) T cell migration into the central nervous system during health and disease: different molecular keys allow access to different central nervous system compartments. Clin Exp Neuroimmunol 1(2):79–93. doi:10.1111/j.1759-1961.2010.009.x
52. Engelhardt B, Kappos L (2008) Natalizumab: targeting alpha4-integrins in multiple sclerosis. Neurodegener Dis 5(1):16–22. doi:10.1159/000109933
53. Engelhardt B, Ransohoff RM (2012) Capture, crawl, cross: the T cell code to breach the blood–brain barriers. Trends Immunol 33(12):579–589. doi:10.1016/j.it.2012.07.004
54. Engelhardt B, Sorokin L (2009) The blood–brain and the blood-cerebrospinal fluid barriers: function and dysfunction. Semin Immunopathol 31(4):497–511. doi:10.1007/s00281-009-0177-0
55. Evans R, Patzak I, Svensson L, De Filippo K, Jones K, McDowall A, Hogg N (2009) Integrins in immunity. J Cell Sci 122(Pt 2):215–225. doi:10.1242/jcs.019117
56. Faulkner M (2015) Risk of progressive multifocal leukoencephalopathy in patients with multiple sclerosis. Expert Opin Drug Saf 14(11):1737–1748. doi:10.1517/14740338.2015.1093620
57. Fife BT, Huffnagle GB, Kuziel WA, Karpus WJ (2000) CC chemokine receptor 2 is critical for induction of experimental autoimmune encephalomyelitis. J Exp Med 192(6):899–905
58. Fife BT, Kennedy KJ, Paniagua MC, Lukacs NW, Kunkel SL, Luster AD, Karpus WJ (2001) CXCL10 (IFN-gamma-inducible protein-10) control of encephalitogenic CD4+ T cell accumulation in the central nervous system during experimental autoimmune encephalomyelitis. J Immunol 166(12):7617–7624
59. Flanagan K, Fitzgerald K, Baker J, Regnstrom K, Gardai S, Bard F, Mocci S, Seto P, You M, Larochelle C, Prat A, Chow S, Li L, Vandevert C, Zago W, Lorenzana C, Nishioka C, Hoffman J, Botelho R, Willits C, Tanaka K, Johnston J, Yednock T (2012) Laminin-411 is a vascular ligand for MCAM and facilitates TH17 cell entry into the CNS. PLoS One 7(7):e40443. doi:10.1371/journal.pone.0040443
60. Gingras AR, Vogel KP, Steinhoff HJ, Ziegler WH, Patel B, Emsley J, Critchley DR, Roberts GC, Barsukov IL (2006) Structural and dynamic characterization of a vinculin binding site in the talin rod. Biochemistry 45(6):1805–1817. doi:10.1021/bi052136l
61. Gotsch U, Borges E, Bosse R, Böggemeyer E, Simon M, Mossmann H, Vestweber D (1997) VE-cadherin antibody accelerates neutrophil recruitment in vivo. J Cell Sci 110:583–588
62. Greenwood J, Heasman SJ, Alvarez JI, Prat A, Lyck R, Engelhardt B (2011) Review: leucocyte-endothelial cell crosstalk at the blood–brain barrier: a prerequisite for successful immune cell entry to the brain. Neuropathol Appl Neurobiol 37(1):24–39. doi:10.1111/j.1365-2990.2010.01140.x
63. Greenwood J, Howes R, Lightman S (1994) The blood-retinal barrier in experimental autoimmune uveoretinitis – leukocyte interactions and functional damage. Lab Invest 70(N1):39–52

64. Guyon A (2014) CXCL12 chemokine and GABA neurotransmitter systems crosstalk and their putative roles. Front Cell Neurosci 5:115. doi:10.3389/fncel.2014.00115
65. Hagman S, Raunio M, Rossi M, Dastidar P, Elovaara I (2011) Disease-associated inflammatory biomarker profiles in blood in different subtypes of multiple sclerosis: prospective clinical and MRI follow-up study. J Neuroimmunol 234(1–2):141–147. doi:10.1016/j.jneuroim.2011.02.009
66. Henninger DD, Panes J, Eppihimer M, Russell J, Gerritsen M, Anderson DC, Granger DN (1997) Cytokine-induced VCAM-1 and ICAM-1 expression in different organs of the mouse. J Immunol 158(4):1825–1832
67. Hickey WF (1999) Leukocyte traffic in the central nervous system: the participants and their roles. Semin Immunol 11(2):125–137
68. Hirata T, Merrill-Skoloff G, Aab M, Yang J, Furie BC, Furie B (2000) P-Selectin glycoprotein ligand 1 (PSGL-1) is a physiological ligand for E-selectin in mediating T helper 1 lymphocyte migration. J Exp Med 192(11):1669–1676
69. Hofman FM, Hinton DR, Johnson K, Merrill JE (1989) Tumor necrosis factor identified in multiple sclerosis brain. J Exp Med 170(2):607–612
70. Holman DW, Klein RS, Ransohoff RM (2011) The blood–brain barrier, chemokines and multiple sclerosis. Biochim Biophys Acta 1812(2):220–230. doi:10.1016/j.bbadis.2010.07.019
71. Huang DR, Wang J, Kivisakk P, Rollins BJ, Ransohoff RM (2001) Absence of monocyte chemoattractant protein 1 in mice leads to decreased local macrophage recruitment and antigen-specific T helper cell type 1 immune response in experimental autoimmune encephalomyelitis. J Exp Med 193(6):713–726
72. Huang MT, Larbi KY, Scheiermann C, Woodfin A, Gerwin N, Haskard DO, Nourshargh S (2006) ICAM-2 mediates neutrophil transmigration in vivo: evidence for stimulus specificity and a role in PECAM-1-independent transmigration. Blood 107:4721–4727
73. Huber JD, Witt KA, Hom S, Egleton RD, Mark KS, Davis TP (2001) Inflammatory pain alters blood–brain barrier permeability and tight junctional protein expression. Am J Physiol Heart Circ Physiol 280(3):H1241–H1248
74. Ifergan I, Kebir H, Terouz S, Alvarez JI, Lecuyer MA, Gendron S, Bourbonniere L, Dunay IR, Bouthillier A, Moumdjian R, Fontana A, Haqqani A, Klopstein A, Prinz M, Lopez-Vales R, Birchler T, Prat A (2011) Role of Ninjurin-1 in the migration of myeloid cells to central nervous system inflammatory lesions. Ann Neurol 70(5):751–763. doi:10.1002/ana.22519
75. Imeri F, Schwalm S, Lyck R, Zivkovic A, Stark H, Engelhardt B, Pfeilschifter J, Huwiler A (2016) Sphingosine kinase 2 deficient mice exhibit reduced experimental autoimmune encephalomyelitis: resistance to FTY720 but not ST-968 treatments. Neuropharmacology 105:341–350. doi:10.1016/j.neuropharm.2016.01.031
76. Izikson L, Klein RS, Charo IF, Weiner HL, Luster AD (2000) Resistance to experimental autoimmune encephalomyelitis in mice lacking the CC chemokine receptor (CCR)2. J Exp Med 192(7):1075–1080
77. Karakose E, Schiller HB, Fassler R (2010) The kindlins at a glance. J Cell Sci 123(Pt 14):2353–2356. doi:10.1242/jcs.064600
78. Kerfoot S, Kubes P (2002) Overlapping roles of P-selectin and alpha 4 integrin to recruit leukocytes to the central nervous system in experimental autoimmune encephalomyelitis. J Immunol 169:1000–1006
79. Kerfoot SM, Norman MU, Lapointe BM, Bonder CS, Zbytnuik L, Kubes P (2006) Reevaluation of P-selectin and alpha 4 integrin as targets for the treatment of experimental autoimmune encephalomyelitis. J Immunol 176:6225–6234
80. Krumbholz M, Theil D, Cepok S, Hemmer B, Kivisakk P, Ransohoff RM, Hofbauer M, Farina C, Derfuss T, Hartle C, Newcombe J, Hohlfeld R, Meinl E (2006) Chemokines in multiple sclerosis: CXCL12 and CXCL13 up-regulation is differentially linked to CNS immune cell recruitment. Brain 129(Pt 1):200–211

81. Kwee L, Baldwin HS, Shen HM, Stewart CL, Buc C, Buch CA, Labow MA (1995) Defective development of the embryonic and extraembryonic circulatory systems in vascular cell adhesion molecule (VCAM-1) deficient mice. Development 121:489–503
82. Lagerstrom MC, Schioth HB (2008) Structural diversity of G protein-coupled receptors and significance for drug discovery. Nat Rev Drug Discov 7(4):339–357. doi:10.1038/nrd2518
83. Lai-Cheong JE, Parsons M, McGrath JA (2010) The role of kindlins in cell biology and relevance to human disease. Int J Biochem Cell Biol 42(5):595–603. doi:10.1016/j.biocel.2009.10.015
84. Lalor PF, Edwards S, McNab G, Salmi M, Jalkanen S, Adams DH (2002) Vascular adhesion protein-1 mediates adhesion and transmigration of lymphocytes on human hepatic endothelial cells. J Immunol 169(2):983–992
85. Lalor SJ, Segal BM (2010) Lymphoid chemokines in the CNS. J Neuroimmunol 224(1–2):56–61. doi:10.1016/j.jneuroim.2010.05.017
86. Larochelle C, Cayrol R, Kebir H, Alvarez JI, Lecuyer MA, Ifergan I, Viel E, Bourbonniere L, Beauseigle D, Terouz S, Hachehouche L, Gendron S, Poirier J, Jobin C, Duquette P, Flanagan K, Yednock T, Arbour N, Prat A (2012) Melanoma cell adhesion molecule identifies encephalitogenic T lymphocytes and promotes their recruitment to the central nervous system. Brain 135(Pt 10):2906–2924. doi:10.1093/brain/aws212
87. Larochelle C, Lecuyer MA, Alvarez JI, Charabati M, Saint-Laurent O, Ghannam S, Kebir H, Flanagan K, Yednock T, Duquette P, Arbour N, Prat A (2015) Melanoma cell adhesion molecule-positive CD8 T lymphocytes mediate central nervous system inflammation. Ann Neurol 78(1):39–53. doi:10.1002/ana.24415
88. Laschinger M, Engelhardt B (2000) Interaction of alpha4-integrin with VCAM-1 is involved in adhesion of encephalitogenic T cell blasts to brain endothelium but not in their transendothelial migration in vitro. J Neuroimmunol 102(1):32–43
89. Lee SJ, Benveniste EN (1999) Adhesion molecule expression and regulation on cells of the central nervous system. J Neuroimmunol 98:77–88
90. Lefort CT, Rossaint J, Moser M, Petrich BG, Zarbock A, Monkley SJ, Critchley DR, Ginsberg MH, Fassler R, Ley K (2012) Distinct roles for talin-1 and kindlin-3 in LFA-1 extension and affinity regulation. Blood 119(18):4275–4282. doi:10.1182/blood-2011-08-373118
91. Ley K (2003) The role of selectins in inflammation and disease. Trends Mol Med 9(6):263–268
92. Ley K, Kansas GS (2004) Selectins in T-cell recruitment to non-lymphoid tissues and sites of inflammation. Nat Rev Immunol 4(5):325–335
93. Liu YJ, Guo DW, Tian L, Shang DS, Zhao WD, Li B, Fang WG, Zhu L, Chen YH (2010) Peripheral T cells derived from Alzheimer's disease patients overexpress CXCR2 contributing to its transendothelial migration, which is microglial TNF-alpha-dependent. Neurobiol Aging 31(2):175–188. doi:10.1016/j.neurobiolaging.2008.03.024
94. Locksley RM, Killeen N, Lenardo MJ (2001) The TNF and TNF receptor superfamilies: integrating mammalian biology. Cell 104(4):487–501
95. Lopes Pinheiro MA, Kooij G, Mizee MR, Kamermans A, Enzmann G, Lyck R, Schwaninger M, Engelhardt B, de Vries HE (2016) Immune cell trafficking across the barriers of the central nervous system in multiple sclerosis and stroke. Biochim Biophys Acta 1862(3):461–471. doi:10.1016/j.bbadis.2015.10.018
96. Lossinsky AS, Badmajew V, Robson JA, Moretz RC, Wisniewski HM (1989) Sites of egress of inflammatory cells and horseradish peroxidase transport across the blood–brain barrier in a murine model of chronic relapsing experimental allergic encephalomyelitis. Acta Neuropathol 78(4):359–371
97. Lossinsky AS, Pluta R, Song MJ, Badmajew V, Moretz RC, Wisniewski HM (1991) Mechanisms of inflammatory cell attachment in chronic relapsing experimental allergic encephalomyelitis: a scanning and high-voltage electron microscopic study of the injured mouse blood–brain barrier. Microvasc Res 41(3):299–310

98. Lowe JB (2002) Glycosylation in the control of selectin counter-receptor structure and function. Immunol Rev 186:19–36
99. Luo BH, Carman CV, Springer TA (2007) Structural basis of integrin regulation and signaling. Annu Rev Immunol 25:619–647
100. Lyck R, Engelhardt B (2012) Going against the tide--how encephalitogenic T cells breach the blood–brain barrier. J Vasc Res 49(6):497–509. doi:10.1159/000341232
101. Lyck R, Enzmann G (2015) The physiological roles of ICAM-1 and ICAM-2 in neutrophil migration into tissues. Curr Opin Hematol 22(1):53–59. doi:10.1097/MOH.0000000000000103
102. Lyck R, Reiss Y, Gerwin N, Greenwood J, Adamson P, Engelhardt B (2003) T-cell interaction with ICAM-1/ICAM-2 double-deficient brain endothelium in vitro: the cytoplasmic tail of endothelial ICAM-1 is necessary for transendothelial migration of T cells. Blood 102(10):3675–3683. doi:10.1182/blood-2003-02-0358
103. Makgoba MW, Sanders ME, Ginther Luce GE, Dustin ML, Springer TA, Clark EA, Mannoni P, Shaw S (1988) ICAM-1 a ligand for LFA-1-dependent adhesion of B, T and myeloid cells. Nature 331(6151):86–88. doi:10.1038/331086a0
104. Malfitano AM, Marasco G, Proto MC, Laezza C, Gazzerro P, Bifulco M (2014) Statins in neurological disorders: an overview and update. Pharmacol Res 88:74–83. doi:10.1016/j.phrs.2014.06.007
105. Man SM, Ma YR, Shang DS, Zhao WD, Li B, Guo DW, Fang WG, Zhu L, Chen YH (2007) Peripheral T cells overexpress MIP-1alpha to enhance its transendothelial migration in Alzheimer's disease. Neurobiol Aging 28(4):485–496. doi:10.1016/j.neurobiolaging.2006.02.013
106. Martinelli R, Gegg M, Longbottom R, Adamson P, Turowski P, Greenwood J (2009) ICAM-1-mediated endothelial nitric oxide synthase activation via calcium and AMP-activated protein kinase is required for transendothelial lymphocyte migration. Mol Biol Cell 20(3):995–1005
107. Martinelli R, Kamci M, Sage PT, Massol R, Varghese L, Sciuto T, Toporsian M, Dvorak AM, Kirchhausen T, Springer TA, Carman CV (2013) Release of cellular tension signals self-restorative ventral lamellipodia to heal barrier micro-wounds. J Cell Biol 201(3):449–465. doi:10.1083/jcb.201209077
108. Martinelli R, Zeiger AS, Whitfield M, Sciuto TE, Dvorak A, Van Vliet KJ, Greenwood J, Carman CV (2014) Probing the biomechanical contribution of the endothelium to lymphocyte migration: diapedesis by the path of least resistance. J Cell Sci 127(Pt 17):3720–3734. doi:10.1242/jcs.148619
109. Martins TB, Rose JW, Jaskowski TD, Wilson AR, Husebye D, Seraj HS, Hill HR (2011) Analysis of proinflammatory and anti-inflammatory cytokine serum concentrations in patients with multiple sclerosis by using a multiplexed immunoassay. Am J Clin Pathol 136(5):696–704. doi:10.1309/AJCP7UBK8IBVMVNR
110. Matsuki T, Nakae S, Sudo K, Horai R, Iwakura Y (2006) Abnormal T cell activation caused by the imbalance of the IL-1/IL-1R antagonist system is responsible for the development of experimental autoimmune encephalomyelitis. Int Immunol 18(2):399–407. doi:10.1093/intimm/dxh379
111. McCandless EE, Piccio L, Woerner BM, Schmidt RE, Rubin JB, Cross AH, Klein RS (2008) Pathological expression of CXCL12 at the blood–brain barrier correlates with severity of multiple sclerosis. Am J Pathol 172(3):799–808. doi:10.2353/ajpath.2008.070918
112. McCandless EE, Wang Q, Woerner BM, Harper JM, Klein RS (2006) CXCL12 limits inflammation by localizing mononuclear infiltrates to the perivascular space during experimental autoimmune encephalomyelitis. J Immunol 177(11):8053–8064
113. McCoy MK, Tansey MG (2008) TNF signaling inhibition in the CNS: implications for normal brain function and neurodegenerative disease. J Neuroinflammation 5:45. doi:10.1186/1742-2094-5-45
114. McEver RP, Zhu C (2010) Rolling cell adhesion. Annu Rev Cell Dev Biol 26:363–396. doi:10.1146/annurev.cellbio.042308.113238

115. McGeer EG, McGeer PL (1998) The importance of inflammatory mechanisms in Alzheimer disease. Exp Gerontol 33(5):371–378
116. Miklossy J, Doudet DD, Schwab C, Yu S, McGeer EG, McGeer PL (2006) Role of ICAM-1 in persisting inflammation in Parkinson disease and MPTP monkeys. Exp Neurol 197(2):275–283. doi:10.1016/j.expneurol.2005.10.034
117. Moretti FA, Moser M, Lyck R, Abadier M, Ruppert R, Engelhardt B, Fassler R (2013) Kindlin-3 regulates integrin activation and adhesion reinforcement of effector T cells. Proc Natl Acad Sci U S A 110(42):17005–17010. doi:10.1073/pnas.1316032110
118. Moser M, Legate KR, Zent R, Fassler R (2009) The tail of integrins, talin, and kindlins. Science (New York NY) 324(5929):895–899. doi:10.1126/science.1163865
119. Nicoletti F, Patti F, Cocuzza C, Zaccone P, Nicoletti A, Di Marco R, Reggio A (1996) Elevated serum levels of interleukin-12 in chronic progressive multiple sclerosis. J Neuroimmunol 70(1):87–90
120. Nourshargh S, Alon R (2014) Leukocyte migration into inflamed tissues. Immunity 41(5):694–707. doi:10.1016/j.immuni.2014.10.008
121. Oldham WM, Hamm HE (2008) Heterotrimeric G protein activation by G-protein-coupled receptors. Nat Rev Mol Cell Biol 9(1):60–71. doi:10.1038/nrm2299
122. Ottum PA, Arellano G, Reyes LI, Iruretagoyena M, Naves R (2015) Opposing roles of interferon-gamma on cells of the central nervous system in autoimmune neuroinflammation. Front Immunol 6:539. doi:10.3389/fimmu.2015.00539
123. Ozaki H, Ishii K, Horiuchi H, Arai H, Kawamoto T, Okawa K, Iwamatsu A, Kita T (1999) Cutting edge: combined treatment of TNF-alpha and IFN-gamma causes redistribution of junctional adhesion molecule in human endothelial cells. J Immunol 163:553–557
124. Pannecoeck R, Serruys D, Benmeridja L, Delanghe JR, van Geel N, Speeckaert R, Speeckaert MM (2015) Vascular adhesion protein-1: role in human pathology and application as a biomarker. Crit Rev Clin Lab Sci 52(6):284–300. doi:10.3109/10408363.2015.1050714
125. Pfeiffer F, Schafer J, Lyck R, Makrides V, Brunner S, Schaeren-Wiemers N, Deutsch U, Engelhardt B (2011) Claudin-1 induced sealing of blood–brain barrier tight junctions ameliorates chronic experimental autoimmune encephalomyelitis. Acta Neuropathol 122(5):601–614. doi:10.1007/s00401-011-0883-2
126. Piccio L, Rossi B, Scarpini E, Laudanna C, Giagulli C, Issekutz AC, Vestweber D, Butcher EC, Constantin G (2002) Molecular mechanisms involved in lymphocyte recruitment in inflamed brain microvessels: critical roles for P-selectin glycoprotein ligand-1 and heterotrimeric G(i)-linked receptors. J Immunol 168(4):1940–1949
127. Planas AM, Gorina R, Chamorro A (2006) Signalling pathways mediating inflammatory responses in brain ischaemia. Biochem Soc Trans 34(Pt 6):1267–1270. doi:10.1042/BST0341267
128. Proudfoot AE, Handel TM, Johnson Z, Lau EK, LiWang P, Clark-Lewis I, Borlat F, Wells TN, Kosco-Vilbois MH (2003) Glycosaminoglycan binding and oligomerization are essential for the in vivo activity of certain chemokines. Proc Natl Acad Sci U S A 100(4):1885–1890. doi:10.1073/pnas.0334864100
129. Quandt J, Dorovini-Zis K (2004) The beta chemokines CCL4 and CCL5 enhance adhesion of specific CD4+ T cell subsets to human brain endothelial cells. J Neuropathol Exp Neurol 63(4):350–362
130. Ramos TN, Bullard DC, Barnum SR (2014) ICAM-1: isoforms and phenotypes. J Immunol 192(10):4469–4474. doi:10.4049/jimmunol.1400135
131. Reitsma S, Slaaf DW, Vink H, van Zandvoort MA, oude Egbrink MG (2007) The endothelial glycocalyx: composition, functions, and visualization. Pflugers Arch 454(3):345–359. doi:10.1007/s00424-007-0212-8
132. Rudolph H, Klopstein A, Gruber I, Blatti C, Lyck R, Engelhardt B (2016) Post-arrest stalling rather than crawling favors CD8+ over CD4+ T-cell migration across the blood–brain barrier under flow in vitro. Eur J Immunol 46:2187–2203. doi:10.1002/eji.201546251
133. Sa-Pereira I, Brites D, Brito MA (2012) Neurovascular unit: a focus on pericytes. Mol Neurobiol 45(2):327–347. doi:10.1007/s12035-012-8244-2

134. Sathiyanadan K, Coisne C, Enzmann G, Deutsch U, Engelhardt B (2014) PSGL-1 and E/P--selectins are essential for T-cell rolling in inflamed CNS microvessels but dispensable for initiation of EAE. Eur J Immunol 44(8):2287–2294. doi:10.1002/eji.201344214
135. Schiffenbauer J, Streit WJ, Butfiloski E, LaBow M, Edwards C 3rd, Moldawer LL (2000) The induction of EAE is only partially dependent on TNF receptor signaling but requires the IL-1 type I receptor. Clin Immunol 95(2):117–123. doi:10.1006/clim.2000.4851
136. Schulte D, Kuppers V, Dartsch N, Broermann A, Li H, Zarbock A, Kamenyeva O, Kiefer F, Khandoga A, Massberg S, Vestweber D (2011) Stabilizing the VE-cadherin-catenin complex blocks leukocyte extravasation and vascular permeability. Embo J 30(20):4157–4170. doi:10.1038/emboj.2011.304
137. Selmaj KW, Raine CS (1995) Experimental autoimmune encephalomyelitis: immunotherapy with anti-tumor necrosis factor antibodies and soluble tumor necrosis factor receptors. Neurology 45(6 Suppl 6):S44–S49
138. Sharief MK, Hentges R, Thomas E (1991) Significance of CSF immunoglobulins in monitoring neurologic disease activity in Behcet's disease. Neurology 41(9):1398–1401
139. Shattil SJ, Kim C, Ginsberg MH (2010) The final steps of integrin activation: the end game. Nat Rev Mol Cell Biol 11(4):288–300. doi:10.1038/nrm2871
140. Shulman Z, Alon R (2012) Real-time analysis of integrin-dependent transendothelial migration and integrin-independent interstitial motility of leukocytes. Methods Mol Biol 757:31–45. doi:10.1007/978-1-61779-166-6_3
141. Shulman Z, Cohen SJ, Roediger B, Kalchenko V, Jain R, Grabovsky V, Klein E, Shinder V, Stoler-Barak L, Feigelson SW, Meshel T, Nurmi SM, Goldstein I, Hartley O, Gahmberg CG, Etzioni A, Weninger W, Ben-Baruch A, Alon R (2012) Transendothelial migration of lymphocytes mediated by intraendothelial vesicle stores rather than by extracellular chemokine depots. Nat Immunol 13(1):67–76. doi:10.1038/ni.2173
142. Sims JE, Smith DE (2010) The IL-1 family: regulators of immunity. Nat Rev Immunol 10(2):89–102. doi:10.1038/nri2691
143. Smith A, Stanley P, Jones K, Svensson L, McDowall A, Hogg N (2007) The role of the integrin LFA-1 in T-lymphocyte migration. Immunol Rev 218:135–146
144. Sobel RA, Mitchell ME, Fondren G (1990) Intercellular Adhesion Molecule-1 (ICAM-1) in cellular immune reactions in the human central nervous system. Am J Pathol 136:1309–1316
145. Song KH, Kwon KW, Choi JC, Jung J, Park Y, Suh KY, Doh J (2014) T cells sense biophysical cues using lamellipodia and filopodia to optimize intraluminal path finding. Integr Biol (Camb) 6(4):450–459. doi:10.1039/c4ib00021h
146. Steffen BJ, Butcher EC, Engelhardt B (1994) Evidence for involvement of ICAM-1 and VCAM-1 in lymphocyte interaction with endothelium in experimental autoimmune encephalomyelitis in the central nervous system in the SJL/J mouse. Am J Pathol 145(1):189–201
147. Steiner O, Coisne C, Cecchelli R, Boscacci R, Deutsch U, Engelhardt B, Lyck R (2010) Differential roles for endothelial ICAM-1, ICAM-2, and VCAM-1 in shear-resistant T cell arrest, polarization, and directed crawling on blood–brain barrier endothelium. J Immunol 185(8):4846–4855. doi:10.4049/jimmunol.0903732
148. Steiner O, Coisne C, Engelhardt B, Lyck R (2011) Comparison of immortalized bEnd5 and primary mouse brain microvascular endothelial cells as in vitro blood–brain barrier models for the study of T cell extravasation. J Cereb Blood Flow Metab 31(1):315–327. doi:10.1038/jcbfm.2010.96
149. Stoler-Barak L, Moussion C, Shezen E, Hatzav M, Sixt M, Alon R (2014) Blood vessels pattern heparan sulfate gradients between their apical and basolateral aspects. PLoS One 9(1):e85699. doi:10.1371/journal.pone.0085699
150. Sutton C, Brereton C, Keogh B, Mills KH, Lavelle EC (2006) A crucial role for interleukin (IL)-1 in the induction of IL-17-producing T cells that mediate autoimmune encephalomyelitis. J Exp Med 203(7):1685–1691. doi:10.1084/jem.20060285

151. Tadokoro S, Shattil SJ, Eto K, Tai V, Liddington RC, de Pereda JM, Ginsberg MH, Calderwood DA (2003) Talin binding to integrin beta tails: a final common step in integrin activation. Science (New York NY) 302(5642):103–106. doi:10.1126/science.1086652
152. Talamonti M, Spallone G, Di Stefani A, Costanzo A, Chimenti S (2011) Efalizumab. Expert Opin Drug Saf 10(2):239–251. doi:10.1517/14740338.2011.524925
153. Te Riet J, Helenius J, Strohmeyer N, Cambi A, Figdor CG, Muller DJ (2014) Dynamic coupling of ALCAM to the actin cortex strengthens cell adhesion to CD6. J Cell Sci 127(Pt 7):1595–1606. doi:10.1242/jcs.141077
154. Thelen M, Stein JV (2008) How chemokines invite leukocytes to dance. Nat Immunol 9(9):953–959. doi:10.1038/ni.f.207
155. Thompson PW, Randi AM, Ridley AJ (2002) Intercellular adhesion molecule (ICAM)-1, but not ICAM-2, activates RhoA and stimulates c-fos and rhoA transcription in endothelial cells. J Immunol 169:1007–1013
156. Tietz S, Engelhardt B (2015) Brain barriers: crosstalk between complex tight junctions and adherens junctions. J Cell Biol 209(4):493–506. doi:10.1083/jcb.201412147
157. Tiruppathi C, Minshall RD, Paria BC, Vogel SM, Malik AB (2002) Role of Ca2+ signaling in the regulation of endothelial permeability. Vascul Pharmacol 39(4–5):173–185
158. Togo T, Akiyama H, Iseki E, Kondo H, Ikeda K, Kato M, Oda T, Tsuchiya K, Kosaka K (2002) Occurrence of T cells in the brain of Alzheimer's disease and other neurological diseases. J Neuroimmunol 124(1–2):83–92
159. Town T, Tan J, Flavell RA, Mullan M (2005) T-cells in Alzheimer's disease. Neuromolecular Med 7(3):255–264. doi:10.1385/NMM:7:3:255
160. Traugott U, Lebon P (1988) Multiple sclerosis: involvement of interferons in lesion pathogenesis. Ann Neurol 24(2):243–251. doi:10.1002/ana.410240211
161. Trenova AG, Manova MG, Kostadinova II, Murdjeva MA, Hristova DR, Vasileva TV, Zahariev ZI (2011) Clinical and laboratory study of pro-inflammatory and antiinflammatory cytokines in women with multiple sclerosis. Folia Med (Plovdiv) 53(2):29–35
162. Tsai HC, Han MH (2016) Sphingosine-1-phosphate (S1P) and S1P signaling pathway: therapeutic targets in autoimmunity and inflammation. Drugs 76:1067–1079. doi:10.1007/s40265-016-0603-2
163. Vacchini A, Locati M, Borroni EM (2016) Overview and potential unifying themes of the atypical chemokine receptor family. J Leukoc Biol 99(6):883–892. doi:10.1189/jlb.2MR1015-477R
164. Vajkoczy P, Laschinger M, Engelhardt B (2001) Alpha4-integrin-VCAM-1 binding mediates G protein-independent capture of encephalitogenic T cell blasts to CNS white matter microvessels. J Clin Invest 108(4):557–565
165. Valignat MP, Theodoly O, Gucciardi A, Hogg N, Lellouch AC (2013) T lymphocytes orient against the direction of fluid flow during LFA-1-mediated migration. Biophys J 104(2):322–331. doi:10.1016/j.bpj.2012.12.007
166. van Buul JD, Allingham MJ, Samson T, Meller J, Boulter E, Garcia-Mata R, Burridge K (2007) RhoG regulates endothelial apical cup assembly downstream from ICAM1 engagement and is involved in leukocyte trans-endothelial migration. J Cell Biol 178(7):1279–1293. doi:10.1083/jcb.200612053
167. van Buul JD, Kanters E, Hordijk PL (2007) Endothelial signaling by Ig-like cell adhesion molecules. Arterioscler Thromb Vasc Biol 27(9):1870–1876. doi:10.1161/ATVBAHA.107.145821
168. Vestweber D, Wessel F, Nottebaum AF (2014) Similarities and differences in the regulation of leukocyte extravasation and vascular permeability. Semin Immunopathol 36(2):177–192. doi:10.1007/s00281-014-0419-7
169. Wallez Y, Cand F, Cruzalegui F, Wernstedt C, Souchelnytskyi S, Vilgrain I, Huber P (2007) Src kinase phosphorylates vascular endothelial-cadherin in response to vascular endothelial growth factor: identification of tyrosine 685 as the unique target site. Oncogene 26(7):1067–1077. doi:10.1038/sj.onc.1209855

170. Watts AO, Verkaar F, van der Lee MM, Timmerman CA, Kuijer M, van Offenbeek J, van Lith LH, Smit MJ, Leurs R, Zaman GJ, Vischer HF (2013) beta-Arrestin recruitment and G protein signaling by the atypical human chemokine decoy receptor CCX-CKR. J Biol Chem 288(10):7169–7181. doi:10.1074/jbc.M112.406108
171. Weiss N, Miller F, Cazaubon S, Couraud PO (2009) The blood–brain barrier in brain homeostasis and neurological diseases. Biochim Biophys Acta 1788(4):842–857. doi:10.1016/j.bbamem.2008.10.022
172. Wen SR, Liu GJ, Feng RN, Gong FC, Zhong H, Duan SR, Bi S (2012) Increased levels of IL-23 and osteopontin in serum and cerebrospinal fluid of multiple sclerosis patients. J Neuroimmunol 244(1–2):94–96. doi:10.1016/j.jneuroim.2011.12.004
173. Wessel F, Winderlich M, Holm M, Frye M, Rivera-Galdos R, Vockel M, Linnepe R, Ipe U, Stadtmann A, Zarbock A, Nottebaum AF, Vestweber D (2014) Leukocyte extravasation and vascular permeability are each controlled in vivo by different tyrosine residues of VE-cadherin. Nat Immunol 15(3):223–230. doi:10.1038/ni.2824
174. Wolburg H, Wolburg-Buchholz K, Engelhardt B (2005) Diapedesis of mononuclear cells across cerebral venules during experimental autoimmune encephalomyelitis leaves tight junctions intact. Acta Neuropathol (Berl) 109(2):181–190
175. Wong RK, Baldwin AL, Heimark RL (1999) Cadherin-5 redistribution at sites of TNF-alpha and IFN-gamma-induced permeability in mesenteric venules. Am J Physiol 276(2 Pt 2):H736–H748
176. Xie H, Lim YC, Luscinskas FW, Lichtman AH (1999) Acquisition of selectin binding and peripheral homing properties by CD4(+) and CD8(+) T cells. J Exp Med 189(11):1765–1776
177. Yang L, Kowalski JR, Yacono P, Bajmoczi M, Shaw SK, Froio RM, Golan DE, Thomas SM, Luscinskas FW (2006) Endothelial cell cortactin coordinates intercellular adhesion molecule-1 clustering and actin cytoskeleton remodeling during polymorphonuclear leukocyte adhesion and transmigration. J Immunol 177(9):6440–6449
178. Yednock TA, Cannon C, Fritz LC, Sanchez-Madrid F, Steinman L, Karin N (1992) Prevention of experimental autoimmune encephalomyelitis by antibodies against alpha 4 beta 1 integrin. Nature 356(6364):63–66
179. Zimmerman AW, Joosten B, Torensma R, Parnes JR, van Leeuwen FN, Figdor CG (2006) Long-term engagement of CD6 and ALCAM is essential for T-cell proliferation induced by dendritic cells. Blood 107(8):3212–3220. doi:10.1182/blood-2005-09-3881

Neuroinflammation in Bacterial Meningitis

Philipp Agyeman, Denis Grandgirard, and Stephen L. Leib

Abstract Under physiologic conditions, the brain is a microbiologically sterile site and is protected from infection by highly specialized barriers, including the hard bony skull, the tough dura mater, and the restrictive blood–brain barrier (BBB). Host defense mechanisms in the central nervous system (CNS) are limited and tightly regulated. Peripheral immune cells and plasma proteins are largely excluded from the brain parenchyma. Once they have breached the protective barriers and entered the CNS, bacteria multiply within the cerebrospinal fluid space (CSF) highly efficiently exhibiting similar kinetics as in vitro and reaching concentrations of up to 10^9 CFU/mL.

In response to the multiplying bacteria and their components, i.e., cell wall fragments, lipopolysaccharides, teichoic and lipoteichoic acids, peptidoglycans, bacterial DNA, and other cytosolic factors, resident cells in the perivascular space and the meninges release pro-inflammatory signaling molecules. Tumor necrosis factor-α, interleukin-1β, and IL-6 are released early on and trigger a cascade of other inflammatory mediators, including a variety of cytokines, chemokines, platelet-activating factor, antimicrobial peptides, prostaglandins, matrix metalloproteinases, nitric oxide, and reactive oxygen species initiating a self-perpetuating inflammatory cascade.

The immediate consequences of the intense inflammatory reaction are a massive influx of leukocytes, the breakdown of the blood–brain barrier with the formation of brain edema, and alterations of the cerebral blood flow. This overshooting inflammatory reaction to the invading pathogens causes damage to the brain parenchyma as collateral damage and is the driving pathophysiologic mechanism of inflammatory inner ear damage, brain cortical ischemic injury, and hippocampal apoptosis, the most frequent histopathological correlates of the neurofunctional sequelae of bacterial meningitis.

P. Agyeman, MD
Institute for Infectious Diseases, University of Bern, Bern 3001, Switzerland

Department of Pediatrics, University Children's Hospital, University of Bern, Bern 3001, Switzerland

D. Grandgirard, PhD • S.L. Leib, MD (✉)
Institute for Infectious Diseases, University of Bern, Bern 3001, Switzerland
e-mail: Stephen.leib@ifik.unibe.ch

R. Lyck, G. Enzmann (eds.), *The Blood Brain Barrier and Inflammation*, Progress in Inflammation Research, DOI 10.1007/978-3-319-45514-3_10

1 Introduction

Meningitis is an inflammation of the membranes enclosing the brain and spinal cord. Meningitis may be caused by viral, fungal, or parasitic infection, but is frequently caused by bacteria capable of penetrating the barriers protecting the brain. The barrier function of the blood–brain barrier (BBB) and the blood–cerebrospinal fluid barrier (BCSFB) largely excludes peripheral immune cells and plasma proteins in normal physiological conditions, which led to the concept of the central nervous system (CNS) as a site of immune privilege. The lack of immediate and sufficient bactericidal activity in the cerebrospinal fluid (CSF) allows bacteria to multiply almost as efficiently as in vitro, to reach titers of up to 10^9 colony-forming units (CFU)/mL and to spread over the entire surface of the brain, the spinal cord, and along penetrating vessels. This explains that prior to the antibiotic era, bacterial meningitis almost always led to death.

To gain access to the CNS, bacteria must (a) colonize the host, (b) invade the bloodstream or cause focal infection in the vicinity of the brain, (c) survive in the bloodstream, and (d) cross the BBB or the BCSFB (Fig. 1). Bacterial meningitis can arise from bacteremia if pathogens colonizing the nasopharyngeal and intestinal mucosa invade the bloodstream or by continuous infection from nearby foci, e.g., sinusitis, mastoiditis. It may also follow a breach of the brain barriers by a foreign body (e.g., cerebrospinal fluid shunt, cochlear implant), a neurosurgical procedure, or trauma.

Once a pathogen has reached the CNS, a cascade of events follows until the full symptomatic manifestations of bacterial meningitis. These include the induction of cytokines and chemokines; activation of inflammatory mediators such as nitric oxide (NO), reactive oxygen species (ROS), or matrix metalloproteinases (MMP); recruitment of white blood cells to the site of infection; and cytotoxic events. In this chapter, the pathogenesis of bacterial meningitis and its neurological sequelae are discussed.

2 Pathogenesis of Bacterial Meningitis

2.1 *Bacterial Invasion of the Host and Penetration of the Blood–Brain Barrier*

Figure 2 shows the pathogenic steps involved in the development of bacterial meningitis.

2.1.1 Bacterial Colonization of the Host

Colonization of the host is a common first step of the major bacterial pathogens that cause meningitis (Table 1). *S. pneumoniae*, *N. meningitidis*, and *H. influenzae* type B colonize the nasopharynx and are transmitted from person to person by the

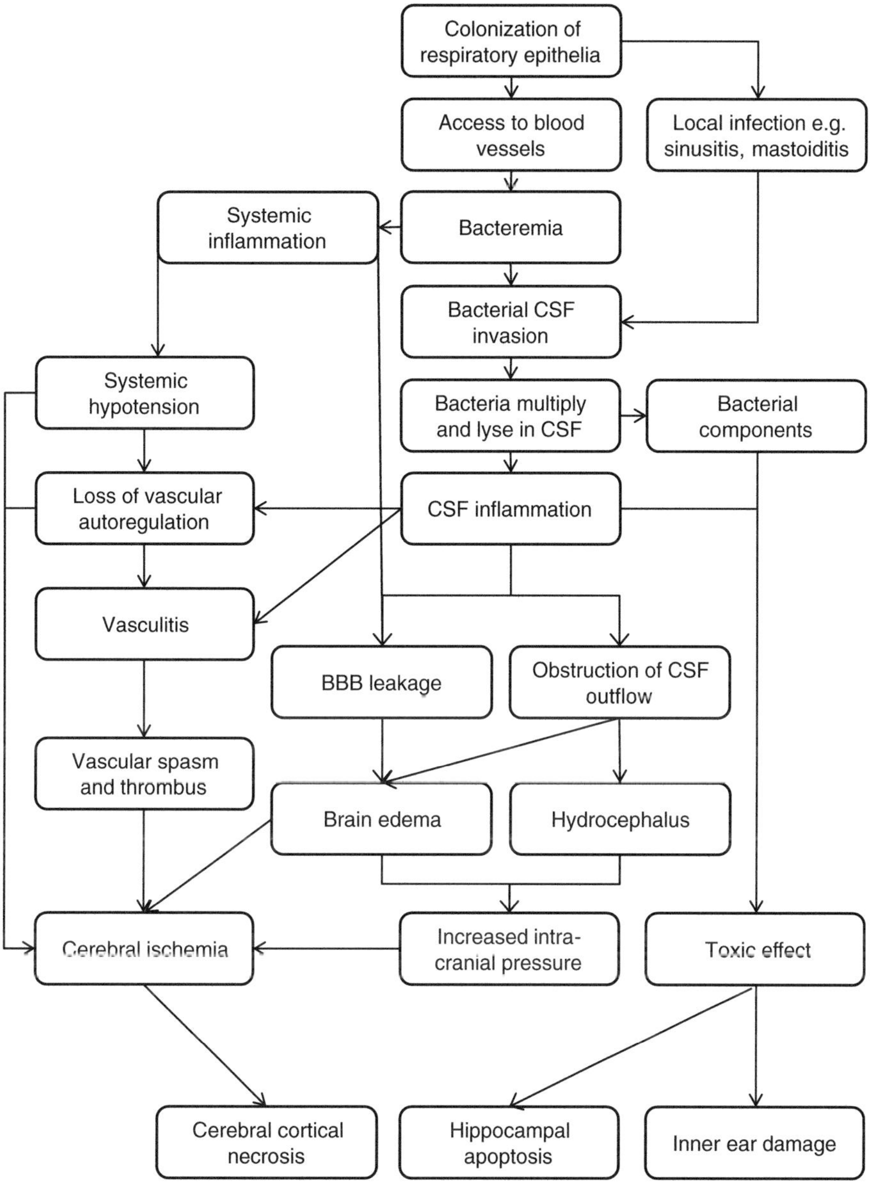

Fig. 1 Selected mechanisms that contribute to the pathogenesis of bacterial meningitis and the development of brain injury

respiratory route. Injury of the functional integrity of the respiratory tract mucosa by viral infections [2, 3] and physical damage to the mucosa [4] may increase the risk of invasive bacterial disease.

To colonize the nasopharyngeal mucosa, a pathogen first has to evade mucosal defense mechanisms like ciliary clearance [5–10], secretory immunoglobulin A

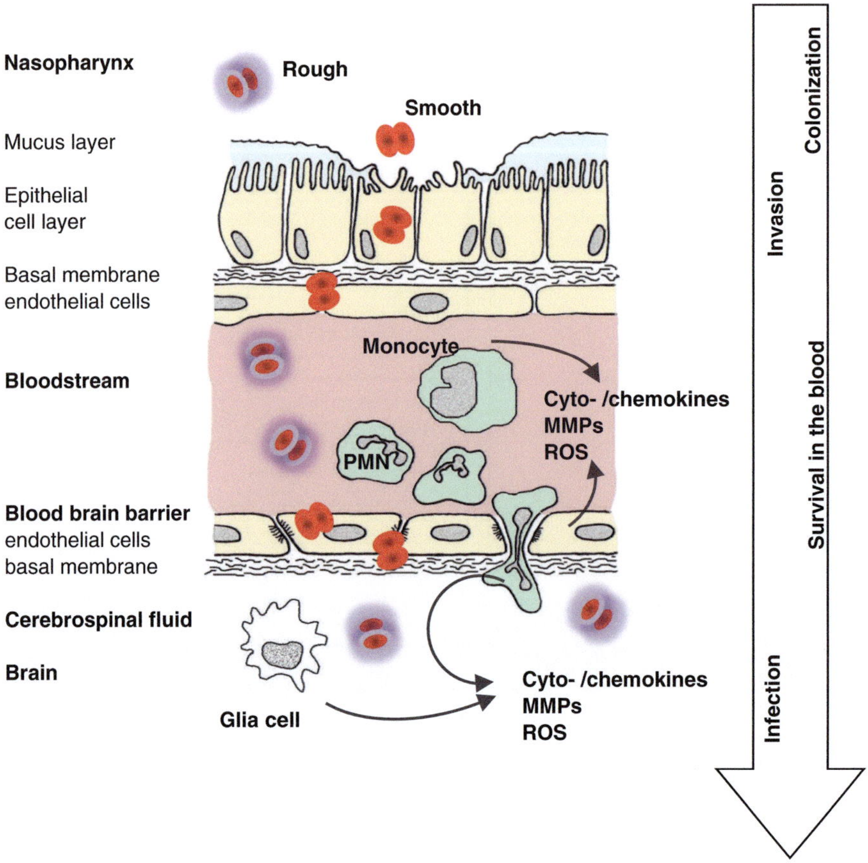

Fig. 2 Pathogenetic steps involved in the development of bacterial meningitis: (i) colonization of the nasopharynx and invasion of blood capillaries via crossing of the mucosal barrier, the epithelia and the vascular endothelium; (ii) bacteremia, survival in the blood and crossing of the blood–brain barrier (BBB); and (iii) bacterial multiplication in the CSF, induction of inflammation, BBB breakdown, and invasion of blood-derived neutrophils into the subarachnoid and ventricular space Adherence of bacteria to the mucosal epithelium and colonization are affected by phase variation. Bacteria cross the mucosal barrier through or between epithelial cells. Once in the bloodstream, expression of a sufficiently thick capsule is necessary to protect the bacteria from circulating antibodies, complement-mediated bacterial killing, and neutrophil phagocytosis. Bacteria penetrate the BBB by (i) paracellular passage, (ii) transcellular passage, or (iii) invasion within white blood cells during diapedesis (intracellular pathogens). Neutrophilic pleocytosis occurs along a chemotactic gradient mainly created by chemokines locally expressed in the brain. Matrix metalloproteinases (MMP), such as MMP-8 and MMP-9, facilitate the process of extravasation by degrading extracellular matrix components of the brain microvasculature

(IgA) [11–13], and lysozyme [14–17]. Secondly, it has to prevail against other organisms in the aerobic environment of the upper respiratory tract [18–20]. Finally, it must adhere to the mucosal epithelium [10, 21–25].

Group B streptococcus (GBS) (*S. agalactiae*), *E. coli*, and *L. monocytogenes* colonize the gastrointestinal tract and are transmitted by the oral, vaginal, or fecal–oral

Table 1 Most common bacterial organisms causing meningitis in humans according to broad age categories. Relative frequency of occurrence in studies performed in different European countries are given as percentage

	Causative organism	Relative frequency of occurrence in selected European countries
Neonatal bacterial meningitis	Group B streptococcus	58 %
	Escherichia coli	21 %
	Listeria monocytogenes	2 %
	Streptococcus pneumoniae	4 %
Pediatric bacterial meningitis beyond the neonatal age	*Neisseria meningitidis*	50 %
	Streptococcus pneumoniae	37 %
	Haemophilus influenzae	5 %
Adult bacterial meningitis	*Streptococcus pneumoniae*	53 %
	Neisseria meningitidis	27 %
	Haemophilus influenzae	3 %
	Listeria monocytogenes	4 %

Adapted from: van de Beek et al. [1]

route. They mainly cause meningitis at the vulnerable extremes of age and in immunocompromised persons. In neonatal meningitis, colonization of the intestinal tract of the infant follows similar principles as outlined for nasopharyngeal colonization [26–34]. The food-borne pathogen *L. monocytogenes* primarily affects immunocompromised patients and pregnant women and is transmitted to the fetus through the placenta or to the newborn during delivery [35, 36].

2.1.2 Invasion of the Bloodstream

To enter the bloodstream at the site of colonization, bacteria penetrate the mucosal barrier formed by epithelial cells and the lining of the blood vessels formed by the vascular endothelium (Fig. 2). Depending on the pathogen, the paracellular or transcellular route across the epithelial and endothelial layers is preferred.

Most known mechanisms by which meningeal pathogens invade human epithelial and endothelial cells via the transcellular route take advantage of the ubiquitous clathrin-mediated cellular endocytosis mechanism [37–39]. This then leads to the uptake into vacuoles that may either recycle back to the cell surface, fuse with lysosomes, or transit to the basolateral cell surface where the bacteria are released out of the cell [40–42].

Actin rearrangements further enhance bacterial internalization. Some transmembrane proteins of the epithelial cell that are targeted by the bacterial pathogen and initiate endocytosis are polymeric immunoglobulin receptor (pIgR) [43, 44], carcinoembryonic antigen-related cell adhesion molecule (CEACAM) proteins [45–47], and epithelial cadherin (E-cadherin) [48, 49].

Paracellular migration of bacteria across the epithelial barrier is facilitated by toll-like receptor (TLR)-dependent downregulation of tight junction protein expression [50] or extracellular degradation of junctional components mediated by bacterial enzymes or bacteria-bound plasmin [51–54].

The final barrier between epithelial cells and mesenchymal cells is the epithelial basement membrane. Meningeal pathogens may degrade the epithelial basement membrane by secreting proteolytic enzymes [51, 55], abuse the host's plasminogen–plasmin system [53, 56–58], and cross the basement membrane in phagocytic cells [59, 60]. Finally, attachment to extracellular matrix (ECM) components is important in bacterial infection pathogenesis, and ECM adhesins have been identified in most meningeal pathogens (e.g., PavA of *S. pneumoniae* [21], Opc of *N. meningitidis* [61], fimbriae of *H. influenzae* [62], and ScpB of GBS [31]).

2.1.3 Intravascular Survival

Once in the bloodstream, meningeal pathogens are immediately exposed to the host's immune defense. Clearance of *S. pneumoniae*, *H. influenzae*, and *N. meningitidis* strongly relies on antibody-mediated opsonization followed by activation of the complement system. *S. pneumoniae* and *H. influenzae* are then taken up by host leukocytes and killed intracellularly, while the membrane attack complex, formed by the terminal complement factors, is important for the killing of *N. meningitidis* [63]. Patients with impaired opsonization or complement activation (e.g., patients with sickle cell disease or complement deficiencies) have a higher susceptibility to invasive infections due to *N. meningitidis* and *S. pneumoniae* [64–67].

The polysaccharide capsule, although impeding adherence to the mucosal surface and entry into epithelial cells [10, 68], is the primary survival factor in the bloodstream [69]. It shields bacteria against circulating antibodies [70, 71], complement-mediated bacterial killing [72], and neutrophil phagocytosis [73]. Meningeal pathogens may also manipulate the complement cascade by enhancing the activity of regulatory proteins [74–79].

Intravascular survival of the intracellular pathogen *L. monocytogenes* depends mainly on its ability to evade intracellular killing in the phagolysosome [80]. Bacterial clearance by phagocytosis is an important early step in listeriosis, but the intracellular survival of *L. monocytogenes* facilitates evasion of innate immune mechanisms ("Trojan-horse" mechanism).

In summary, the intricate defense mechanisms of bacteria against the host immune system only need to protect few organisms, which may then cause meningitis [81].

2.1.4 Meningeal Invasion

The CNS is a tightly controlled environment. The boundary to the blood is provided by the endothelial BBB, the epithelial BCSFB, and the arachnoid [82]. Blood-borne pathogens must cross the BBB or the BCSFB to cause bacterial meningitis. Clinical observations and experimental studies link the magnitude of bacteremia with the risk of meningitis development for most meningeal pathogens [65, 83, 84]. High bacterial load in the blood alone, however, is not sufficient for the development of bacterial meningitis. The expression of selected bacterial adhesion and invasion factors is necessary for the invasion of the CNS [85].

Meningeal pathogens may migrate into the CSF by transcellular or paracellular migration [85]. The BCSFB may be more vulnerable to bacterial migration via the paracellular route because the tight junctions between the epithelial cells have a lower electrical resistance [82]. For *L. monocytogenes*, the previously described Trojan-horse mechanism has also been implicated in the invasion of the CNS [86–88].

Direct invasion of endothelial and epithelial cells by meningeal pathogens involves ligand–receptor interactions between bacteria and host cells followed by rearrangements of the actin cytoskeleton. This ultimately leads to the uptake of the pathogen in a vacuole and transcellular transport [42, 85, 89]. Attachment of meningeal pathogens to the tight endothelium of the BBB and the fenestrated endothelium of the BCSFB is facilitated by receptors on the apical side of the endothelial cells, some of which may be upregulated by inflammation. The 37/67-kDa laminin receptor is a shared target of cytotoxic necrotizing factor 1 (CNF1)-expressing *E. coli* K1, *S. pneumoniae*, *N. meningitidis*, and *H. influenzae* [90–94]. Bacterial adhesins mediate adhesion to the 37/67-kDa laminin receptor, including CNF1 in *E. coli*, choline-binding protein A (CbpA) in *S. pneumoniae*, secretin PilQ and porin PorA in *N. meningitidis*, and porin OmpP2 in *H. influenzae* [92, 93, 95]. *S. pneumoniae* may also bind to the cell adhesion molecule platelet endothelial cell adhesion molecule-1 (PECAM-1) and pIgR [96, 97].

Fimbriae and pili, hair-like structures frequently found on the surface of bacteria, are instrumental to bacterial adhesion. Type IV pili mediate attachment, bacterial aggregation, and persistence of *N. meningitidis* on brain endothelial cells [98]. Binding of *E. coli* K1 to brain endothelial cells is facilitated by FimH on type 1 fimbriae CD48 and the association of the outer membrane protein A (OmpA) with glycoproteins on the cell surface [85]. Attachment of GBS to brain endothelial cells is facilitated by several adhesins including fibrinogen receptor FbsA, laminin receptor Lmb, and the pilus tip adhesins PilA and Srr1 [99–103].

The exact mechanisms by which *S. pneumoniae* invades the BBB are not yet known. Partially conflicting in vitro and in vivo data indicate that binding of *S. pneumoniae* to transmembrane receptors pIgR and PAFr may trigger clathrin-mediated cellular endocytosis [41, 97, 104]. Specifically, the interaction of phosphorylcholine with PAFr and expression of NanA seem to be necessary for the invasion of the BBB by *S. pneumoniae* [40, 105–107].

E. coli invades the BBB following the induction of cytoskeletal rearrangements and actin condensations through the activation of the Rho family GTPases [85, 108, 109]. The interaction of several *E. coli* proteins with the BBB has been shown to facilitate cytoskeletal rearrangements culminating in the uptake of *E. coli* into a vacuole [94, 110–117].

For GBS, experimental data support an important role of the fibronectin-binding protein SfbA and pili in BBB invasion [101, 118]. The expression of fibronectin-binding protein OpC has also been shown to be instrumental to the invasion of the BBB by *N. meningitidis* [119]. *N. meningitidis* also form bacterial microcolonies on the luminal face of brain endothelial cells followed by the reorganization of the intracellular cytoskeleton. This may lead to the opening of intercellular junctions followed by paracellular penetration of the BBB [98]. Receptor-mediated signaling, as well as local inflammation with the recruitment of leucocytes, may further modify tight junctions and allow paracellular penetration [98, 101].

Conflicting evidence exists, whether the transcellular invasion of brain endothelial cells by *L. monocytogenes* may be mediated by InlA and InlB in a similar fashion as in the intestine and placenta [86, 120].

Bacteremia-independent invasion of the CNS may follow focal infections of structures close to the brain (e.g., otitis media, mastoiditis, and sinusitis) or after disruption of the integrity of the skull and meninges (e.g., malformations, trauma, neurosurgery). In a case series of 87 patients with pneumococcal meningitis, more than half of the patients had radiological signs consistent with otitis or sinusitis [121]. In a large observational study, otitis and sinusitis were reported as predisposing conditions in 25 % of patients [122]. This is supported by an experimental model of *S. pneumoniae* meningitis, in which a *galU* mutant and its parent pneumococcal strain both caused meningitis following otitis media infection in gerbils, despite the mutant's impaired ability to disseminate to the bloodstream following infection [123].

For *L. monocytogenes*, *S. pneumoniae* and *N. meningitidis* retrograde access to the CNS via the neural route has also been documented [86, 124, 125]. For *L. monocytogenes*, however, this mechanism is likely only important in ruminants. These observations support the notion that meningeal pathogens can gain access to the CNS by several routes. The relative importance of the different routes of infection in bacterial meningitis is not exactly known.

3 Induction and Modulation of Inflammation

Our understanding of the cascade of events leading to brain inflammation during meningitis is mostly based on experimental models, in which bacteria are inoculated intranasally, but more often intracisternally or intracerebrally. These models do not fully replicate the development of meningitis as observed in patients, especially when considering the initial anatomical progression from the nasopharynx.

Early recognition of the pathogens involves brain vascular endothelial cells or perivascular macrophages [126], as well as ependymal cells. These may be aided by a trafficking population of central memory T cells which interact with local antigen-presenting cells and initiate inflammation. The very recent discovery of functional lymphatic vessels in the CNS, however, may considerably change the perception on the role of adaptive immunity in the brain [127]. Leukocytes, mainly neutrophils in the case of bacterial meningitis, are initially excluded from the CNS, being recruited later during infection [124–126]. Most pathogens responsible for bacterial meningitis have been shown to induce inflammation in brain microvascular endothelial cells in vitro, which facilitates their invasion through the blood–brain barrier. In experimental GBS meningitis, the early induction of tumor necrosis factor-α (TNF-α) and interleukin (IL)-1β mRNA expression has been documented in the ependyma and the meninges [127]. Little is known about the role of perivascular macrophages, but their depletion has been shown to worsen the clinical symptoms, to increase bacterial counts in the CSF, and to decrease leukocytes extravasation. However, since higher levels of CSF inflammation were still detected in depleted animals, a local production of inflammatory mediators in the parenchyma was also suggested [128].

During disease progression, cells in the parenchyma, especially next to the site of the initial inflammation, also express inflammatory factors, including TNF-α and IL-1β [127]. Microglia and astrocytes are sentinels in the brain parenchyma that are able to detect invading pathogens. It is postulated that, as the infection/inflammation progresses, diffusible bacterial products or host-derived factors penetrate the parenchyma, triggering a secondary response.

How the innate immune system recognizes invading pathogens is mostly determined by its ability to "sense" the bacterial surface. Bacterial pathogens causing meningitis usually are encapsulated microorganisms. The polysaccharide capsule is a virulence factor which is necessary for establishing invasive disease by preventing opsonophagocytosis (see above). In vivo, the pneumococcal capsule could also impair recognition by the innate immune system, in particular the toll-like receptor-mediated pathway [129]. The capsule can partially cover subcapsular bacterial components, depending on its thickness. For gram-positive bacteria, immunogenic components comprise teichoic (TA) and lipoteichoic acids (LTA) and peptidoglycan. For gram-negative bacteria, the endotoxin lipopolysaccharide (LPS) is recognized by the innate immune system. These components are released either by autolysis, when bacteria reach sufficient high density, or by the effect of bacteriolytic antibiotics and trigger the inflammatory reaction in bacterial meningitis. Specialized extracellular (i.e., toll-like receptors, TLRs) or intracytoplasmic (i.e., NOD-like receptors, NLRs) receptors can sense the major meningeal pathogens [130–133]. Most cells of the brain express these receptors, with microglia expressing the full repertoire of TLRs [134]. Brain endothelial cells and macrophages are also known to express these receptors. Negative outcome during bacterial meningitis has been associated with genetic variations in these pattern recognition receptors (PRR) [135–137]. TLR2 detects cell wall components of gram-positive bacteria, while TLR4 recognizes those of gram-negative bacteria. TLR9 recognizes bacterial DNA and other components. Scavenger receptors (SRs), initially described for their role in the identification and removal of modified lipoproteins, are found at the surface of perivascular and meningeal macrophages [138]. SRA-1 (CD204) and SRA-2 (MARCO) can detect *N. meningitidis* or *S. pneumoniae* [138]. The G-coupled receptor protein formyl peptide receptor-like-1 (FPRL-1), which increases the expression of antimicrobial peptide CRAMP, recognizes *S. pneumoniae*. Of note, it seems that the resident immune system in the brain parenchyma may already be activated during the initial presence of pneumococci in the blood in a bacteremia-derived meningitis model, even before the development of meningitis and the recruitment of neutrophils [139].

3.1 Cytokines and Chemokines

An increase in cytokines and chemokines expression levels has been documented in the brain parenchyma or the CSF of patients with bacterial meningitis and experimental models [140–143]. Cytokine and chemokine profiles vary in vitro and in experimental infection models in response to different meningeal pathogens. These differences are presumably related to pathogen-specific activation of PRR [144–147] and may in part explain the differences in mortality and morbidity observed in patients affected by the different pathogens.

The early release of TNF-α, IL-1β, and IL-6 in the course of bacterial meningitis is part of the initial innate immune answer. The expression of TNF-α and IL-1β is detected first in the ependyma and the meninges and later in the parenchyma in experimental meningitis [127]. The initial immune reaction induces a cascade of inflammatory mediators, including other cytokines and chemokines, antimicrobial peptides, prostaglandins, MMPs, NO, and ROS.

During bacterial meningitis, TNF-α activity is kept at a sustained high level in the CSF due to the continuous release of bacterial products into the CSF or by a positive feedback loop in the inflammatory cascade [148]. Similar to PPR engagement, binding of TNF-α to the TNF-α receptor leads to NF-kB activation resulting in the expression of further mediators described hereafter. Experimental administration of TNF-α into the CSF of rabbits or rats induces BBB breakdown, neutrophil influx, and increased cerebral blood flow (CBF), recapitulating pathophysiologic changes characteristic of bacterial meningitis in humans [149–154].

IL-1β is released by mononuclear phagocytes, glial cells, and endothelial cells in the CNS. Caspase-1 in the inflammasome complex activates the precursors of IL-1 or IL-18 [132, 155–157]. The CSF concentration of IL-1β significantly correlates with other inflammatory parameters and is a predictor of adverse disease outcome [158]. When experimentally injected into the CSF, IL-1 triggers a meningeal inflammation without detectable TNF-α activity, whereas the concomitant injection of TNF-α and IL-1 results in a synergistic increase in leukocyte influx into the CSF [159].

IL-6 is produced by monocytes, endothelial cells, and astrocytes, essentially in response to IL-1β and TNF-α. Its presence in the CSF of patients with bacterial meningitis is not correlated with any of the indices of meningeal inflammation or with disease severity [160]. IL-6 efficiently induces the expression of acute-phase proteins, fever, leukocytosis, and activation of the complement and clotting cascades [161]. IL-6 also possesses anti-inflammatory properties, by inhibiting TNF-α and IL-1β production in vitro and inducing IL-1 receptor antagonist [162].

Chemokine levels, including IL-8 (CXCL8), CXC5 (ENA-78), CXCL1 (GROα), and monocyte chemoattractant protein-1 (MCP-1, CCL2), MIP-1 (CCL3), and MIP-1 (CCL3), are elevated in the CSF of patients with bacterial meningitis [163, 164]. Cells shown to produce these chemokines upon stimulation include monocyte macrophages, polymorphonuclear leukocytes, endothelial cells, astrocytes, microglia, and neurons. Chemokines primarily activate and attract leukocytes to the site of inflammation. By enhancing neutrophil adhesion to endothelial cells, IL-8 regulates the migration of neutrophils toward the CNS during bacterial meningitis [142].

IL-10 is an anti-inflammatory cytokine that inhibits the production of TNF-α, IL-1, IL-6, and IL-8 in vitro. High levels of IL-10 have been found in the CSF of patients with bacterial meningitis. In experimental meningitis, application of IL-10 decreased brain edema and down-modulated subarachnoid space inflammation [165]. In addition to IL-10, transforming growth factor-β (TGF-β) also possesses anti-inflammatory properties, as demonstrated in experimental meningitis, where its application reduced cerebral edema, intracranial pressure, and CBF [166]. Similarly, IL-19 is produced by astrocytes in response to *S. pneumoniae* infection and acts as an anti-inflammatory modulator [167].

In bacterial meningitis, especially caused by *S. pneumoniae*, IFN-γ levels are increased in the CSF [157, 168]. IFN-γ stimulates nonspecific defense mechanisms such as phagocytosis or cytokine release by macrophages and polymorphonuclear leukocytes. In experimental pneumococcal meningitis in mice, the inflammasome complex initiates the production of IFN-γ [157]. IFN-γ inactivation results in increased survival, attenuation of pro-inflammatory cytokines in the CSF, less brain injury, and improved neurofunctional outcome [169].

3.2 *Matrix Metalloproteinases and Related Proteins*

MMPs are important role players to the pathogenesis of bacterial meningitis, participating in the breakdown of the BBB and the intrathecal production of cytokines, therefore contributing to the pleocytosis of neutrophils in the CSF [170, 171]. As endopeptidases are able to degrade ECM components, MMPs facilitate cell migration, tissue remodeling, and cytotoxicity. MMPs also modulate cytokine production and release. MMPs and related metalloproteinases, e.g., TNF-α converting enzymes (TACE/ADAM-17), transform membrane-bound cytokines, cytokine receptors, and adhesion molecules to their soluble forms by virtue of their sheddase activity. In return, cytokines such as TNF-α, IL-1, and IL-2 modulate the expression and regulation of MMPs. Most MMPs are not constitutively expressed but produced in response to increased levels of cytokines, eicosanoids, growth factors, and the presence of pathogens. Negative regulation of MMP activity is mediated by tissue inhibitors of metalloproteinases (TIMP), the specific endogenous inhibitors of MMPs. These inhibit MMPs by the formation of complexes with pro-peptide containing inactive and cleaved activated forms of MMPs. Similar to MMPs, TIMPs are also regulated in response to changing levels of a variety of signaling molecules [172].

Levels of MMP-1, MMP-3, MMP-7, MMP-8, MMP-9, and MMP-10 have been found to be elevated in the CSF of patients with bacterial meningitis. Modest increases in TIMP-1 and TIMP-2 levels have also been observed while TIMP-4 was concomitantly downregulated in the CSF of the same patient cohort [173–175]. In the brain tissue of patients with purulent meningoencephalitis, endothelial cells and infiltrating leukocytes have been shown to produce MMP-9, TIMP-1, and TIMP-2 [176].

In pediatric patients with bacterial meningitis, MMP-9 levels in the CSF fell by 90 % within 1 week, while TIMP-1 levels continued to rise by 51 % [177]. Persistent high levels of MMP-9 may be associated with a negative neurological outcome in childhood bacterial meningitis [177, 178].

In experimental pneumococcal meningitis, changes in MMP-9/TIMP-1 ratio and an increase in collagen degradation were observed, resulting in cortical brain damage [179]. MMP inhibitors prevented this effect [180, 181]. Also, combined inhibition of MMP and TACE led to a reduction of hippocampal apoptosis and preserved learning capacity of animals that recovered from bacterial meningitis [148, 182].

3.2.1 Neutrophil Invasion

Marked neutrophil pleocytosis in the CSF is a distinctive feature of bacterial meningitis. In response to chemotactic stimuli, neutrophils extravasate across the fenestrated endothelium followed by an inverse transmigration across the tight epithelial ependyma to accumulate in the CSF where they contribute to the deleterious effects of inflammation in the brain [183–185].

Inflammation induces E-selectin and P-selectin expression on the surface of endothelial cells. This allows the binding of neutrophils through P-selectin glycoprotein ligand-1 and other glycosylated ligands [186]. Deficiency in E-selectin or P-selectin expression resulted in almost complete inhibition of neutrophil influx when recombinant IL-1β and TNF-α were directly injected into the CSF of mice [187]. Blocking of selectins by the polysaccharide fucoidan also attenuates CSF neutrophil pleocytosis in experimental pneumococcal meningitis [188–190]. The firm adhesion of chemokine-activated leukocytes to the endothelium is mediated by the binding of integrins to members of the immunoglobulin-like superfamily on the endothelium, including ICAM-1 and vascular cell adhesion molecule-1 (VCAM-1). Antibody-mediated blocking of integrins or ICAM-1 decreases leukocyte influx in experimental meningitis models, confirming an important role for these molecules in assisting leukocyte entry into the inflamed neural tissue [191–193].

Bacteria-induced transmigration of leukocytes across the BBB or the BCSFB may occur by the paracellular or transcellular route as determined in in vitro models [194–197]. Intracellular signaling induced by the interaction of integrins on the leukocytes with endothelial cell adhesion molecules leads to structural changes in endothelial cells including remodeling of the actin cytoskeleton. This facilitates migration of leukocytes either transcellularly or paracellularly [198]. Interestingly, the early presence of leukocytes in the CSF does not affect bacterial multiplication in experimental meningitis models [199, 200].

3.2.2 Resolution of the Inflammation

Apart from the production of anti-inflammatory activity of IL-6, IL-10, and TGF-β, different mechanisms are initiated later during the disease course and participate in the resolution of inflammation. The clearance of invading neutrophils by apoptosis is necessary to avoid the excess release of neurotoxic molecules. Experimental inhibition of apoptosis in neutrophils by overexpression of the antiapoptotic factor *bcl-2* prolonged the expression of pro-inflammatory cytokines in experimental pneumococcal meningitis in mice. On the contrary, treatment with roscovitine, a specific inducer of apoptosis in neutrophils, reduced damage and improved survival in vivo in the same model [201]. Similarly, TNF-related apoptosis-inducing ligand (TRAIL) released by microglia triggers leukocyte apoptosis. Elevated levels of TRAIL are documented in the CSF of patients with pneumococcal meningitis. The administration of recombinant TRAIL increased leukocyte apoptosis and decreased inflammation and neuronal apoptosis in experimental pneumococcal meningitis, while TRAIL knockout animals displayed a worse outcome than wild-type mice [202].

4 Sites of Damage in Meningitis

Derived from observations in patients and experimental models of bacterial meningitis, the cerebral vasculature, the brain parenchyma (cortex or hippocampus), and the inner ear are primarily injured by the disease. The morbidity observed in survivors reflects damage to these structures.

Figure 3 shows the histopathology of experimental pneumococcal meningitis and group B streptococcal meningitis.

4.1 Cerebral Vasculature

The cerebral vasculature is one of the primary sites involved in the development of meningitis (Fig. 3e). The early activation of endothelial cells by inflammatory mediators, the disruption of endothelial function by meningeal pathogens, and the neutrophil influx into the CSF lead to several pathological changes. These include the disruption of the BBB, brain edema, loss of CBF autoregulation, and focal and global changes of CBF resulting in cerebral ischemia. Extensive cerebral infarction, resulting from vasculitis and coagulation disturbances, and circulatory failure due to septic shock may lead to acute death in bacterial meningitis [203].

4.1.1 Pathology

Subarachnoid space inflammation is a characteristic feature of acute bacterial meningitis. Arterial narrowing has been shown to be a predominant finding in patients with arterial complication. Vasculitis or vasospasm may lead to brain ischemia. Additionally, venous thrombosis has been observed in adult patients with pneumococcal meningitis [121, 204, 205]. In fatal cases of neonatal meningitis, inflammatory vasculitis is uniformly present, indicating that the cerebral vasculature of the

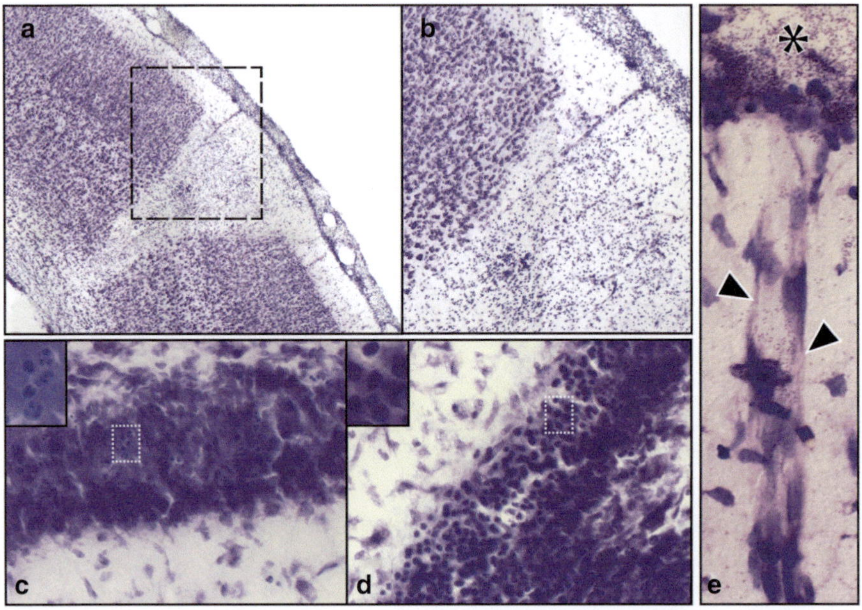

Fig. 3 Histopathology of experimental pneumococcal meningitis (**a**–**c**, **e**) or group B streptococcal meningitis (**d**). (**a**) Extensive cortical injury consisting of areas of cortical necrosis. A wedge-shaped distribution of reduced neuronal density, suggestive of ischemic damage, can be distinguished from neighboring more healthy cortical tissue (*cresyl violet*, magnification × 10). (**b**) Cortical neuronal loss on the right contains necrotic neurons characterized by cell swelling and fading of cytoarchitecture and is sharply demarcated from preserved neurons on the upper left (*cresyl violet*, magnification × 40). (**c**) Hippocampal dentate gyrus histology of an infant rat suffering from pneumococcal meningitis at 42 h after infection. Apoptotic cells are characterized by the presence of round or oval apoptotic bodies consisting of regularly shaped, often fragmented, dark chromatin clumps (*upper left inset*). Apoptotic cells are mostly observed in the inner rim (i.e., the subgranular zone) of the dentate gyrus (*cresyl violet*, magnification × 40). (**d**) Dentate gyrus of the hippocampus of an infant rat infected with group B streptococcus. Large clusters of cells with uniformly dense and shrunken (pyknotic) nuclear morphology (*upper left inset*) are observed throughout the entire blade. (**e**) Subarachnoid space inflammation consisting of bacteria and inflammatory cells (*asterisk*) extending into the Virchow-Robin space (*arrowheads*) around the penetrating cortical vasculature (*cresyl violet*, magnification × 63)

neonate may be particularly susceptible to inflammatory damage, leading to notably severe structural damage to the neonatal brain [206].

4.1.2 Disruption of the Blood–Brain Barrier

In meningitis, the permeability of the BBB and BCSFB increases as the result of functionally relevant alterations induced by the disease process. Mainly paracellular leakage has been studied in this respect. The remodeling of the actin cytoskeleton, the reorganization of tight junctions, and the enzymatic degradation of tight junctions and basement membrane components participate in the permeabilization of the BBB [198, 207].

The meningeal pathogens are able to induce direct damage to endothelial or epithelial cells. In vitro cytotoxic damage to brain endothelial cells has been seen with *S. pneumoniae*, mainly mediated by pneumolysin [208–210], *N. meningitidis*, primarily mediated by NO [211], and *E. coli* K1 [212]. Pneumolysin was also shown to lead to astrocyte cytoskeleton remodeling in vitro [213], which might result in BBB disruption. In experimental GBS and *S. pneumoniae* meningitis, increased expression of the transcriptional repressor Snail1 has been found, leading to downregulation of tight junction components, like ZO-1, claudin-5, and occludin [214].

Signaling and effector molecules released during the inflammation process and changes caused by the interaction of inflammatory cells with the brain barriers also alter the function of cells constituting the BBB. A variety of cytokines and chemokines have been shown to induce tight junction and cytoskeleton rearrangements leading to BBB disruption [215–218].

MMPs increase BBB permeability by proteolytic breakdown of the basement membrane and tight junctions. In experimental meningococcal meningitis in rats, disruption of the BBB, increased intracranial pressure, and CSF pleocytosis were observed in parallel to an increase of MMP-9 activity in the CSF 6 h after infection [170]. Experimental inhibition of MMPs reduces occurrence of intracerebral bleeding and BBB breakdown in mice with meningococcal meningitis [219].

ROS and NO also contribute to BBB disruption via lipid peroxidation and oxidative damage to cell membrane proteins [220]. Neutralization of ROS and NO by radical scavengers in experimental meningitis prevented cerebral damage including BBB breakdown [221–224]. Additionally, ROS and NO lead to tight junction and cytoskeletal reorganization and MMP activation [225].

As a consequence of the increased BBB/BCSFB permeability, potentially harmful molecules found at high concentrations in the blood can accumulate in the CSF, unfavorably modifying the brain's microenvironment. On the other hand, CSF penetration of antibiotics is facilitated by BBB opening, therefore improving the efficiency of therapy by increasing antibiotic CSF/serum ratio in comparison to noninflammatory conditions [226].

4.1.3 Alterations of Cerebral Blood Flow

Marked changes in CBF are observed during bacterial meningitis. Brain edema, obstructive hydrocephalus, meningitis-associated cerebritis, cerebral infarction and cerebral venous thrombosis, and status epilepticus may all lead to intracerebral hypertension [227, 228]. In the advanced disease stages, CBF is usually reduced [204, 229, 230]. Focal changes in the vasculature, loss of CBF autoregulation, intracranial pressure, and systemic hypotension perturb CBF in bacterial meningitis.

In humans, cerebrovascular changes in bacterial meningitis include segmental narrowing of vessels, irregularities of vessel walls, and arterial and venous thromboses [121, 231–233]. The middle and anterior cerebral arteries or the basilar artery are frequently affected by stroke associated with bacterial meningitis. Vessel narrowing results in an increased velocity of cerebral blood, an early predictor of cerebrovascular complications [121, 204]. Sequelae of vascular complications may result in hemiparesis or quadriparesis [206, 234, 235].

Disturbance of CBF autoregulation, combined with systemic hypotension, may lead to cerebral ischemia [236–240]. Alternatively, systemic hypertension may augment vasogenic edema and intracranial pressure [240, 241], ultimately limiting CBF. In addition to global cerebral hypoperfusion, vasculitis of large and small arteries traversing the inflamed subarachnoid space leads to regional hypoperfusion. This form of ischemic damage (Fig. 3a, b) may be responsible for permanent neurologic sequelae following bacterial meningitis [235, 236].

NO plays a crucial albeit complex role in modulating CBF during meningitis. Early in the disease, the vasodilatory effect of NO contributes to the hyperemia induced by the subarachnoid space inflammation. Later, NO generated in the vasculature provides some protection against ischemia by attenuating the effects of the progressive decline in CBF due to the production of vasoconstrictive factors (see later discussion).

ROS play a critical role in modulating CBF during meningitis [221, 242–244]. Generation of ROS is localized primarily to the cells constituting the subarachnoid and ventricular inflammation and the cerebral vasculature, as shown in infant rats with experimental meningitis [221]. In this model, the cerebral vasculature showed evidence of marked oxidative alterations, whereas oxidative damage to the brain parenchyma itself was not documented conclusively [245]. Antioxidant treatment prevents oxidative vascular damage, hyperemia, CBF decline, and rise of ICP in experimental pneumococcal and GBS meningitis [221, 222, 243, 244].

The reaction of NO with ROS produces peroxynitrite, a strong oxidant that exerts cytotoxic effects [220, 246]. Nitrotyrosine residues, a reaction product of peroxynitrite with proteins, were detected in the meninges, the cortical blood vessels, and inflammatory cells in the brains of patients with bacterial meningitis and corresponding animal models [220, 247]. Treatment with urate, a peroxynitrite scavenger, reduced the meningeal inflammation, BBB disruption, and intracranial hypertension [247].

Increased endothelin levels are found in the CSF of patients with bacterial meningitis [248]. This potent vasoconstrictive peptide is produced in the CNS by vascular endothelial cells, astrocytes, and neurons and participates in CBF regulation. Endothelin synthesis is triggered by cytokines, i.e., TNF-α [249–251], and inhibited by NO [252]. In experimental pneumococcal meningitis, an endothelin antagonist (bosentan) significantly prevented the reduction of CBF and attenuated the extent of cerebral ischemia [253].

4.2 Inner Ear

Unilateral or bilateral sensorineural hearing loss (SNHL) is the most common neurologic sequelae following bacterial meningitis and is found in 5–30 % of survivors [254–257], with *S. pneumoniae* causing the highest rate of sensorineural hearing loss [258–261]. Hearing loss during bacterial meningitis is progressive rather than abrupt, and its magnitude depends on the duration of untreated infection [255]. Hearing loss is the direct result of inflammation in the inner ear during the acute phase of the disease. Magnetic resonance imaging (MRI) studies in humans with meningitis have confirmed the inflammatory involvement of the inner ear [262]. In survivors of meningitis, progressive cochlear ossification and spiral ganglion loss are observed years after the disease [263, 264].

4.2.1 Pathology

During the acute stage of meningitis, suppurative labyrinthitis is usually observed in the human temporal bone of patients, often accompanied by purulent infiltrate in the perilymphatic duct [265]. Studies in experimental meningitis have shown that bacteria and inflammatory cells are present in the cochlea in the earliest stages of pneumococcal infection [266, 267]. Inflammatory infiltration of the cochlea progresses via the cochlear aqueduct to the perilymphatic space (scala tympani) and via the spiral ligament to the endolymphatic space [267–269]. Damage to the blood–labyrinth barrier, the hair cells, and the spiral ganglion is observed first at the base and later at the apex of the cochlea, corresponding to hearing loss first at high then at low frequencies [270]. The first signs of damage and hearing loss are observed for hair cells with a partial reversibility of associated hearing loss [270]. Later during the course of the disease, irreparable ultrastructural inner ear damage, especially the loss of spiral ganglion neuronal cells, is associated with severe to profound deafness [266, 271]. Toxic effects of the meningeal pathogen (e.g., pneumolysin from *S. pneumoniae*) and inflammatory mediators appear to be responsible for the cytopathic effects [272, 273]. In particular, the production of ROS and NO has been involved in the pathogenesis of cochlear damage and hearing loss in bacterial meningitis [270]. Treatment with peroxynitrite scavengers or antioxidants attenuated hearing loss and protected spiral ganglion neuronal cells [274, 275].

4.3 Central Nervous Tissue

4.3.1 Neurologic Sequelae

Bacterial meningitis causes damage to both cortical and subcortical brain structures. The neurologic sequelae resulting from brain damage include hearing impairment (see previous discussion), obstructive hydrocephalus, focal sensory-motor deficits, mental retardation, seizure disorders, and cortical blindness. Behavioral and cognitive sequelae in children and adults after bacterial meningitis are common [276–281]. Morbidity in patients surviving bacterial meningitis is usually higher with *S. pneumoniae* as causative pathogen [122, 254, 282].

Follow-up studies revealed that neurologic sequelae of childhood meningitis persist for more than 10 years, impacting the school performance of affected children [277,

279, 283, 284]. In adults tested 0.5–13.5 years after acute bacterial meningitis, 32 % of 155 meningitis survivors suffered from relevant impairment of psychomotor performance, speed of cognitive processes, and concentration and memory functions [280].

4.3.2 Pathology

Histological changes are often seen in a focal pattern (Fig. 3a, b). Infarctions in the frontal, temporal, and parietal lobes, the basal ganglia, the thalamus, as well as periventricular white matter injuries are found in newborns or children affected by bacterial meningitis, leading to cerebral atrophy and hydrocephalus [205, 285–287]. Structural changes were also seen in the temporal lobe and the limbic system of adult patients [281]. Neuronal loss is associated with a marked activation of astrocytes and microglia [288], as well as axonal injury [289, 290]. Furthermore, meningitis induced by gram-negative bacteria is often characterized by the development of brain abscesses [291, 292].

Apoptotic cell death has been observed in the dentate gyrus of the hippocampus during meningitis [221, 293, 294] (Fig. 3c) with the presence of condensed, fragmented nuclei (Fig. 3c, inset), and the detection of fragmented DNA and active caspase-3, an effector enzyme involved in the execution of programmed cell death [295]. A second form of hippocampal neuronal damage with morphologically distinct features (uniformly shrunken nuclei, clusters of damaged cells) may also be found in the lower blade of the dentate gyrus cell, spanning the entire width of the dentate gyrus band (Fig. 3d). This form of hippocampal damage is reminiscent of ischemia-related neuronal damage [296] and is the preferential pattern of neuronal injury observed in meningitis caused by GBS [216, 294].

The neuronal injury in the hippocampus is of particular significance because experimental data suggest that it is related to learning impairment following meningitis [148, 297, 298]. The observation of cell death primarily of immature progenitor cells in the subgranular zone of the dentate gyrus, cells necessary for the acquisition of new memory, strengthens this hypothesis [294, 295, 299]. Cognitive impairment and learning disabilities following meningitis may thus be reflected by damage to the dentate gyrus of the hippocampus [148, 297, 298, 300].

Confirming the findings in animal models, brain sections of patients who died from bacterial meningitis showed apoptotic neurons with cells staining for the active form of caspase-3 in the dentate gyrus [301]. Also, volumetric measurements of the hippocampus by MRI techniques showed unilateral and bilateral hippocampal atrophy in patients surviving meningitis [302], potentially reflecting the apoptotic loss of neurons observed by histopathology.

4.3.3 Brain Edema and Cerebral Herniation

Cerebral edema in bacterial meningitis may be the combined result of vasogenic, cytotoxic, interstitial, or osmotic edema [296]. Vasogenic cerebral edema is the consequence of increased BBB permeability, with the extravasation of plasma proteins into the brain parenchyma (see earlier discussion) and mainly affects the white matter [296]. Cytotoxic edema is an increase in intracellular water due to the intracellular accumulation of osmotically effective ions (i.e., sodium, potassium, or glutamate). Cytotoxic mechanisms include ischemia and the effect of excitatory amino acids (EAA) [296, 303, 304]. Cytotoxic edema affects the gray and white matter. The increase of CSF influx due to a more sustained production (increased blood flow in the choroid plexus) or a decreased resorption (increased CSF outflow resistance across the arachnoid villi system of the sagittal sinus) is the cause of interstitial edema [305]. Finally, osmotic edema may be caused by inappropriate secretion of antidiuretic hormone during the course of bacterial meningitis. The observed hypoosmolality of serum results in a net influx of water into the brain [296, 306].

Brain edema is an important contributor to the fatal outcome of bacterial meningitis [121, 307, 308]. Extensive brain edema leads to increased intracranial pressure culminating in the herniation of brain tissue and compression of the brainstem.

Aquaporins (AQP) are pore-forming membrane proteins regulating water homeostasis of the brain [309]. In cytotoxic edema, the transcellular influx of water across the BBB is facilitated by AQP-4, while in vasogenic and interstitial brain edema, the absence of AQP-4 leads to more severe brain edema by limiting outflow of CSF [296, 310]. In a mouse model of pneumococcal meningitis, AQP-4 was significantly upregulated and associated with increased cytotoxic brain edema [311]. AQP-4 upregulation was also found in the brain of a patient with bacterial meningitis [312], but not in the CSF of patients with bacterial meningitis [313].

4.3.4 Mediators of Cell Death in Neuronal Tissue

EAAs including glutamate induce neuronal apoptosis and necrosis and appear to mediate cellular injury in a variety of brain disorders. In bacterial meningitis, ischemia and EAA may cause direct neuronal toxicity as suggested from experimental models or clinical studies [288, 304]. In patients and experimental pneumococcal meningitis in rabbits, CSF glutamate concentrations were significantly elevated and correlated with the severity of the disease [314, 315]. The cause for increased concentrations of EAA in the brain during meningitis is still poorly understood, but may be related to BBB disruption. Alternatively, pore-forming cytolysins, like pneumolysin, may cause the release of glutamate from astrocytes, leading to subsequent glutamate-dependent synaptic damage [316]. Kynurenic acid, a nonselective inhibitor of the neurotoxic effect of EAA acting on NMDA receptors, significantly attenuated brain injury, both in the cortex and in the hippocampus in an infant rat model of neonatal meningitis [288, 304].

The intensity and duration of the inflammatory response in bacterial meningitis correlates with the development of neuronal damage. Meningeal pathogens interfere with the programmed cell death of immune cells, possibly prolonging inflammation. For example, *N. meningitidis* inhibits apoptosis in neutrophils [317] or macrophages [318]. Similarly GBS interfere with TLR-2-mediated programmed cell death in macrophages and microglia [319].

The brain is one of the most metabolically active organ in the body, therefore requiring a high rate of oxygen consumption. But because it possesses a high level of polyunsaturated fatty acid prone to lipid peroxidation, the brain is particularly vulnerable to oxidative damage caused by ROS. This is also exacerbated by the high abundance of redox-active metals (iron or copper) and the relative low levels of endogenous antioxidant [320]. The high concentration of mitochondria in cerebrovascular endothelial cells might also account for the sensitivity of the BBB to oxidant stressors [321]. The prevention of BBB disruption, attenuation of lipid peroxidation, reduction of ischemia in the brain parenchyma, and the reduced neuronal injury observed with the use of radical scavengers in experimental meningitis underline the role of ROS [322]. Despite these observations, a direct neurotoxic effect of ROS in bacterial meningitis has not been documented conclusively. Most of the beneficial effects of ROS scavengers in experimental meningitis may be primarily due to a beneficial effect at the level of the cerebral vasculature [245] (see previous discussion).

NO may be either neuroprotective or neurotoxic in bacterial meningitis. In its oxidized form, it can inactivate glutamate receptors, therefore reducing EAA-mediated excitotoxicity. In its reduced form, it reacts with superoxide and forms peroxynitrite, which is highly reactive and causes damage to DNA, lipid membranes, and proteins [323]. Caspase-3-dependent hippocampal apoptosis was reduced during experimental pneumococcal meningitis in mice lacking iNOS expression [324]. In contrast, iNOS inhibition by aminoguanidine exacerbates neuronal injury in experimental GBS meningitis [325].

MMPs may exert direct neurotoxic effects by degrading perineuronal components of the ECM, like laminin [326, 327], or the neural cell adhesion molecule NCAM [328]. Indirectly, the shedding of death receptors ligands (FasL, TNF-α) by MMPs can induce death in cells that express these ligands, including neurons [329]. Given as adjuvant therapy, MMP inhibitors reduced the extent of cortical damage, and combined inhibition of MMP and TACE led to a reduction in hippocampal apoptosis [148, 180, 182]. MMP-TACE inhibitors also improved learning capacity after experimental pneumococcal meningitis in infant rats [148].

Bacteria themselves may also directly damage neurons. Pneumococci have been shown to induce cell death in microglia or neurons in vitro [330]. Pneumolysin and superoxide produced by *S. pneumoniae* have been identified as the responsible neurotoxic mediators. In vitro, these toxins cause neuronal death via damage to mitochondria with the subsequent release of the proapoptotic mitochondrial factors cytochrome *c* and AIF [331, 332]. In experimental pneumococcal meningitis in rabbits, pneumolysin co-localized with apoptotic neurons of the hippocampus, and infection with pneumococcal mutants deficient for pneumolysin and superoxide production caused significantly less damage [331]. Pneumolysin was also shown to induce cochlear hair cell death in the rat [333] and endothelial cell damage in vitro [209, 210]. Other bacterial pore-forming toxins such as the beta hemolysin/cytolysin of GBS [334] are able to directly induce neuronal damage.

5 Conclusion

The pathophysiology of bacterial meningitis involves a complex interplay between the invasive pathogen, the anatomical barriers of the CNS, and the immune defense system of the host. In that context, the BCSFB and BBB play a primordial role, by restricting access to the brain. Meningeal pathogens have evolved sophisticated strategies to cross the otherwise hard-to-reach CNS compartment. Doing so, the successfully invading pathogens initiate a catastrophic cascade of events, involving the permeabilization of the barrier membranes, the massive recruitment of neutrophils, and a deleterious inflammatory reaction. Protecting barrier integrity by targeting pathophysiological processes that lead to its impairment, as demonstrated from experimental evidence using pharmacologic interventions including, e.g., radical scavengers or MMP inhibitors, is a promising strategy with the potential to reduce neurofunctional impairment in survivors of the disease.

Acknowledgments This work was supported by grant no. 162583 (to SLL) from the Swiss National Science Foundation.

Bibliography

1. van de Beek, Cabellos et al (2016) ESCMID guideline: diagnosis and treatment of bacterial meningitis. Clin Microbiol Infect 22 Suppl 3:S37–S62. doi: 10.1016/j.cmi.2016.01.007
2. McCullers JA (2006) Insights into the interaction between influenza virus and pneumococcus. Clin Microbiol Rev 19(3):571–582. doi:10.1128/CMR.00058-05, 19/3/571 [pii]
3. Avadhanula V, Rodriguez CA, Devincenzo JP, Wang Y, Webby RJ, Ulett GC, Adderson EE (2006) Respiratory viruses augment the adhesion of bacterial pathogens to respiratory epithelium in a viral species- and cell type-dependent manner. J Virol 80(4):1629–1636. doi:10.1128/JVI.80.4.1629-1636.2006, 80/4/1629 [pii]
4. Greenwood BM (1987) The epidemiology of acute bacterial meningitis in tropical Africa. In: Bacterial meningitis. Academic press, London, p 61–91
5. Mook-Kanamori BB, Geldhoff M, van der Poll T, van de Beek D (2011) Pathogenesis and pathophysiology of pneumococcal meningitis. Clin Microbiol Rev 24(3):557–591. doi:10.1128/CMR.00008-11, 24/3/557 [pii]
6. Steinfort C, Wilson R, Mitchell T, Feldman C, Rutman A, Todd H, Sykes D, Walker J, Saunders K, Andrew PW et al (1989) Effect of Streptococcus pneumoniae on human respiratory epithelium in vitro. Infect Immun 57(7):2006–2013
7. Hirst RA, Sikand KS, Rutman A, Mitchell TJ, Andrew PW, O'Callaghan C (2000) Relative roles of pneumolysin and hydrogen peroxide from Streptococcus pneumoniae in inhibition of ependymal ciliary beat frequency. Infect Immun 68(3):1557–1562
8. Kadioglu A, Taylor S, Iannelli F, Pozzi G, Mitchell TJ, Andrew PW (2002) Upper and lower respiratory tract infection by Streptococcus pneumoniae is affected by pneumolysin deficiency and differences in capsule type. Infect Immun 70(6):2886–2890
9. Rao VK, Krasan GP, Hendrixson DR, Dawid S, St Geme JW 3rd (1999) Molecular determinants of the pathogenesis of disease due to non-typable Haemophilus influenzae. FEMS Microbiol Rev 23(2):99–129, S0168-6445(98)00039-4 [pii]
10. Merz AJ, So M (2000) Interactions of pathogenic neisseriae with epithelial cell membranes. Annu Rev Cell Dev Biol 16:423–457

11. Wani JH, Gilbert JV, Plaut AG, Weiser JN (1996) Identification, cloning, and sequencing of the immunoglobulin A1 protease gene of Streptococcus pneumoniae. Infect Immun 64:3967–3974
12. Poulsen K, Reinholdt J, Jespersgaard C, Boye K, Brown TA, Hauge M, Kilian M (1998) A comprehensive genetic study of streptococcal immunoglobulin A1 proteases: evidence for recombination within and between species. Infect Immun 66(1):181–190
13. Vitovski S, Read RC, Sayers JR (1999) Invasive isolates of Neisseria meningitidis possess enhanced immunoglobulin A1 protease activity compared to colonizing strains. FASEB J 13(2):331–337
14. Callewaert L, Van Herreweghe JM, Vanderkelen L, Leysen S, Voet A, Michiels CW (2012) Guards of the great wall: bacterial lysozyme inhibitors. Trends Microbiol 20(10):501–510. doi:10.1016/j.tim.2012.06.005, S0966-842X(12)00116-3 [pii]
15. Boneca IG, Dussurget O, Cabanes D, Nahori MA, Sousa S, Lecuit M, Psylinakis E, Bouriotis V, Hugot JP, Giovannini M, Coyle A, Bertin J, Namane A, Rousselle JC, Cayet N, Prevost MC, Balloy V, Chignard M, Philpott DJ, Cossart P, Girardin SE (2007) A critical role for peptidoglycan N-deacetylation in Listeria evasion from the host innate immune system. Proc Natl Acad Sci U S A 104(3):997–1002. doi:10.1073/pnas.0609672104, 0609672104 [pii]
16. Davis KM, Akinbi HT, Standish AJ, Weiser JN (2008) Resistance to mucosal lysozyme compensates for the fitness deficit of peptidoglycan modifications by Streptococcus pneumoniae. PLoS Pathog 4(12):e1000241. doi:10.1371/journal.ppat.1000241
17. Davis KM, Weiser JN (2011) Modifications to the peptidoglycan backbone help bacteria to establish infection. Infect Immun 79(2):562–570. doi:10.1128/IAI.00651-10, IAI.00651-10 [pii]
18. Shakhnovich EA, King SJ, Weiser JN (2002) Neuraminidase expressed by Streptococcus pneumoniae desialylates the lipopolysaccharide of Neisseria meningitidis and Haemophilus influenzae: a paradigm for interbacterial competition among pathogens of the human respiratory tract. Infect Immun 70(12):7161–7164
19. Bogaardt C, van Tonder AJ, Brueggemann AB (2015) Genomic analyses of pneumococci reveal a wide diversity of bacteriocins – including pneumocyclicin, a novel circular bacteriocin. BMC Genomics 16(1):554. doi:10.1186/s12864-015-1729-4
20. Margolis E, Yates A, Levin BR (2010) The ecology of nasal colonization of Streptococcus pneumoniae, Haemophilus influenzae and Staphylococcus aureus: the role of competition and interactions with host's immune response. BMC Microbiol 10:59. doi:10.1186/1471-2180-10-59
21. Hammerschmidt S (2006) Adherence molecules of pathogenic pneumococci. Curr Opin Microbiol 9(1):12–20. doi:10.1016/j.mib.2005.11.001, S1369-5274(05)00193-1 [pii]
22. Anderton JM, Rajam G, Romero-Steiner S, Summer S, Kowalczyk AP, Carlone GM, Sampson JS, Ades EW (2007) E-cadherin is a receptor for the common protein pneumococcal surface adhesin A (PsaA) of Streptococcus pneumoniae. Microb Pathog 42(5–6):225–236. doi:10.1016/j.micpath.2007.02.003, S0882-4010(07)00021-6 [pii]
23. Rowe HA, Griffiths NJ, Hill DJ, Virji M (2007) Co-ordinate action of bacterial adhesins and human carcinoembryonic antigen receptors in enhanced cellular invasion by capsulate serum resistant Neisseria meningitidis. Cell Microbiol 9(1):154–168. doi:10.1111/j.1462-5822.2006.00775.x, CMI775 [pii]
24. Weber A, Harris K, Lohrke S, Forney L, Smith AL (1991) Inability to express fimbriae results in impaired ability of Haemophilus influenzae b to colonize the nasopharynx. Infect Immun 59(12):4724–4728
25. Cotter SE, Yeo HJ, Juehne T, St Geme JW 3rd (2005) Architecture and adhesive activity of the Haemophilus influenzae Hsf adhesin. J Bacteriol 187(13):4656–4664. doi:10.1128/JB.187.13.4656-4664.2005, 187/13/4656 [pii]
26. Le Bouguenec C (2005) Adhesins and invasins of pathogenic Escherichia coli. Int J Med Microbiol 295(6–7):471–478
27. Sokurenko EV, Chesnokova V, Dykhuizen DE, Ofek I, Wu XR, Krogfelt KA, Struve C, Schembri MA, Hasty DL (1998) Pathogenic adaptation of Escherichia coli by natural variation of the FimH adhesin. Proc Natl Acad Sci U S A 95(15):8922–8926

28. Antao EM, Wieler LH, Ewers C (2009) Adhesive threads of extraintestinal pathogenic Escherichia coli. Gut Pathog 1(1):22. doi:10.1186/1757-4749-1-22, 1757-4749-1-22 [pii]
29. Pizarro-Cerda J, Cossart P (2006) Bacterial adhesion and entry into host cells. Cell 124(4):715–727. doi:10.1016/j.cell.2006.02.012, S0092-8674(06)00187-5 [pii]
30. Tazi A, Disson O, Bellais S, Bouaboud A, Dmytruk N, Dramsi S, Mistou MY, Khun H, Mechler C, Tardieux I, Trieu-Cuot P, Lecuit M, Poyart C (2010) The surface protein HvgA mediates group B streptococcus hypervirulence and meningeal tropism in neonates. J Exp Med 207(11):2313–2322. doi:10.1084/jem.20092594, jem.20092594 [pii]
31. Maisey HC, Doran KS, Nizet V (2008) Recent advances in understanding the molecular basis of group B Streptococcus virulence. Expert Rev Mol Med 10:e27. doi:10.1017/S1462399408000811, S1462399408000811 [pii]
32. Rinaudo CD, Rosini R, Galeotti CL, Berti F, Necchi F, Reguzzi V, Ghezzo C, Telford JL, Grandi G, Maione D (2010) Specific involvement of pilus type 2a in biofilm formation in group B Streptococcus. PLoS One 5(2):e9216. doi:10.1371/journal.pone.0009216
33. Mengaud J, Ohayon H, Gounon P, Mege RM, Cossart P (1996) E-cadherin is the receptor for internalin, a surface protein required for entry of L. monocytogenes into epithelial cells. Cell 84(6):923–932, S0092-8674(00)81070-3 [pii]
34. Lecuit M (2005) Understanding how Listeria monocytogenes targets and crosses host barriers. Clin Microbiol Infect 11(6):430–436. doi:10.1111/j.1469-0691.2005.01146.x, CLM1146 [pii]
35. Posfay-Barbe KM, Wald ER (2009) Listeriosis. Semin Fetal Neonatal Med 14(4):228–233. doi:10.1016/j.siny.2009.01.006, S1744-165X(09)00006-7 [pii]
36. Lecuit M, Nelson DM, Smith SD, Khun H, Huerre M, Vacher-Lavenu MC, Gordon JI, Cossart P (2004) Targeting and crossing of the human maternofetal barrier by Listeria monocytogenes: role of internalin interaction with trophoblast E-cadherin. Proc Natl Acad Sci U S A 101(16):6152–6157. doi:10.1073/pnas.0401434101, 0401434101 [pii]
37. Conner SD, Schmid SL (2003) Regulated portals of entry into the cell. Nature 422(6927):37–44. doi:10.1038/nature01451, nature01451 [pii]
38. Veiga E, Guttman JA, Bonazzi M, Boucrot E, Toledo-Arana A, Lin AE, Enninga J, Pizarro-Cerda J, Finlay BB, Kirchhausen T, Cossart P (2007) Invasive and adherent bacterial pathogens co-Opt host clathrin for infection. Cell Host Microbe 2(5):340–351. doi:10.1016/j.chom.2007.10.001, S1931-3128(07)00247-8 [pii]
39. Gradstedt H, Iovino F, Bijlsma JJ (2013) Streptococcus pneumoniae invades endothelial host cells via multiple pathways and is killed in a lysosome dependent manner. PLoS One 8(6):e65626. doi:10.1371/journal.pone.0065626
40. Ring A, Weiser JN, Tuomanen EI (1998) Pneumococcal trafficking across the blood–brain barrier. Molecular analysis of a novel bidirectional pathway. J Clin Invest 102(2):347–360. doi:10.1172/JCI2406
41. Radin JN, Orihuela CJ, Murti G, Guglielmo C, Murray PJ, Tuomanen EI (2005) Beta-Arrestin 1 participates in platelet-activating factor receptor-mediated endocytosis of Streptococcus pneumoniae. Infect Immun 73(12):7827–7835. doi:10.1128/IAI.73.12.7827-7835.2005, 73/12/7827 [pii]
42. Kim KJ, Elliott SJ, Di Cello F, Stins MF, Kim KS (2003) The K1 capsule modulates trafficking of E. coli-containing vacuoles and enhances intracellular bacterial survival in human brain microvascular endothelial cells. Cell Microbiol 5(4):245–252, 271 [pii]
43. Zhang JR, Mostov KE, Lamm ME, Nanno M, Shimida S, Ohwaki M, Tuomanen E (2000) The polymeric immunoglobulin receptor translocates pneumococci across human nasopharyngeal epithelial cells. Cell 102(6):827–837
44. Elm C, Braathen R, Bergmann S, Frank R, Vaerman JP, Kaetzel CS, Chhatwal GS, Johansen FE, Hammerschmidt S (2004) Ectodomains 3 and 4 of human polymeric Immunoglobulin receptor (hpIgR) mediate invasion of Streptococcus pneumoniae into the epithelium. J Biol Chem 279(8):6296–6304. doi:10.1074/jbc.M310528200, M310528200 [pii]
45. Gray-Owen SD, Blumberg RS (2006) CEACAM1: contact-dependent control of immunity. Nat Rev Immunol 6(6):433–446. doi:10.1038/nri1864, nri1864 [pii]

46. Sadarangani M, Pollard AJ, Gray-Owen SD (2011) Opa proteins and CEACAMs: pathways of immune engagement for pathogenic Neisseria. FEMS Microbiol Rev 35(3):498–514. doi:10.1111/j.1574-6976.2010.00260.x
47. Griffiths NJ, Bradley CJ, Heyderman RS, Virji M (2007) IFN-gamma amplifies NFkappaB-dependent Neisseria meningitidis invasion of epithelial cells via specific upregulation of CEA-related cell adhesion molecule 1. Cell Microbiol 9(12):2968–2983. doi:10.1111/j.1462-5822.2007.01038.x, CMI1038 [pii]
48. Nikitas G, Deschamps C, Disson O, Niault T, Cossart P, Lecuit M (2011) Transcytosis of Listeria monocytogenes across the intestinal barrier upon specific targeting of goblet cell accessible E-cadherin. J Exp Med 208(11):2263–2277. doi:10.1084/jem.20110560, jem.20110560 [pii]
49. Disson O, Grayo S, Huillet E, Nikitas G, Langa-Vives F, Dussurget O, Ragon M, Le Monnier A, Babinet C, Cossart P, Lecuit M (2008) Conjugated action of two species-specific invasion proteins for fetoplacental listeriosis. Nature 455(7216):1114–1118. doi:10.1038/nature07303, nature07303 [pii]
50. Clarke TB, Francella N, Huegel A, Weiser JN (2011) Invasive bacterial pathogens exploit TLR-mediated downregulation of tight junction components to facilitate translocation across the epithelium. Cell Host Microbe 9(5):404–414. doi:10.1016/j.chom.2011.04.012, S1931-3128(11)00134-X [pii]
51. Zwijnenburg PJ, van der Poll T, Florquin S, van Deventer SJ, Roord JJ, van Furth AM (2001) Experimental pneumococcal meningitis in mice: a model of intranasal infection. J Infect Dis 183(7):1143–1146
52. Hirst RA, Kadioglu A, O'Callaghan C, Andrew PW (2004) The role of pneumolysin in pneumococcal pneumonia and meningitis. Clin Exp Immunol 138(2):195–201
53. Attali C, Durmort C, Vernet T, Di Guilmi AM (2008) The interaction of Streptococcus pneumoniae with plasmin mediates transmigration across endothelial and epithelial monolayers by intercellular junction cleavage. Infect Immun 76(11):5350–5356. doi:10.1128/IAI.00184-08, IAI.00184-08 [pii]
54. Soriani M, Santi I, Taddei A, Rappuoli R, Grandi G, Telford JL (2006) Group B Streptococcus crosses human epithelial cells by a paracellular route. J Infect Dis 193(2):241–250. doi:10.1086/498982, JID35189 [pii]
55. Kostyukova NN, Volkova MO, Ivanova VV, Kvetnaya AS (1995) A study of pathogenic factors of Streptococcus pneumoniae strains causing meningitis. FEMS Immunol Med Microbiol 10(2):133–137
56. Knaust A, Weber MV, Hammerschmidt S, Bergmann S, Frosch M, Kurzai O (2007) Cytosolic proteins contribute to surface plasminogen recruitment of Neisseria meningitidis. J Bacteriol 189(8):3246–3255. doi:10.1128/JB.01966-06, JB.01966-06 [pii]
57. Virkola R, Lahteenmaki K, Eberhard T, Kuusela P, van Alphen L, Ullberg M, Korhonen TK (1996) Interaction of Haemophilus influenzae with the mammalian extracellular matrix. J Infect Dis 173(5):1137–1147
58. Bergmann S, Rohde M, Preissner KT, Hammerschmidt S (2005) The nine residue plasminogen-binding motif of the pneumococcal enolase is the major cofactor of plasmin-mediated degradation of extracellular matrix, dissolution of fibrin and transmigration. Thromb Haemost 94(2):304–311. doi:10.1267/THRO05020304, 05080304 [pii]
59. Steukers L, Glorieux S, Vandekerckhove AP, Favoreel HW, Nauwynck HJ (2012) Diverse microbial interactions with the basement membrane barrier. Trends Microbiol 20(3):147–155. doi:10.1016/j.tim.2012.01.001, S0966-842X(12)00002-9 [pii]
60. Pron B, Boumaila C, Jaubert F, Berche P, Milon G, Geissmann F, Gaillard JL (2001) Dendritic cells are early cellular targets of Listeria monocytogenes after intestinal delivery and are involved in bacterial spread in the host. Cell Microbiol 3(5):331–340, cmi120 [pii]
61. Carbonnelle E, Hill DJ, Morand P, Griffiths NJ, Bourdoulous S, Murillo I, Nassif X, Virji M (2009) Meningococcal interactions with the host. Vaccine 27(Suppl 2):B78–B89. doi:10.1016/j.vaccine.2009.04.069, S0264-410X(09)00636-7 [pii]

62. Virkola R, Brummer M, Rauvala H, van Alphen L, Korhonen TK (2000) Interaction of fimbriae of Haemophilus influenzae type B with heparin-binding extracellular matrix proteins. Infect Immun 68(10):5696–5701
63. Walport MJ (2001) Complement. First of two parts. N Engl J Med 344(14):1058–1066. doi:10.1056/NEJM200104053441406
64. Overturf GD (2003) Indications for the immunological evaluation of patients with meningitis. Clin Infect Dis 36(2):189–194
65. Dietzman DE, Fischer GW, Schoenknecht FD (1974) Neonatal Escherichia coli septicemia – bacterial counts in blood. J Pediatr 85(1):128–130
66. Adriani KS, Brouwer MC, Geldhoff M, Baas F, Zwinderman AH, Paul Morgan B, Harris CL, van der Ende A, van de Beek D (2013) Common polymorphisms in the complement system and susceptiblity to bacterial meningitis. J Infect 66:255–262. doi:10.1016/j.jinf.2012.10.008, S0163-4453(12)00294-0 [pii]
67. Brouwer MC, de Gans J, Heckenberg SG, Zwinderman AH, van der Poll T, van de Beek D (2009) Host genetic susceptibility to pneumococcal and meningococcal disease: a systematic review and meta-analysis. Lancet Infect Dis 9(1):31–44. doi:10.1016/S1473-3099(08)70261-5, S1473-3099(08)70261-5 [pii]
68. Albiger B, Johansson L, Jonsson AB (2003) Lipooligosaccharide-deficient Neisseria meningitidis shows altered pilus-associated characteristics. Infect Immun 71(1):155–162
69. Kim KS, Itabashi H, Gemski P, Sadoff J, Warren RL, Cross AS (1992) The K1 capsule is the critical determinant in the development of Escherichia coli meningitis in the rat. J Clin Invest 90(3):897–905
70. Hyams C, Camberlein E, Cohen JM, Bax K, Brown JS (2010) The Streptococcus pneumoniae capsule inhibits complement activity and neutrophil phagocytosis by multiple mechanisms. Infect Immun 78(2):704–715. doi:10.1128/IAI.00881-09, IAI.00881-09 [pii]
71. Kilian M, Reinholdt J, Lomholt H, Poulsen K, Frandsen EV (1996) Biological significance of IgA1 proteases in bacterial colonization and pathogenesis: critical evaluation of experimental evidence. APMIS 104(5):321–338
72. Vogel U, Hammerschmidt S, Frosch M (1996) Sialic acids of both the capsule and the sialylated lipooligosaccharide of Neisseria meningitis serogroup B are prerequisites for virulence of meningococci in the infant rat. Med Microbiol Immunol (Berl) 185(2):81–87
73. Peppoloni S, Ricci S, Orsi CF, Colombari B, De Santi MM, Messino M, Fabio G, Zanardi A, Righi E, Braione V, Tripodi S, Chiavolini D, Cintorino M, Zoli M, Oggioni MR, Blasi E, Pozzi G (2010) The encapsulated strain TIGR4 of Streptococcus pneumoniae is phagocytosed but is resistant to intracellular killing by mouse microglia. Microbes Infect 12(12–13):990–1001. doi:10.1016/j.micinf.2010.06.010
74. Agarwal V, Hammerschmidt S, Malm S, Bergmann S, Riesbeck K, Blom AM (2012) Enolase of Streptococcus pneumoniae binds human complement inhibitor C4b-binding protein and contributes to complement evasion. J Immunol 189(7):3575–3584. doi:10.4049/jimmunol.1102934, jimmunol.1102934 [pii]
75. Santi I, Scarselli M, Mariani M, Pezzicoli A, Masignani V, Taddei A, Grandi G, Telford JL, Soriani M (2007) BibA: a novel immunogenic bacterial adhesin contributing to group B Streptococcus survival in human blood. Mol Microbiol 63(3):754–767. doi:10.1111/j.1365-2958.2006.05555.x, MMI5555 [pii]
76. Zipfel PF, Skerka C, Hellwage J, Jokiranta ST, Meri S, Brade V, Kraiczy P, Noris M, Remuzzi G (2002) Factor H family proteins: on complement, microbes and human diseases. Biochem Soc Trans 30(Pt 6):971–978. doi:10.1042/bst0300971
77. Maruvada R, Prasadarao NV, Rubens CE (2009) Acquisition of factor H by a novel surface protein on group B Streptococcus promotes complement degradation. Faseb J 23(11):3967–3977. doi:10.1096/fj.09-138149, fj.09-138149 [pii]
78. Lewis LA, Carter M, Ram S (2012) The relative roles of factor H binding protein, neisserial surface protein A, and lipooligosaccharide sialylation in regulation of the alternative

pathway of complement on meningococci. J Immunol 188(10):5063–5072. doi:10.4049/jimmunol.1103748, jimmunol.1103748 [pii]
79. Singh B, Su YC, Riesbeck K (2010) Vitronectin in bacterial pathogenesis: a host protein used in complement escape and cellular invasion. Mol Microbiol 78(3):545–560. doi:10.1111/j.1365-2958.2010.07373.x
80. Stavru F, Archambaud C, Cossart P (2011) Cell biology and immunology of Listeria monocytogenes infections: novel insights. Immunol Rev 240(1):160–184. doi:10.1111/j.1600-065X.2010.00993.x
81. Moxon ER, Murphy PA (1978) Haemophilus influenzae bacteremia and meningitis resulting from survival of a single organism. Proc Natl Acad Sci U S A 75(3):1534–1536
82. Redzic Z (2011) Molecular biology of the blood–brain and the blood-cerebrospinal fluid barriers: similarities and differences. Fluids Barriers CNS 8(1):3. doi:10.1186/2045-8118-8-3, 2045-8118-8-3 [pii]
83. Kim KS (2002) Strategy of Escherichia coli for crossing the blood–brain barrier. J Infect Dis 186(Suppl 2):S220–S224
84. Moxon ER, Ostrow PT (1977) Haemophilus influenzae meningitis in infant rats: role of bacteremia in pathogenesis of age-dependent inflammatory responses in cerebrospinal fluid. J Infect Dis 135(2):303–307
85. Kim KS (2008) Mechanisms of microbial traversal of the blood–brain barrier. Nat Rev Microbiol 6(8):625–634. doi:10.1038/nrmicro1952, nrmicro1952 [pii]
86. Drevets DA, Leenen PJ, Greenfield RA (2004) Invasion of the central nervous system by intracellular bacteria. Clin Microbiol Rev 17(2):323–347
87. Drevets DA, Jelinek TA, Freitag NE (2001) Listeria monocytogenes-infected phagocytes can initiate central nervous system infection in mice. Infect Immun 69(3):1344–1350. doi:10.1128/IAI.69.3.1344-1350.2001
88. Drevets DA, Dillon MJ, Schawang JS, Van Rooijen N, Ehrchen J, Sunderkotter C, Leenen PJ (2004) The Ly-6Chigh monocyte subpopulation transports Listeria monocytogenes into the brain during systemic infection of mice. J Immunol 172(7):4418–4424
89. Nassif X, Bourdoulous S, Eugene E, Couraud PO (2002) How do extracellular pathogens cross the blood–brain barrier? Trends Microbiol 10(5):227–232
90. Wang MH, Kim KS (2013) Cytotoxic necrotizing factor 1 contributes to Escherichia coli meningitis. Toxins (Basel) 5(11):2270–2280. doi:10.3390/toxins5112270
91. Kim BY, Kang J, Kim KS (2005) Invasion processes of pathogenic Escherichia coli. Int J Med Microbiol 295(6–7):463–470
92. Orihuela CJ, Mahdavi J, Thornton J, Mann B, Wooldridge KG, Abouseada N, Oldfield NJ, Self T, Ala'Aldeen DA, Tuomanen EI (2009) Laminin receptor initiates bacterial contact with the blood brain barrier in experimental meningitis models. J Clin Invest 119(6):1638–1646. doi:10.1172/JCI36759, 36759 [pii]
93. Abouseada NM, Assafi MS, Mahdavi J, Oldfield NJ, Wheldon LM, Wooldridge KG, Ala'Aldeen DA (2012) Mapping the laminin receptor binding domains of Neisseria meningitidis PorA and Haemophilus influenzae OmpP2. PLoS One 7(9):e46233. doi:10.1371/journal.pone.0046233
94. Chung JW, Hong SJ, Kim KJ, Goti D, Stins MF, Shin S, Dawson VL, Dawson TM, Kim KS (2003) 37-kDa laminin receptor precursor modulates cytotoxic necrotizing factor 1-mediated RhoA activation and bacterial uptake. J Biol Chem 278(19):16857–16862. doi:10.1074/jbc.M301028200
95. Kim KJ, Chung JW, Kim KS (2005) 67-kDa laminin receptor promotes internalization of cytotoxic necrotizing factor 1-expressing Escherichia coli K1 into human brain microvascular endothelial cells. J Biol Chem 280(2):1360–1368. doi:10.1074/jbc.M410176200
96. Iovino F, Molema G, Bijlsma JJ (2014) Platelet endothelial cell adhesion molecule-1, a putative receptor for the adhesion of Streptococcus pneumoniae to the vascular endothelium of the blood–brain barrier. Infect Immun 82(9):3555–3566. doi:10.1128/IAI.00046-14

97. Iovino F, Molema G, Bijlsma JJ (2014) Streptococcus pneumoniae interacts with pIgR expressed by the brain microvascular endothelium but does not co-localize with PAF receptor. PLoS One 9(5):e97914. doi:10.1371/journal.pone.0097914
98. Coureuil M, Join-Lambert O, Lecuyer H, Bourdoulous S, Marullo S, Nassif X (2012) Mechanism of meningeal invasion by Neisseria meningitidis. Virulence 3(2):164–172. doi:10.4161/viru.18639, 18639 [pii]
99. Maisey HC, Hensler M, Nizet V, Doran KS (2007) Group B streptococcal pilus proteins contribute to adherence to and invasion of brain microvascular endothelial cells. J Bacteriol 189(4):1464–1467. doi:10.1128/JB.01153-06, JB.01153-06 [pii]
100. Tenenbaum T, Bloier C, Adam R, Reinscheid DJ, Schroten H (2005) Adherence to and invasion of human brain microvascular endothelial cells are promoted by fibrinogen-binding protein FbsA of Streptococcus agalactiae. Infect Immun 73(7):4404–4409. doi:10.1128/IAI.73.7.4404-4409.2005, 73/7/4404 [pii]
101. Banerjee A, Kim BJ, Carmona EM, Cutting AS, Gurney MA, Carlos C, Feuer R, Prasadarao NV, Doran KS (2011) Bacterial Pili exploit integrin machinery to promote immune activation and efficient blood–brain barrier penetration. Nat Commun 2:462. doi:10.1038/ncomms1474, ncomms1474 [pii]
102. Seo HS, Mu R, Kim BJ, Doran KS, Sullam PM (2012) Binding of glycoprotein Srr1 of Streptococcus agalactiae to fibrinogen promotes attachment to brain endothelium and the development of meningitis. PLoS Pathog 8(10):e1002947. doi:10.1371/journal.ppat.1002947
103. Tenenbaum T, Spellerberg B, Adam R, Vogel M, Kim KS, Schroten H (2007) Streptococcus agalactiae invasion of human brain microvascular endothelial cells is promoted by the laminin-binding protein Lmb. Microbes Infect 9(6):714–720. doi:10.1016/j.micinf.2007.02.015
104. Asmat TM, Agarwal V, Saleh M, Hammerschmidt S (2014) Endocytosis of Streptococcus pneumoniae via the polymeric immunoglobulin receptor of epithelial cells relies on clathrin and caveolin dependent mechanisms. Int J Med Microbiol 304(8):1233–1246. doi:10.1016/j.ijmm.2014.10.001
105. Uchiyama S, Carlin AF, Khosravi A, Weiman S, Banerjee A, Quach D, Hightower G, Mitchell TJ, Doran KS, Nizet V (2009) The surface-anchored NanA protein promotes pneumococcal brain endothelial cell invasion. J Exp Med 206(9):1845–1852. doi:10.1084/jem.20090386, jem.20090386 [pii]
106. Banerjee A, Van Sorge NM, Sheen TR, Uchiyama S, Mitchell TJ, Doran KS (2010) Activation of brain endothelium by pneumococcal neuraminidase NanA promotes bacterial internalization. Cell Microbiol 12(11):1576–1588. doi:10.1111/j.1462-5822.2010.01490.x
107. Cundell DR, Gerard NP, Gerard C, Idanpaan-Heikkila I, Tuomanen EI (1995) Streptococcus pneumoniae anchor to activated human cells by the receptor for platelet-activating factor. Nature 377(6548):435–438
108. Reddy MA, Wass CA, Kim KS, Schlaepfer DD, Prasadarao NV (2000) Involvement of focal adhesion kinase in Escherichia coli invasion of human brain microvascular endothelial cells. Infect Immun 68(11):6423–6430
109. Prasadarao NV, Wass CA, Stins MF, Shimada H, Kim KS (1999) Outer membrane protein A-promoted actin condensation of brain microvascular endothelial cells is required for Escherichia coli invasion. Infect Immun 67(11):5775–5783
110. Khan NA, Kim Y, Shin S, Kim KS (2007) FimH-mediated Escherichia coli K1 invasion of human brain microvascular endothelial cells. Cell Microbiol 9(1):169–178. doi:10.1111/j.1462-5822.2006.00779.x
111. Maruvada R, Kim KS (2012) IbeA and OmpA of Escherichia coli K1 exploit Rac1 activation for invasion of human brain microvascular endothelial cells. Infect Immun 80(6):2035–2041. doi:10.1128/IAI.06320-11
112. Zou Y, He L, Huang SH (2006) Identification of a surface protein on human brain microvascular endothelial cells as vimentin interacting with Escherichia coli invasion protein IbeA. Biochem Biophys Res Commun 351(3):625–630. doi:10.1016/j.bbrc.2006.10.091

113. Zou Y, He L, Wu CH, Cao H, Xie ZH, Ouyang Y, Wang Y, Jong A, Huang SH (2007) PSF is an IbeA-binding protein contributing to meningitic Escherichia coli K1 invasion of human brain microvascular endothelial cells. Med Microbiol Immunol 196(3):135–143. doi:10.1007/s00430-006-0034-x
114. Teng CH, Tseng YT, Maruvada R, Pearce D, Xie Y, Paul-Satyaseela M, Kim KS (2010) NlpI contributes to Escherichia coli K1 strain RS218 interaction with human brain microvascular endothelial cells. Infect Immun 78(7):3090–3096. doi:10.1128/IAI.00034-10
115. Prasadarao NV, Wass CA, Huang SH, Kim KS (1999) Identification and characterization of a novel Ibe10 binding protein that contributes to Escherichia coli invasion of brain microvascular endothelial cells. Infect Immun 67(3):1131–1138
116. Hoffman JA, Badger JL, Zhang Y, Huang SH, Kim KS (2000) Escherichia coli K1 aslA contributes to invasion of brain microvascular endothelial cells in vitro and in vivo. Infect Immun 68(9):5062–5067
117. Parthasarathy G, Yao Y, Kim KS (2007) Flagella promote Escherichia coli K1 association with and invasion of human brain microvascular endothelial cells. Infect Immun 75(6):2937–2945. doi:10.1128/IAI.01543-06
118. Mu R, Kim BJ, Paco C, Del Rosario Y, Courtney HS, Doran KS (2014) Identification of a group B streptococcal fibronectin binding protein, SfbA, that contributes to invasion of brain endothelium and development of meningitis. Infect Immun 82(6):2276–2286. doi:10.1128/IAI.01559-13
119. Unkmeir A, Latsch K, Dietrich G, Wintermeyer E, Schinke B, Schwender S, Kim KS, Eigenthaler M, Frosch M (2002) Fibronectin mediates Opc-dependent internalization of Neisseria meningitidis in human brain microvascular endothelial cells. Mol Microbiol 46(4):933–946
120. Disson O, Lecuit M (2012) Targeting of the central nervous system by Listeria monocytogenes. Virulence 3(2):213–221. doi:10.4161/viru.19586, 19586 [pii]
121. Kastenbauer S, Pfister HW (2003) Pneumococcal meningitis in adults: spectrum of complications and prognostic factors in a series of 87 cases. Brain 126(Pt 5):1015–1025
122. van de Beek D, de Gans J, Spanjaard L, Weisfelt M, Reitsma JB, Vermeulen M (2004) Clinical features and prognostic factors in adults with bacterial meningitis. N Engl J Med 351(18):1849–1859
123. Marra A, Brigham D (2001) Streptococcus pneumoniae causes experimental meningitis following intranasal and otitis media infections via a nonhematogenous route. Infect Immun 69(12):7318–7325
124. van Ginkel FW, McGhee JR, Watt JM, Campos-Torres A, Parish LA, Briles DE (2003) Pneumococcal carriage results in ganglioside-mediated olfactory tissue infection. Proc Natl Acad Sci U S A 100(24):14363–14367. doi:10.1073/pnas.2235844100, 2235844100 [pii]
125. Sjolinder H, Jonsson AB (2010) Olfactory nerve – a novel invasion route of Neisseria meningitidis to reach the meninges. PLoS One 5(11):e14034. doi:10.1371/journal.pone.0014034
126. Konsman JP, Drukarch B, Van Dam AM (2007) (Peri)vascular production and action of pro-inflammatory cytokines in brain pathology. Clin Sci 112(1):1–25. doi:10.1042/CS20060043
127. Kim YS, Honkaniemi J, Sharp FR, Täuber MG (2004) Expression of proinflammatory cytokines tumor necrosis factor-alpha and interleukin-1beta in the brain during experimental group B streptococcal meningitis. Brain Res Mol Brain Res 128(1):95–102. doi:10.1016/j.molbrainres.2004.06.009
128. Polfliet MM, Zwijnenburg PJ, van Furth AM, van der Poll T, Dopp EA, Renardel de Lavalette C, van Kesteren-Hendrikx EM, van Rooijen N, Dijkstra CD, van den Berg TK (2001) Meningeal and perivascular macrophages of the central nervous system play a protective role during bacterial meningitis. J Immunol 167(8):4644–4650
129. de Vos AF, Dessing MC, Lammers AJ, de Porto AP, Florquin S, de Boer OJ, de Beer R, Terpstra S, Bootsma HJ, Hermans PW, van 't Veer C, van der Poll T (2015) The polysaccharide capsule of Streptococcus pneumonia partially impedes MyD88-mediated immunity during pneumonia in mice. PLoS One 10(2):e0118181. doi:10.1371/journal.pone.0118181

130. Hanamsagar R, Hanke ML, Kielian T (2012) Toll-like receptor (TLR) and inflammasome actions in the central nervous system. Trends Immunol 33(7):333–342. doi:10.1016/j.it.2012.03.001, S1471-4906(12)00051-8 [pii]
131. Koedel U (2009) Toll-like receptors in bacterial meningitis. Curr Top Microbiol Immunol 336:15–40. doi:10.1007/978-3-642-00549-7_2
132. Hoegen T, Tremel N, Klein M, Angele B, Wagner H, Kirschning C, Pfister HW, Fontana A, Hammerschmidt S, Koedel U (2011) The NLRP3 inflammasome contributes to brain injury in pneumococcal meningitis and is activated through ATP-dependent lysosomal cathepsin B release. J Immunol 187(10):5440–5451. doi:10.4049/jimmunol.1100790, jimmunol.1100790 [pii]
133. Neal JW, Gasque P (2013) How does the brain limit the severity of inflammation and tissue injury during bacterial meningitis? J Neuropathol Exp Neurol 72(5):370–385. doi:10.1097/NEN.0b013e3182909f2f
134. Kigerl KA, de Rivero Vaccari JP, Dietrich WD, Popovich PG, Keane RW (2014) Pattern recognition receptors and central nervous system repair. Exp Neurol 258:5–16. doi:10.1016/j.expneurol.2014.01.001
135. Geldhoff M, Mook-Kanamori BB, Brouwer MC, Valls Seron M, Baas F, van der Ende A, van de Beek D (2013) Genetic variation in inflammasome genes is associated with outcome in bacterial meningitis. Immunogenetics 65(1):9–16. doi:10.1007/s00251-012-0653-x
136. Sanders MS, van Well GT, Ouburg S, Morre SA, van Furth AM (2012) Toll-like receptor 9 polymorphisms are associated with severity variables in a cohort of meningococcal meningitis survivors. BMC Infect Dis 12:112. doi:10.1186/1471-2334-12-112, 1471-2334-12-112 [pii]
137. van Well GT, Sanders MS, Ouburg S, van Furth AM, Morre SA (2012) Polymorphisms in toll-like receptors 2, 4, and 9 are highly associated with hearing loss in survivors of bacterial meningitis. PLoS One 7(5):e35837. doi:10.1371/journal.pone.0035837, PONE-D-11-19179 [pii]
138. Canton J, Neculai D, Grinstein S (2013) Scavenger receptors in homeostasis and immunity. Nat Rev Immunol 13(9):621–634. doi:10.1038/nri3515
139. Iovino F, Orihuela CJ, Moorlag HE, Molema G, Bijlsma JJ (2013) Interactions between blood-borne Streptococcus pneumoniae and the blood–brain barrier preceding meningitis. PLoS One 8(7):e68408. doi:10.1371/journal.pone.0068408
140. Täuber MG, Moser B (1999) Cytokines and chemokines in meningeal inflammation: biology and clinical implications. Clin Infect Dis 28(1):1–11; quiz 12
141. Kastenbauer S, Angele B, Sporer B, Pfister HW, Koedel U (2005) Patterns of protein expression in infectious meningitis: a cerebrospinal fluid protein array analysis. J Neuroimmunol 164(1–2):134–139
142. Zwijnenburg PJ, van der Poll T, Roord JJ, van Furth AM (2006) Chemotactic factors in cerebrospinal fluid during bacterial meningitis. Infect Immun 74(3):1445–1451. doi:10.1128/IAI.74.3.1445-1451.2006, 74/3/1445 [pii]
143. Klein M, Paul R, Angele B, Popp B, Pfister HW, Koedel U (2006) Protein expression pattern in experimental pneumococcal meningitis. Microbes Infect 8(4):974–983. doi:10.1016/j.micinf.2005.10.013, S1286-4579(05)00381-3 [pii]
144. Mogensen TH, Paludan SR, Kilian M, Ostergaard L (2006) Live Streptococcus pneumoniae, Haemophilus influenzae, and Neisseria meningitidis activate the inflammatory response through toll-like receptors 2, 4, and 9 in species-specific patterns. J Leukoc Biol 80(2):267–277. doi:10.1189/jlb.1105626, jlb.1105626 [pii]
145. Fowler MI, Weller RO, Heckels JE, Christodoulides M (2004) Different meningitis-causing bacteria induce distinct inflammatory responses on interaction with cells of the human meninges. Cell Microbiol 6(6):555–567
146. Tietze K, Dalpke A, Morath S, Mutters R, Heeg K, Nonnenmacher C (2006) Differences in innate immune responses upon stimulation with gram-positive and gram-negative bacteria. J Periodontal Res 41(5):447–454. doi:10.1111/j.1600-0765.2006.00890.x, JRE890 [pii]

147. Diab A, Zhu J, Linquist L, Wretlind B, Bakhiet M, Link H (1997) Haemophilus influenzae and streptococcus pneumoniae induce different intracerebral mRNA cytokine patterns during the course of experimental bacterial meningitis. Clin Exp Immunol 109:233–241
148. Leib SL, Clements JM, Lindberg RL, Heimgartner C, Loeffler JM, Pfister LA, Tauber MG, Leppert D (2001) Inhibition of matrix metalloproteinases and tumour necrosis factor alpha converting enzyme as adjuvant therapy in pneumococcal meningitis. Brain 124(Pt 9):1734–1742
149. Tureen J (1995) Effect of recombinant human tumor necrosis factor-alpha on cerebral oxygen uptake, cerebrospinal fluid lactate, and cerebral blood flow in the rabbit: role of nitric oxide. J Clin Invest 95(3):1086–1091
150. Rosenberg GA, Estrada EY, Dencoff JE, Stetler-Stevenson WG (1995) Tumor necrosis factor-alpha-induced gelatinase B causes delayed opening of the blood–brain barrier: an expanded therapeutic window. Brain Res 703(1–2):151–155
151. Andersson PB, Perry VH, Gordon S (1992) Intracerebral injection of proinflammatory cytokines or leukocyte chemotaxins induces minimal myelomonocytic cell recruitment to the parenchyma of the central nervous system. J Exp Med 176(1):255–259
152. Liu T, Clark RK, McDonnell PC, Young PR, White RF, Barone FC, Feuerstein GZ (1994) Tumor necrosis factor-alpha expression in ischemic neurons. Stroke 25(7):1481–1488
153. Glabinski A, Krajewski S, Rafalowska J (1998) Tumor necrosis factor-alpha induced pathology in the rat brain: characterization of stereotaxic injection model. Folia Neuropathol 36(1):52–62
154. Candelario-Jalil E, Taheri S, Yang Y, Sood R, Grossetete M, Estrada EY, Fiebich BL, Rosenberg GA (2007) Cyclooxygenase inhibition limits blood–brain barrier disruption following intracerebral injection of tumor necrosis factor-alpha in the rat. J Pharmacol Exp Ther 323(2):488–498. doi:10.1124/jpet.107.127035, jpet.107.127035 [pii]
155. Costa A, Gupta R, Signorino G, Malara A, Cardile F, Biondo C, Midiri A, Galbo R, Trieu-Cuot P, Papasergi S, Teti G, Henneke P, Mancuso G, Golenbock DT, Beninati C (2012) Activation of the NLRP3 inflammasome by group B streptococci. J Immunol 188(4):1953–1960. doi:10.4049/jimmunol.1102543, jimmunol.1102543 [pii]
156. Opitz B, Puschel A, Schmeck B, Hocke AC, Rosseau S, Hammerschmidt S, Schumann RR, Suttorp N, Hippenstiel S (2004) Nucleotide-binding oligomerization domain proteins are innate immune receptors for internalized Streptococcus pneumoniae. J Biol Chem 279(35):36426–36432
157. Mitchell AJ, Yau B, McQuillan JA, Ball HJ, Too LK, Abtin A, Hertzog P, Leib SL, Jones CA, Gerega SK, Weninger W, Hunt NH (2012) Inflammasome-dependent IFN-gamma drives pathogenesis in Streptococcus pneumoniae meningitis. J Immunol 189(10):4970–4980. doi:10.4049/jimmunol.1201687, jimmunol.1201687 [pii]
158. Waage A, Halstensen A, Shalaby R (1998) Local production of tumor necrosis factor, interleukin-l, and interleukin-6 in meningococcal meningitis. J Exp Med 170:1859–1864
159. Ramilo O, Saez-Llorens X, Mertsola J, Jafari H, Olsen KD, Hansen EJ, Yoshinaga M, Ohkawara S, Nariuchi H, McCracken GH Jr (1990) Tumor necrosis factor alpha/cachectin and interleukin 1 beta initiate meningeal inflammation. J Exp Med 172(2):497–507
160. Rusconi F, Parizzi F, Garlaschi L, Assael BM, Sironi M, Ghezzi P, Mantovani A (1991) Interleukin 6 activity in infants and children with bacterial meningitis. The collaborative study on meningitis. Pediatr Infect Dis J 10(2):117–121
161. Hirano T, Akira S, Taga T, Kishimoto T (1990) Biological and clinical aspects of interleukin-6. Immunol Today 11:443–445
162. Cohen MC, Cohen S (1996) Cytokine function, a study in biologic diversity. Am J Clin Pathol 105:589–598
163. Spanaus KS, Nadal D, Pfister HW, Seebach J, Widmer U, Frei K, Gloor S, Fontana A (1997) C-X-C and C-C chemokines are expressed in the cerebrospinal fluid in bacterial meningitis and mediate chemotactic activity on peripheral blood-derived polymorphonuclear and mononuclear cells in vitro. J Immunol 158(4):1956–1964

164. Zwijnenburg PJ, de Bie HM, Roord JJ, van der Poll T, van Furth AM (2003) Chemotactic activity of CXCL5 in cerebrospinal fluid of children with bacterial meningitis. J Neuroimmunol 145(1–2):148–153
165. Paris MM, Hickey SM, Trujillo M, Ahmed A, Olsen K, McCracken GH Jr (1997) The effect of interleukin-10 on meningeal inflammation in experimental bacterial meningitis. J Infect Dis 176(5):1239–1246
166. Pfister HW, Frei K, Ottnad B, Koedel U, Tomasz A, Fontana A (1992) Transforming growth factor beta 2 inhibits cerebrovascular changes and brain edema formation in the tumor necrosis factor alpha-independent early phase of experimental pneumococcal meningitis. J Exp Med 176(1):265–268
167. Cooley ID, Chauhan VS, Donneyz MA, Marriott I (2014) Astrocytes produce IL-19 in response to bacterial challenge and are sensitive to the immunosuppressive effects of this IL-10 family member. Glia 62(5):818–828. doi:10.1002/glia.22644
168. Grandgirard D, Gaumann R, Coulibaly B, Dangy JP, Sie A, Junghanss T, Schudel H, Pluschke G, Leib SL (2013) The causative pathogen determines the inflammatory profile in cerebrospinal fluid and outcome in patients with bacterial meningitis. Mediators Inflamm 2013:312476. doi:10.1155/2013/312476
169. Too LK, Ball HJ, McGregor IS, Hunt NH (2014) The pro-inflammatory cytokine interferon-gamma is an important driver of neuropathology and behavioural sequelae in experimental pneumococcal meningitis. Brain Behav Immun 40:252–268. doi:10.1016/j.bbi.2014.02.020
170. Kieseier BC, Paul R, Koedel U, Seifert T, Clements JM, Gearing AJ, Pfister HW, Hartung HP (1999) Differential expression of matrix metalloproteinases in bacterial meningitis. Brain 122(Pt 8):1579–1587
171. Paul R, Lorenzl S, Koedel U, Sporer B, Vogel U, Frosch M, Pfister HW (1998) Matrix metalloproteinases contribute to the blood–brain barrier disruption during bacterial meningitis. Ann Neurol 44(4):592–600
172. Leppert D, Lindberg RL, Kappos L, Leib SL (2001) Matrix metalloproteinases: multifunctional effectors of inflammation in multiple sclerosis and bacterial meningitis. Brain Res Brain Res Rev 36(2–3):249–257
173. Green JA, Thi Hong Chau T, Farrar JJ, Friedland JS, Thwaites GE (2011) CNS infection, CSF matrix metalloproteinase concentrations, and clinical/laboratory features. Neurology 76(6):577–579. doi:10.1212/WNL.0b013e31820b7600, 76/6/577 [pii]
174. Tsai HC, Shi MH, Lee SS, Wann SR, Tai MH, Chen YS (2011) Expression of matrix metalloproteinases and their tissue inhibitors in the serum and cerebrospinal fluid of patients with meningitis. Clin Microbiol Infect 17(5):780–784. doi:10.1111/j.1469-0691.2010.03393.x
175. Lindberg RL, Sorsa T, Tervahartiala T, Hoffmann F, Mellanen L, Kappos L, Schaad UB, Leib SL, Leppert D (2006) Gelatinase B [matrix metalloproteinase (MMP)-9] and collagenases (MMP-8/-13) are upregulated in cerebrospinal fluid during aseptic and bacterial meningitis in children. Neuropathol Appl Neurobiol 32(3):304–317
176. Sulik A, Chyczewski L (2008) Immunohistochemical analysis of MMP-9, MMP-2 and TIMP-1, TIMP-2 expression in the central nervous system following infection with viral and bacterial meningitis. Folia Histochem Cytobiol 46(4):437–442. doi:10.2478/v10042-008-0058-8, 468Q020105H97207 [pii]
177. Roine I, Pelkonen T, Lauhio A, Lappalainen M, Cruzeiro ML, Bernardino L, Tervahartiala T, Sorsa T, Peltola H (2015) Changes in MMP-9 and TIMP-1 concentrations in cerebrospinal fluid after 1 week of treatment of childhood bacterial meningitis. J Clin Microbiol 53(7):2340–2342. doi:10.1128/JCM.00714-15
178. Leppert D, Leib SL, Grygar C, Miller KM, Schaad UB, Hollander GA (2000) Matrix metalloproteinase (MMP)-8 and MMP-9 in cerebrospinal fluid during bacterial meningitis: association with blood–brain barrier damage and neurological sequelae. Clin Infect Dis 31(1):80–84
179. Sellner J, Leib SL (2006) In bacterial meningitis cortical brain damage is associated with changes in parenchymal MMP-9/TIMP-1 ratio and increased collagen type IV degradation. Neurobiol Dis 21(3):647–656

180. Meli DN, Loeffler JM, Baumann P, Neumann U, Buhl T, Leppert D, Leib SL (2004) In pneumococcal meningitis a novel water-soluble inhibitor of matrix metalloproteinases and TNF-alpha converting enzyme attenuates seizures and injury of the cerebral cortex. J Neuroimmunol 151(1–2):6–11. doi:10.1016/j.jneuroim.2004.01.026
181. Liechti FD, Bachtold F, Grandgirard D, Leppert D, Leib SL (2015) The matrix metalloproteinase inhibitor RS-130830 attenuates brain injury in experimental pneumococcal meningitis. J Neuroinflammation 12:43. doi:10.1186/s12974-015-0257-0
182. Liechti FD, Grandgirard D, Leppert D, Leib SL (2014) Matrix metalloproteinase inhibition lowers mortality and brain injury in experimental pneumococcal meningitis. Infect Immun 82(4):1710–1718. doi:10.1128/IAI.00073-14
183. Tunkel AR, Scheld WM (1993) Pathogenesis and pathophysiology of bacterial meningitis. Clin Microbiol Rev 6(2):118–136
184. Korthuis RJ, Anderson DC, Granger DN (1994) Role of neutrophil-endothelial cell adhesion in inflammatory disorders. J Crit Care 9(1):47–71
185. Täuber MG, Sande MA (1984) Pathogenesis of bacterial meningitis: contributions by experimental models in rabbits. Infection 12(Suppl 1):S3–S10
186. Ley K, Laudanna C, Cybulsky MI, Nourshargh S (2007) Getting to the site of inflammation: the leukocyte adhesion cascade updated. Nat Rev Immunol 7(9):678–689. doi:10.1038/nri2156, nri2156 [pii]
187. Tang T, Frenette PS, Hynes RO, Wagner DD, Mayadas TN (1996) Cytokine-induced meningitis is dramatically attenuated in mice deficient in endothelial selectins. J Clin Invest 97(11):2485–2490
188. Granert C, Raud J, Xie X, Lindquist L, Lindbom L (1994) Inhibition of leukocyte rolling with polysaccharide fucoidan prevents pleocytosis in experimental meningitis in the rabbit. J Clin Invest 93(3):929–936
189. Ostergaard C, Yieng-Kow RV, Benfield T, Frimodt-Moller N, Espersen F, Lundgren JD (2000) Inhibition of leukocyte entry into the brain by the selectin blocker fucoidan decreases interleukin-1 (IL-1) levels but increases IL-8 levels in cerebrospinal fluid during experimental pneumococcal meningitis in rabbits. Infect Immun 68(6):3153–3157
190. Brandt CT, Lundgren JD, Frimodt-Moller N, Christensen T, Benfield T, Espersen F, Hougaard DM, Ostergaard C (2005) Blocking of leukocyte accumulation in the cerebrospinal fluid augments bacteremia and increases lethality in experimental pneumococcal meningitis. J Neuroimmunol 166(1–2):126–131
191. Weber JR, Angstwurm K, Burger W, Einhaupl KM, Dirnagl U (1995) Anti ICAM-1 (CD 54) monoclonal antibody reduces inflammatory changes in experimental bacterial meningitis. J Neuroimmunol 63(1):63–68
192. Saez-Llorens X, Jafari HS, Severien C, Parras F, Olsen KD, Hansen EJ, Singer II, McCracken GH Jr (1991) Enhanced attenuation of meningeal inflammation and brain edema by concomitant administration of anti-CD18 monoclonal antibodies and dexamethasone in experimental Haemophilus meningitis. J Clin Invest 88(6):2003–2011
193. Tuomanen EI, Prasad SM, George JS, Hoepelman AI, Ibsen P, Heron I, Starzyk RM (1993) Reversible opening of the blood–brain barrier by anti-bacterial antibodies. Proc Natl Acad Sci U S A 90(16):7824–7828
194. Steinmann U, Borkowski J, Wolburg H, Schroppel B, Findeisen P, Weiss C, Ishikawa H, Schwerk C, Schroten H, Tenenbaum T (2013) Transmigration of polymorphonuclear neutrophils and monocytes through the human blood-cerebrospinal fluid barrier after bacterial infection in vitro. J Neuroinflammation 10:31. doi:10.1186/1742-2094-10-31
195. Wewer C, Seibt A, Wolburg H, Greune L, Schmidt MA, Berger J, Galla HJ, Quitsch U, Schwerk C, Schroten H, Tenenbaum T (2011) Transcellular migration of neutrophil granulocytes through the blood-cerebrospinal fluid barrier after infection with Streptococcus suis. J Neuroinflammation 8:51. doi:10.1186/1742-2094-8-51
196. Che X, Chi F, Wang L, Jong TD, Wu CH, Wang X, Huang SH (2011) Involvement of IbeA in meningitic Escherichia coli K1-induced polymorphonuclear leukocyte transmigration across brain endothelial cells. Brain Pathol 21(4):389–404. doi:10.1111/j.1750-3639.2010.00463.x

197. von Wedel-Parlow M, Schrot S, Lemmen J, Treeratanapiboon L, Wegener J, Galla HJ (2011) Neutrophils cross the BBB primarily on transcellular pathways: an in vitro study. Brain Res 1367:62–76. doi:10.1016/j.brainres.2010.09.076, S0006-8993(10)02142-6 [pii]
198. Wittchen ES (2009) Endothelial signaling in paracellular and transcellular leukocyte transmigration. Front Biosci 14:2522–2545, 3395 [pii]
199. Ernst JD, Decazes JM, Sande MA (1983) Experimental pneumococcal meningitis: role of leukocytes in pathogenesis. Infect Immun 41(1):275–279
200. Tuomanen EI, Saukkonen K, Sande S, Cioffe C, Wright SD (1989) Reduction of inflammation, tissue damage, and mortality in bacterial meningitis in rabbits treated with monoclonal antibodies against adhesion-promoting receptors of leukocytes. J Exp Med 170(3):959–969
201. Koedel U, Frankenberg T, Kirschnek S, Obermaier B, Hacker H, Paul R, Hacker G (2009) Apoptosis is essential for neutrophil functional shutdown and determines tissue damage in experimental pneumococcal meningitis. PLoS Pathog 5(5):e1000461. doi:10.1371/journal.ppat.1000461
202. Hoffmann O, Priller J, Prozorovski T, Schulze-Topphoff U, Baeva N, Lunemann JD, Aktas O, Mahrhofer C, Stricker S, Zipp F, Weber JR (2007) TRAIL limits excessive host immune responses in bacterial meningitis. J Clin Invest 117(7):2004–2013. doi:10.1172/JCI30356
203. McMillan DA, Lin CY, Aronin SI, Quagliarello VJ (2001) Community-acquired bacterial meningitis in adults: categorization of causes and timing of death. Clin Infect Dis 33(7):969–975
204. Klein M, Koedel U, Pfefferkorn T, Zeller G, Woehrl B, Pfister HW (2011) Arterial cerebrovascular complications in 94 adults with acute bacterial meningitis. Crit Care 15(6):R281. doi:10.1186/cc10565, cc10565 [pii]
205. Jan W, Zimmerman RA, Bilaniuk LT, Hunter JV, Simon EM, Haselgrove J (2003) Diffusion-weighted imaging in acute bacterial meningitis in infancy. Neuroradiology 45(9):634–639. doi:10.1007/s00234-003-1035-8
206. Takeoka M, Takahashi T (2002) Infectious and inflammatory disorders of the circulatory system and stroke in childhood. Curr Opin Neurol 15(2):159–164
207. Weiss N, Miller F, Cazaubon S, Couraud PO (2009) The blood–brain barrier in brain homeostasis and neurological diseases. Biochim Biophys Acta 1788(4):842–857. doi:10.1016/j.bbamem.2008.10.022, S0005-2736(08)00348-9 [pii]
208. Zysk G, Schneider-Wald BK, Hwang JH, Bejo L, Kim KS, Mitchell TJ, Hakenbeck R, Heinz HP (2001) Pneumolysin is the main inducer of cytotoxicity to brain microvascular endothelial cells caused by Streptococcus pneumoniae. Infect Immun 69(2):845–852
209. N'Guessan PD, Schmeck B, Ayim A, Hocke AC, Brell B, Hammerschmidt S, Rosseau S, Suttorp N, Hippenstiel S (2005) Streptococcus pneumoniae R6x induced p38 MAPK and JNK-mediated caspase-dependent apoptosis in human endothelial cells. Thromb Haemost 94(2):295–303. doi:10.1267/THRO05020295, 05080295 [pii]
210. Bermpohl D, Halle A, Freyer D, Dagand E, Braun JS, Bechmann I, Schroder NW, Weber JR (2005) Bacterial programmed cell death of cerebral endothelial cells involves dual death pathways. J Clin Invest 115(6):1607–1615. doi:10.1172/JCI23223
211. Constantin D, Ala'Aldeent D, Murphy S (2002) Transcriptional activation of nitric oxide synthase-2, and NO-induced cell death, in mouse cerebrovascular endothelium exposed to Neisseria meningitidis. J Neurochem 81(2):270–276
212. Khan NA, Iqbal J, Siddiqui R (2012) Escherichia coli K1-induced cytopathogenicity of human brain microvascular endothelial cells. Microb Pathog 53(5–6):269–275. doi:10.1016/j.micpath.2012.07.001
213. Hupp S, Heimeroth V, Wippel C, Fortsch C, Ma J, Mitchell TJ, Iliev AI (2012) Astrocytic tissue remodeling by the meningitis neurotoxin pneumolysin facilitates pathogen tissue penetration and produces interstitial brain edema. Glia 60(1):137–146. doi:10.1002/glia.21256
214. Kim BJ, Hancock BM, Bermudez A, Del Cid N, Reyes E, van Sorge NM, Lauth X, Smurthwaite CA, Hilton BJ, Stotland A, Banerjee A, Buchanan J, Wolkowicz R, Traver D, Doran KS (2015) Bacterial induction of Snail1 contributes to blood–brain barrier disruption. J Clin Invest 125(6):2473–2483. doi:10.1172/JCI74159

215. Daneman R (2012) The blood–brain barrier in health and disease. Ann Neurol 72(5):648–672. doi:10.1002/ana.23648
216. Stamatovic SM, Keep RF, Andjelkovic AV (2008) Brain endothelial cell-cell junctions: how to "open" the blood brain barrier. Curr Neuropharmacol 6(3):179–192. doi:10.2174/157015908785777210
217. Paul R, Koedel U, Winkler F, Kieseier BC, Fontana A, Kopf M, Hartung HP, Pfister HW (2003) Lack of IL-6 augments inflammatory response but decreases vascular permeability in bacterial meningitis. Brain 126(Pt 8):1873–1882
218. Stamatovic SM, Dimitrijevic OB, Keep RF, Andjelkovic AV (2006) Inflammation and brain edema: new insights into the role of chemokines and their receptors. Acta Neurochir Suppl 96:444–450
219. Ricci S, Grandgirard D, Wenzel M, Braccini T, Salvatore P, Oggioni MR, Leib SL, Koedel U (2014) Inhibition of matrix metalloproteinases attenuates brain damage in experimental meningococcal meningitis. BMC Infect Dis 14:726. doi:10.1186/s12879-014-0726-6
220. Kastenbauer S, Koedel U, Becker BF, Pfister HW (2002) Oxidative stress in bacterial meningitis in humans. Neurology 58(2):186–191
221. Leib SL, Kim YS, Chow LL, Sheldon RA, Tauber MG (1996) Reactive oxygen intermediates contribute to necrotic and apoptotic neuronal injury in an infant rat model of bacterial meningitis due to group B streptococci. J Clin Invest 98(11):2632–2639. doi:10.1172/JCI119084
222. Auer M, Pfister LA, Leppert D, Tauber MG, Leib SL (2000) Effects of clinically used antioxidants in experimental pneumococcal meningitis. J Infect Dis 182(1):347–350
223. Koedel U, Pfister HW (1997) Protective effect of the antioxidant N-acetyl-L-cysteine in pneumococcal meningitis in the rat. Neurosci Lett 225(1):33–36
224. Kastenbauer S, Koedel U, Becker BF, Pfister HW (2001) Experimental meningitis in the rat: protection by uric acid at human physiological blood concentrations. Eur J Pharmacol 425(2):149–152
225. Pun PB, Lu J, Moochhala S (2009) Involvement of ROS in BBB dysfunction. Free Radic Res 43(4):348–364. doi:10.1080/10715760902751902, 909050919 [pii]
226. Owens HM, Destache CJ, Dash AK (1999) Simple liquid chromatographic method for the analysis of the blood brain barrier permeability characteristics of ceftriaxone in an experimental rabbit meningitis model. J Chromatogr B Biomed Sci Appl 728(1):97–105
227. Fishman RA (1975) Brain edema. N Engl J Med 293(14):706–711. doi:10.1056/NEJM197510022931407
228. Brown LW, Feigin RD (1994) Bacterial meningitis: fluid balance and therapy. Pediatr Ann 23(2):93–98
229. Lu CH, Chang HW, Lui CC, Huang CR, Chang WN (2006) Cerebral haemodynamics in acute bacterial meningitis in adults. QJM 99(12):863–869. doi:10.1093/qjmed/hcl119, 99/12/863 [pii]
230. Ries S, Schminke U, Fassbender K, Daffertshofer M, Steinke W, Hennerici M (1997) Cerebrovascular involvement in the acute phase of bacterial meningitis. J Neurol 244(1):51–55
231. Pfister HW, Borasio GD, Dirnagl U, Bauer M, Einhaupl KM (1992) Cerebrovascular complications of bacterial meningitis in adults. Neurology 42(8):1497–1504
232. Pfister HW, Feiden W, Einhaupl KM (1993) Spectrum of complications during bacterial meningitis in adults. Results of a prospective clinical study. Arch Neurol 50(6):575–581
233. Vergouwen MD, Schut ES, Troost D, van de Beek D (2010) Diffuse cerebral intravascular coagulation and cerebral infarction in pneumococcal meningitis. Neurocrit Care 13(2):217–227. doi:10.1007/s12028-010-9387-5
234. Perry JR, Bilbao JM, Gray T (1992) Fatal basilar vasculopathy complicating bacterial meningitis. Stroke 23(8):1175–1178
235. Ment LR, Ehrenkranz RA, Duncan CC (1986) Bacterial meningitis as an etiology of perinatal cerebral infarction. Pediatr Neurol 2(5):276–279
236. Koedel U, Bernatowicz A, Paul R, Frei K, Fontana A, Pfister HW (1995) Experimental pneumococcal meningitis: cerebrovascular alterations, brain edema, and meningeal inflammation are linked to the production of nitric oxide. Ann Neurol 37(3):313–323

237. Tureen J, Liu Q, Chow L (1996) Near-infrared spectroscopy in experimental pneumococcal meningitis in the rabbit: cerebral hemodynamics and metabolism. Pediatr Res 40(5):759–763
238. Täuber MG (1989) Brain edema, intracranial pressure and cerebral blood flow in bacterial meningitis. Pediatr Infect Dis J 8(12):915–917
239. Pedersen M, Brandt CT, Knudsen GM, Ostergaard C, Skinhoj P, Frimodt-Moller N, Moller K (2007) Cerebral blood flow autoregulation in early experimental S. pneumoniae meningitis. J Appl Physiol 102(1):72–78. doi:10.1152/japplphysiol.00697.2006, 00697.2006 [pii]
240. Tureen JH, Dworkin RJ, Kennedy SL, Sachdeva M, Sande MA (1990) Loss of cerebrovascular autoregulation in experimental meningitis in rabbits. J Clin Invest 85(2):577–581
241. Pfister HW, Koedel U, Haberl RL, Dirnagl U, Feiden W, Ruckdeschel G, Einhaupl KM (1990) Microvascular changes during the early phase of experimental bacterial meningitis. J Cereb Blood Flow Metab 10(6):914–922
242. Lorenzl S, Koedel U, Frei K, Bernatowicz A, Fontana A, Pfister HW (1995) Protective effect of a 21-aminosteroid during experimental pneumococcal meningitis. J Infect Dis 172(1):113–118
243. Pfister HW, Kodel U, Dirnagl U, Haberl RL, Ruckdeschel G, Einhaupl KM (1992) Effect of catalase on regional cerebral blood flow and brain edema during the early phase of experimental pneumococcal meningitis. J Infect Dis 166(6):1442–1445
244. Hoffmann OM, Becker D, Weber JR (2007) Bacterial hydrogen peroxide contributes to cerebral hyperemia during early stages of experimental pneumococcal meningitis. J Cereb Blood Flow Metab 27(11):1792–1797. doi:10.1038/sj.jcbfm.9600474, 9600474 [pii]
245. Schaper M, Gergely S, Lykkesfeldt J, Zbaren J, Leib SL, Tauber MG, Christen S (2002) Cerebral vasculature is the major target of oxidative protein alterations in bacterial meningitis. J Neuropathol Exp Neurol 61(7):605–613
246. Szabo C (2003) Multiple pathways of peroxynitrite cytotoxicity. Toxicol Lett 140–141:105–112
247. Kastenbauer S, Koedel U, Becker BF, Pfister HW (2002) Pneumococcal meningitis in the rat: evaluation of peroxynitrite scavengers for adjunctive therapy. Eur J Pharmacol 449(1–2):177–181
248. Koedel U, Gorriz C, Lorenzl S, Pfister HW (1997) Increased endothelin levels in cerebrospinal fluid samples from adults with bacterial meningitis. Clin Infect Dis 25(2):329–330
249. Skopal J, Turbucz P, Vastag M, Bori Z, Pek M, deChatel R, Nagy Z, Toth M, Karadi I (1998) Regulation of endothelin release from human brain microvessel endothelial cells. J Cardiovasc Pharmacol 31(Suppl 1):S370–S372
250. Sibson NR, Blamire AM, Perry VH, Gauldie J, Styles P, Anthony DC (2002) TNF-alpha reduces cerebral blood volume and disrupts tissue homeostasis via an endothelin- and TNFR2-dependent pathway. Brain 125(Pt 11):2446–2459
251. Sury MD, Frese-Schaper M, Muhlemann MK, Schulthess FT, Blasig IE, Tauber MG, Shaw SG, Christen S (2006) Evidence that N-acetylcysteine inhibits TNF-alpha-induced cerebrovascular endothelin-1 upregulation via inhibition of mitogen- and stress-activated protein kinase. Free Radic Biol Med 41(9):1372–1383. doi:10.1016/j.freeradbiomed.2006.07.016, S0891-5849(06)00469-2 [pii]
252. Brunner F, Bras-Silva C, Cerdeira AS, Leite-Moreira AF (2006) Cardiovascular endothelins: essential regulators of cardiovascular homeostasis. Pharmacol Ther 111(2):508–531. doi:10.1016/j.pharmthera.2005.11.001, S0163-7258(05)00262-7 [pii]
253. Pfister LA, Tureen JH, Shaw S, Christen S, Ferriero DM, Tauber MG, Leib SL (2000) Endothelin inhibition improves cerebral blood flow and is neuroprotective in pneumococcal meningitis. Ann Neurol 47(3):329–335
254. Edmond K, Clark A, Korczak VS, Sanderson C, Griffiths UK, Rudan I (2010) Global and regional risk of disabling sequelae from bacterial meningitis: a systematic review and meta-analysis. Lancet Infect Dis 10(5):317–328. doi:10.1016/S1473-3099(10)70048-7, S1473-3099(10)70048-7 [pii]
255. Kesser BW, Hashisaki GT, Spindel JH, Ruth RA, Scheld WM (1999) Time course of hearing loss in an animal model of pneumococcal meningitis. Otolaryngol Head Neck Surg 120(5):628–637

256. Woolley AL, Kirk KA, Neumann AM Jr, McWilliams SM, Murray J, Freind D, Wiatrak BJ (1999) Risk factors for hearing loss from meningitis in children: the Children's Hospital experience. Arch Otolaryngol Head Neck Surg 125(5):509–514
257. Oostenbrink R, Maas M, Moons KG, Moll HA (2002) Sequelae after bacterial meningitis in childhood. Scand J Infect Dis 34(5):379–382
258. Edmond K, Dieye Y, Griffiths UK, Fleming J, Ba O, Diallo N, Mulholland K (2010) Prospective cohort study of disabling sequelae and quality of life in children with bacterial meningitis in urban Senegal. Pediatr Infect Dis J 29(11):1023–1029. doi:10.1097/INF.0b013e3181e598ea
259. Adachi N, Ito K, Sakata H (2010) Risk factors for hearing loss after pediatric meningitis in Japan. Ann Otol Rhinol Laryngol 119(5):294–296
260. Douglas SA, Sanli H, Gibson WP (2008) Meningitis resulting in hearing loss and labyrinthitis ossificans – does the causative organism matter? Cochlear Implants Int 9(2):90–96. doi:10.1002/cii.344
261. Kutz JW, Simon LM, Chennupati SK, Giannoni CM, Manolidis S (2006) Clinical predictors for hearing loss in children with bacterial meningitis. Arch Otolaryngol Head Neck Surg 132(9):941–945. doi:10.1001/archotol.132.9.941, 132/9/941 [pii]
262. Dichgans M, Jager L, Mayer T, Schorn K, Pfister HW (1999) Bacterial meningitis in adults: demonstration of inner ear involvement using high-resolution MRI. Neurology 52(5):1003–1009
263. Caye-Thomasen P, Dam MS, Omland SH, Mantoni M (2012) Cochlear ossification in patients with profound hearing loss following bacterial meningitis. Acta Otolaryngol 132(7):720–725. doi:10.3109/00016489.2012.656323
264. Nadol JB Jr, Hsu WC (1991) Histopathologic correlation of spiral ganglion cell count and new bone formation in the cochlea following meningogenic labyrinthitis and deafness. Ann Otol Rhinol Laryngol 100(9 Pt 1):712–716
265. Merchant SN, Gopen Q (1996) A human temporal bone study of acute bacterial meningogenic labyrinthitis. Am J Otol 17(3):375–385
266. Klein M, Koedel U, Pfister HW, Kastenbauer S (2003) Morphological correlates of acute and permanent hearing loss during experimental pneumococcal meningitis. Brain Pathol 13(2):123–132
267. Caye-Thomasen P, Worsoe L, Brandt CT, Miyazaki H, Ostergaard C, Frimodt-Moller N, Thomsen J (2009) Routes, dynamics, and correlates of cochlear inflammation in terminal and recovering experimental meningitis. Laryngoscope 119(8):1560–1570. doi:10.1002/lary.20260
268. Blank AL, Davis GL, VanDeWater TR, Ruben RJ (1994) Acute Streptococcus pneumoniae meningogenic labyrinthitis. An experimental guinea pig model and literature review. Arch Otolaryngol Head Neck Surg 120(12):1342–1346
269. Bhatt S, Halpin C, Hsu W, Thedinger BA, Levine RA, Tuomanen E, Nadol JB Jr (1991) Hearing loss and pneumococcal meningitis: an animal model. Laryngoscope 101(12 Pt 1):1285–1292
270. Klein M, Koedel U, Kastenbauer S, Pfister HW (2008) Nitrogen and oxygen molecules in meningitis-associated labyrinthitis and hearing impairment. Infection 36(1):2–14. doi:10.1007/s15010-007-7153-1
271. Rappaport JM, Bhatt SM, Kimura RS, Lauretano AM, Levine RA (1999) Electron microscopic temporal bone histopathology in experimental pneumococcal meningitis. Ann Otol Rhinol Laryngol 108(6):537–547
272. Brandt CT, Caye-Thomasen P, Lund SP, Worsoe L, Ostergaard C, Frimodt-Moller N, Espersen F, Thomsen J, Lundgren JD (2006) Hearing loss and cochlear damage in experimental pneumococcal meningitis, with special reference to the role of neutrophil granulocytes. Neurobiol Dis 23(2):300–311. doi:10.1016/j.nbd.2006.03.006, S0969-9961(06)00060-X [pii]
273. Winter AJ, Comis SD, Osborne MP, Tarlow MJ, Stephen J, Andrew PW, Hill J, Mitchell TJ (1997) A role for pneumolysin but not neuraminidase in the hearing loss and cochlear damage induced by experimental pneumococcal meningitis in guinea pigs. Infect Immun 65(11):4411–4418

274. Klein M, Koedel U, Pfister HW, Kastenbauer S (2003) Meningitis-associated hearing loss: protection by adjunctive antioxidant therapy. Ann Neurol 54(4):451–458. doi:10.1002/ana.10684
275. Ge NN, Brodie HA, Tinling SP (2008) Long-term hearing loss in gerbils with bacterial meningitis treated with superoxide dismutase. Otol Neurotol 29(8):1061–1067. doi:10.1097/MAO.0b013e31818b6479
276. Bedford H, de Louvois J, Halket S, Peckham C, Hurley R, Harvey D (2001) Meningitis in infancy in England and Wales: follow up at age 5 years. BMJ 323(7312):533–536
277. Anderson V, Anderson P, Grimwood K, Nolan T (2004) Cognitive and executive function 12 years after childhood bacterial meningitis: effect of acute neurologic complications and age of onset. J Pediatr Psychol 29(2):67–81
278. Schmidt H, Heimann B, Djukic M, Mazurek C, Fels C, Wallesch CW, Nau R (2006) Neuropsychological sequelae of bacterial and viral meningitis. Brain 129(Pt 2):333–345
279. Grimwood K, Anderson P, Anderson V, Tan L, Nolan T (2000) Twelve year outcomes following bacterial meningitis: further evidence for persisting effects. Arch Dis Child 83(2):111–116
280. Hoogman M, van de Beek D, Weisfelt M, de Gans J, Schmand B (2007) Cognitive outcome in adults after bacterial meningitis. J Neurol Neurosurg Psychiatry 78(10):1092–1096. doi:10.1136/jnnp.2006.110023, jnnp.2006.110023 [pii]
281. Focke NK, Kallenberg K, Mohr A, Djukic M, Nau R, Schmidt H (2013) Distributed, limbic gray matter atrophy in patients after bacterial meningitis. AJNR Am J Neuroradiol 34:1164–1167. doi:10.3174/ajnr.A3351, ajnr.A3351 [pii]
282. Grimwood K, Nolan TM, Bond L, Anderson VA, Catroppa C, Keir EH (1996) Risk factors for adverse outcomes of bacterial meningitis. J Paediatr Child Health 32(5):457–462
283. Libster R, Edwards KM, Levent F, Edwards MS, Rench MA, Castagnini LA, Cooper T, Sparks RC, Baker CJ, Shah PE (2012) Long-term outcomes of group B streptococcal meningitis. Pediatrics 130(1):e8–e15. doi:10.1542/peds.2011-3453, peds.2011-3453 [pii]
284. Bargui F, D'Agostino I, Mariani-Kurkdjian P, Alberti C, Doit C, Bellier N, Morin L, Galli Gibertini G, Smail A, Zanin A, Lorrot M, Dauger S, Neve M, Faye A, Armoogum P, Bourrillon A, Bingen E, Mercier JC, Bonacorsi S, Nigrovic LE, Titomanlio L (2012) Factors influencing neurological outcome of children with bacterial meningitis at the emergency department. Eur J Pediatr 171(9):1365–1371. doi:10.1007/s00431-012-1733-5
285. Hernandez MI, Sandoval CC, Tapia JL, Mesa T, Escobar R, Huete I, Wei XC, Kirton A (2011) Stroke patterns in neonatal group B streptococcal meningitis. Pediatr Neurol 44(4):282–288. doi:10.1016/j.pediatrneurol.2010.11.002, S0887-8994(10)00491-1 [pii]
286. Jorens PG, Parizel PM, Wojciechowski M, Laridon A, De Weerdt A, Mertens G, Ceulemans B (2008) Streptococcus pneumoniae meningoencephalitis with unusual and widespread white matter lesions. Eur J Paediatr Neurol 12(2):127–132. doi:10.1016/j.ejpn.2007.06.007, S1090-3798(07)00112-2 [pii]
287. Shah DK, Daley AJ, Hunt RW, Volpe JJ, Inder TE (2005) Cerebral white matter injury in the newborn following Escherichia coli meningitis. Eur J Paediatr Neurol 9(1):13–17. doi:10.1016/j.ejpn.2004.09.002, S1090-3798(04)00189-8 [pii]
288. Täuber MG, Kim YS, Leib SL (1997) Neuronal injury in meningitis. In: Peterson PK, Remington JS (eds) In defense of the brain. Blackwell Science, Malden, MA, pp 124–143
289. Nau R, Gerber J, Bunkowski S, Bruck W (2004) Axonal injury, a neglected cause of CNS damage in bacterial meningitis. Neurology 62(3):509–511
290. Gerber J, Seitz RC, Bunkowski S, Bruck W, Nau R (2009) Evidence for frequent focal and diffuse acute axonal injury in human bacterial meningitis. Clin Neuropathol 28(1):33–39, 5350 [pii]
291. Renier D, Flandin C, Hirsch E, Hirsch JF (1988) Brain abscesses in neonates. A study of 30 cases. J Neurosurg 69(6):877–882. doi:10.3171/jns.1988.69.6.0877
292. Cone LA, Leung MM, Byrd RG, Annunziata GM, Lam RY, Herman BK (2003) Multiple cerebral abscesses because of Listeria monocytogenes: three case reports and a literature review of supratentorial listerial brain abscess(es). Surg Neurol 59(4):320–328, S0090301903000569 [pii]

293. Zysk G, Bruck W, Gerber J, Bruck Y, Prange HW, Nau R (1996) Anti-inflammatory treatment influences neuronal apoptotic cell death in the dentate gyrus in experimental pneumococcal meningitis. J Neuropathol Exp Neurol 55(6):722–728
294. Bifrare YD, Gianinazzi C, Imboden H, Leib SL, Tauber MG (2003) Bacterial meningitis causes two distinct forms of cellular damage in the hippocampal dentate gyrus in infant rats. Hippocampus 13(4):481–488
295. Gianinazzi C, Grandgirard D, Imboden H, Egger L, Meli DN, Bifrare YD, Joss PC, Tauber MG, Borner C, Leib SL (2003) Caspase-3 mediates hippocampal apoptosis in pneumococcal meningitis. Acta Neuropathol 105(5):499–507. doi:10.1007/s00401-003-0672-7
296. Nag S, Manias JL, Stewart DJ (2009) Pathology and new players in the pathogenesis of brain edema. Acta Neuropathol 118(2):197–217. doi:10.1007/s00401-009-0541-0
297. Loeffler JM, Ringer R, Hablutzel M, Tauber MG, Leib SL (2001) The free radical scavenger alpha-phenyl-tert-butyl nitrone aggravates hippocampal apoptosis and learning deficits in experimental pneumococcal meningitis. J Infect Dis 183(2):247–252. doi:10.1086/317921
298. Wellmer A, Noeske C, Gerber J, Munzel U, Nau R (2000) Spatial memory and learning deficits after experimental pneumococcal meningitis in mice. Neurosci Lett 296(2–3):137–140
299. Grandgirard D, Bifrare YD, Pleasure SJ, Kummer J, Leib SL, Tauber MG (2007) Pneumococcal meningitis induces apoptosis in recently postmitotic immature neurons in the dentate gyrus of neonatal rats. Dev Neurosci 29(1–2):134–142
300. Leib SL, Heimgartner C, Bifrare YD, Loeffler JM, Taauber MG (2003) Dexamethasone aggravates hippocampal apoptosis and learning deficiency in pneumococcal meningitis in infant rats. Pediatr Res 54(3):353–357. doi:10.1203/01.PDR.0000079185.67878.72
301. Nau R, Soto A, Bruck W (1999) Apoptosis of neurons in the dentate gyrus in humans suffering from bacterial meningitis. J Neuropathol Exp Neurol 58(3):265–274
302. Free SL, Li LM, Fish DR, Shorvon SD, Stevens JM (1996) Bilateral hippocampal volume loss in patients with a history of encephalitis or meningitis. Epilepsia 37(4):400–405
303. Derugin N, Wendland M, Muramatsu K, Roberts TP, Gregory G, Ferriero DM, Vexler ZS (2000) Evolution of brain injury after transient middle cerebral artery occlusion in neonatal rats. Stroke 31(7):1752–1761
304. Leib SL, Kim YS, Ferriero DM, Täuber MG (1996) Neuroprotective effect of excitatory amino acid antagonist kynurenic acid in experimental bacterial meningitis. J Infect Dis 173(1):166–171
305. Scheld WM, Dacey RG, Winn HR, Welsh JE, Jane JA, Sande MA (1980) Cerebrospinal fluid outflow resistance in rabbits with experimental meningitis. Alterations with penicillin and methylprednisolone. J Clin Invest 66:243–253
306. Kaplan SL, Feigin RD (1978) The syndrome of inappropriate secretion of antidiuretic hormone in children with bacterial meningitis. J Pediatr 92:758–761
307. Winkler F, Kastenbauer S, Yousry TA, Maerz U, Pfister HW (2002) Discrepancies between brain CT imaging and severely raised intracranial pressure proven by ventriculostomy in adults with pneumococcal meningitis. J Neurol 249(9):1292–1297
308. Lindvall P, Ahlm C, Ericsson M, Gothefors L, Naredi S, Koskinen LO (2004) Reducing intracranial pressure may increase survival among patients with bacterial meningitis. Clin Infect Dis 38(3):384–390. doi:10.1086/380970, CID31898 [pii]
309. Arcienega II, Brunet JF, Bloch J, Badaut J (2010) Cell locations for AQP1, AQP4 and 9 in the non-human primate brain. Neuroscience 167(4):1103–1114. doi:10.1016/j.neuroscience.2010.02.059, S0306-4522(10)00287-3 [pii]
310. Saadoun S, Papadopoulos MC (2010) Aquaporin-4 in brain and spinal cord oedema. Neuroscience 168(4):1036–1046. doi:10.1016/j.neuroscience.2009.08.019, S0306-4522(09)01339-6 [pii]
311. Papadopoulos MC, Verkman AS (2005) Aquaporin-4 gene disruption in mice reduces brain swelling and mortality in pneumococcal meningitis. J Biol Chem 280(14):13906–13912. doi:10.1074/jbc.M413627200, M413627200 [pii]
312. Saadoun S, Papadopoulos MC, Krishna S (2003) Water transport becomes uncoupled from K+ siphoning in brain contusion, bacterial meningitis, and brain tumours: immunohistochemical case review. J Clin Pathol 56(12):972–975

313. Blocher J, Eckert I, Elster J, Wiefek J, Eiffert H, Schmidt H (2011) Aquaporins AQP1 and AQP4 in the cerebrospinal fluid of bacterial meningitis patients. Neurosci Lett 504(1):23–27. doi:10.1016/j.neulet.2011.08.049, S0304-3940(11)01234-1 [pii]
314. Spranger M, Krempien S, Schwab S, Maiwald M, Bruno K, Hacke W (1996) Excess glutamate in the cerebrospinal fluid in bacterial meningitis. J Neurol Sci 143(1–2):126–131
315. Guerra-Romero L, Tauber MG, Fournier MA, Tureen JH (1992) Lactate and glucose concentrations in brain interstitial fluid, cerebrospinal fluid, and serum during experimental pneumococcal meningitis. J Infect Dis 166(3):546–550
316. Wippel C, Maurer J, Fortsch C, Hupp S, Bohl A, Ma J, Mitchell TJ, Bunkowski S, Bruck W, Nau R, Iliev AI (2013) Bacterial cytolysin during meningitis disrupts the regulation of glutamate in the brain, leading to synaptic damage. PLoS Pathog 9(6):e1003380. doi:10.1371/journal.ppat.1003380
317. Recher M, Malipiero U, Schaer DJ, Koedel U, Pfister HW, Birchler T, Petrausch U, Claus H, Gast H, Fontana A (2013) Inhibition of meningitis-associated neutrophil apoptosis by TNF-alpha depends on functional PI3-kinase in monocytes. J Leukoc Biol 93(2):259–266. doi:10.1189/jlb.0511218, jlb.0511218 [pii]
318. Tunbridge AJ, Stevanin TM, Lee M, Marriott HM, Moir JW, Read RC, Dockrell DH (2006) Inhibition of macrophage apoptosis by Neisseria meningitidis requires nitric oxide detoxification mechanisms. Infect Immun 74(1):729–733
319. Wennekamp J, Henneke P (2008) Induction and termination of inflammatory signaling in group B streptococcal sepsis. Immunol Rev 225:114–127. doi:10.1111/j.1600-065X.2008.00673.x, IMR673 [pii]
320. Kim GH, Kim JE, Rhie SJ, Yoon S (2015) The role of oxidative stress in neurodegenerative diseases. Exp Neurobiol 24(4):325–340. doi:10.5607/en.2015.24.4.325
321. Grammas P, Martinez J, Miller B (2011) Cerebral microvascular endothelium and the pathogenesis of neurodegenerative diseases. Expert Rev Mol Med 13:e19. doi:10.1017/S1462399411001918
322. Klein M, Koedel U, Pfister HW (2006) Oxidative stress in pneumococcal meningitis: a future target for adjunctive therapy? Prog Neurobiol 80(6):269–280. doi:10.1016/j.pneurobio.2006.11.008, S0301-0082(06)00143-2 [pii]
323. Simon RP, Beckman JS (2002) Why pus is bad for the brain. Neurology 58(2):167–168
324. Braun J (2009) Inducible nitric oxide synthase mediates hippocampal caspase-3 activation in pneumococcal meningitis. Int J Neurosci 119(4):455–459. doi:10.1080/00207450802479970, 908868698 [pii]
325. Leib SL, Kim YS, Black SM, Tureen JH, Tauber MG (1998) Inducible nitric oxide synthase and the effect of aminoguanidine in experimental neonatal meningitis. J Infect Dis 177(3):692–700
326. Guo Z, Sun X, He Z, Jiang Y, Zhang X (2010) Role of matrix metalloproteinase-9 in apoptosis of hippocampal neurons in rats during early brain injury after subarachnoid hemorrhage. Neurol Sci 31(2):143–149. doi:10.1007/s10072-009-0192-x
327. Murase S, McKay RD (2012) Matrix metalloproteinase-9 regulates survival of neurons in newborn hippocampus. J Biol Chem 287(15):12184–12194. doi:10.1074/jbc.M111.297671, M111.297671 [pii]
328. Fujita-Hamabe W, Tokuyama S (2012) The involvement of cleavage of neural cell adhesion molecule in neuronal death under oxidative stress conditions in cultured cortical neurons. Biol Pharm Bull 35(4):624–628, JST.JSTAGE/bpb/35.624 [pii]
329. Rosenberg GA (2009) Matrix metalloproteinases and their multiple roles in neurodegenerative diseases. Lancet Neurol 8(2):205–216. doi:10.1016/S1474-4422(09)70016-X, S1474-4422(09)70016-X [pii]
330. Braun JS, Novak R, Murray PJ, Eischen CM, Susin SA, Kroemer G, Halle A, Weber JR, Tuomanen EI, Cleveland JL (2001) Apoptosis-inducing factor mediates microglial and neuronal apoptosis caused by pneumococcus. J Infect Dis 184(10):1300–1309
331. Braun JS, Sublett JE, Freyer D, Mitchell TJ, Cleveland JL, Tuomanen EI, Weber JR (2002) Pneumococcal pneumolysin and H(2)O(2) mediate brain cell apoptosis during meningitis. J Clin Invest 109(1):19–27

332. Braun JS, Hoffmann O, Schickhaus M, Freyer D, Dagand E, Bermpohl D, Mitchell TJ, Bechmann I, Weber JR (2007) Pneumolysin causes neuronal cell death through mitochondrial damage. Infect Immun 75(9):4245–4254. doi:10.1128/IAI.00031-07, IAI.00031-07 [pii]
333. Beurg M, Hafidi A, Skinner L, Cowan G, Hondarrague Y, Mitchell TJ, Dulon D (2005) The mechanism of pneumolysin-induced cochlear hair cell death in the rat. J Physiol 568(Pt 1):211–227. doi:10.1113/jphysiol.2005.092478, jphysiol.2005.092478 [pii]
334. Reiss A, Braun JS, Jager K, Freyer D, Laube G, Buhrer C, Felderhoff-Muser U, Stadelmann C, Nizet V, Weber JR (2011) Bacterial pore-forming cytolysins induce neuronal damage in a rat model of neonatal meningitis. J Infect Dis 203(3):393–400. doi:10.1093/infdis/jiq047, jiq047 [pii]

Blood Vessels in the Brain: A Signaling Hub in Brain Tumor Inflammation

Sylvaine Guerit and Stefan Liebner

Abstract Malignant brain tumors are associated with an extremely poor prognosis. Currently, only limited treatment options are available, resulting in a disappointing benefit in patient survival. In adults, primary brain tumors, mostly belonging to glioblastoma World Health Organization grade IV (GBM; the most malignant glial tumor) are characterized by infiltrative growth, necroses, and extensive vascular proliferations. GBM vessels are poorly differentiated, largely lacking regular blood–brain barrier properties. The vascular abnormalities generate a tumor microenvironment that permits sustained tumor growth by oxidative stress, metabolic changes, and augmented inflammation. Rudolf Virchow acknowledged the importance of inflammation for tumor growth as early as the middle of the 19th century. In addition to a plethora of cytokines and pro-angiogenic factors such as vascular endothelial growth factor, tumor-associated macrophages have recently been shown to secrete Wnt growth factors that were proposed to be a trigger of the so-called angiogenic switch of tumors. The Wnt/β-catenin pathway was implicated in many aspects of angiogenesis and vascular remodeling. Moreover, it has been proven to be crucial for brain angiogenesis and endothelial barrier formation during development and in brain tumors.

Here, we summarize the current concept that the tumor stroma, including blood vessels and inflammatory cells, is the major driver of cancer progression and discuss common and conflicting findings in the light of glioma in the central nervous system, thereby focusing on the Wnt pathway and the endothelium as a signaling hub.

S. Guerit
Institute of Neurology (Edinger Institute), Goethe University Medical School, Heinrich-Hoffmann-Strasse 7, 60528 Frankfurt, Germany

S. Liebner (✉)
German Cancer Consortium (DKTK), Partner Site Frankfurt/Mainz, Frankfurt, Germany

German Cancer Research Center (DKFZ), Heidelberg, Germany

German Centre of Cardiovascular Research (DZHK), Partner Site RheinMain, Frankfurt, Germany

Institute of Neurology (Edinger Institute), Goethe University Medical School, Heinrich-Hoffmann-Strasse 7, 60528 Frankfurt, Germany
e-mail: stefan.liebner@kgu.de

R. Lyck, G. Enzmann (eds.), *The Blood Brain Barrier and Inflammation*, Progress in Inflammation Research, DOI 10.1007/978-3-319-45514-3_11

1 Introduction

The vascular system in the central nervous system (CNS) confers oxygen and a supply of nutrients for the demanding metabolic needs of neuronal function. Moreover, the majority of endothelial cells (ECs) in the CNS exhibit so-called blood–brain barrier (BBB) characteristics, which prevent the free diffusion of hydrophilic substances between the blood and the brain [1, 32]. BBB properties in ECs are dependent on the cellular communication between ECs, pericytes (PCs), astrocytes (ACs), neurons, and perivascular microglia in the neurovascular unit (NVU).

Under pathological conditions, such as trauma, stroke, tumor, and inflammation, and in neurodegenerative diseases, the organization of the BBB and NVU is hampered at different levels and to different extents [32, 42].

It is widely believed that the Wnt/β-catenin pathway is responsible for CNS angiogenesis and for the induction of the BBB phenotype [31, 32, 87, 141]. Wnt signaling is highly conserved in all metazoan organisms and is involved in a wide variety of developmental and adult processes, conferring diverse effects ranging from proliferation, apoptosis, migration, and polarization to stem cell maintenance and differentiation [85]. Currently, 19 different Wnt ligands have been identified that can potentially bind and signal through ten members of the seven transmembrane-spanning, G-protein-coupled receptors of the Frizzled (Fzd) family, together with the co-receptors low-density lipoprotein receptor-related protein (LRP) 5/6. At least three divergent Wnt signaling pathways can be distinguished. Two of them, the Ca^{2+} and the planar cell polarity pathways, do not require β-catenin as a co-transcription factor, and therefore are considered to be "noncanonical" Wnt pathways [47, 113, 130]. Compared with the noncanonical Wnt pathways, which are less well characterized, the canonical Wnt pathway, also called the Wnt/β-catenin pathway, is understood in more detail. The central player of this pathway is the protein β-catenin, which was first characterized as a scaffold protein, linking the cytoplasmic tail of classical cadherins – in ECs vascular endothelial (VE)-cadherin (Cdh5) and neuronal (N)-cadherin (Cdh2) – via α-catenin to the actin cytoskeleton. Without Wnt stimulation, cytoplasmic β-catenin levels are kept low by a degradation complex consisting of axin, adenomatous polyposis coli, casein kinase 1α, and glycogen synthase kinase-3β. In the case of Wnt stimulation, β-catenin can accumulate in the cytoplasm and translocate to the nucleus, where it interacts with members of the T-cell factor/lymphoid enhancer factor family to activate the transcription of target genes. Pathway complexity is further increased by several secreted and soluble Wnt pathway inhibitors called secreted Fzd-related proteins (sFRPs) and Wnt inhibitory factor 1. These proteins directly bind and sequester Wnt ligands, thus blocking the canonical and noncanonical Wnt signaling output [25]. Another class of inhibitors, the Dickkopf (Dkk) protein family, includes the members Dkk1, 2, 3, 4, with Dkk1 and 4 being accepted as pure inhibitors of Wnt/β-catenin signaling [104]. For a comprehensive description of the canonical Wnt pathway, the reader is referred to Duchartre et al. [38].

The characteristics of the BBB have been shown to be induced during development by activation of the Wnt/β-catenin pathway in ECs, driven on the molecular level by Wnt7a/b released from neural precursor cells and by the non-Wnt ligand norrin from ACs [31, 141]. In adult vertebrates, canonical Wnt activity is crucial for the maintenance of the endothelial barrier in CNS vessels throughout lifetime, which becomes apparent in loss-of-function genetic mouse models for Wnt pathway components [152, 172]. Indeed, deletion of downstream Wnt signaling components specifically in ECs leads to downregulation of the tight junction (TJ) protein claudin-5 and concomitant upregulation of the permeability-related protein plasmalemma vesicle-associated protein [87, 173]. The orphan G-protein coupled receptor 124 (Gpr124) was recently shown to specifically function as an additional Fzd4-co-receptor for Wnt7a/b [3, 30, 122, 150, 172].

It is well known that aberrant activation of the Wnt pathway can lead to the formation of various neoplasias, and is best described for colorectal cancer, breast cancer, and the Wnt-type of medulloblastoma in the brain [38, 49]. Although mutations in genes of the Wnt pathways leading to its constitutive activation play a fundamental role in the neoplastic development, the Wnt pathways also have a considerable impact on the so-called tumor stroma, consisting of blood vessels, extracellular matrix, inflammatory cells, and mesenchymal cells (not in the CNS) [58, 120].

In the last decade, the tumor stroma has drawn considerable attention as a therapeutic target, as tumor angiogenesis and the status of tumor inflammation turned out to contribute to tumor formation, progression, and metastasis [34, 50, 59]. This also applies to brain tumors and specifically to human gliomas, of which the *glioblastoma multiforme* (GBM; astrocytoma World Health Organization grade IV) is the most frequent brain tumor in adult humans with very poor prognosis [115].

The GBM is characterized by its high level of vessel recruitment, which are of highly aberrant structures and are consequently largely devoid of BBB characteristics, leading to clinically relevant edema formation and poor delivery of systemic chemotherapeutic agents [90]. Seemingly predisposed to respond to anti-angiogenic and vessel-normalizing therapy, so far clinical trials have been disillusioning [9].

Moreover, a high degree of tumor inflammation has also been observed in GBM, which is in line with the long-standing hypothesis first formulated by Rudolph Virchow that tumor initiation and possibly growth is associated with a chronic inflammatory reaction [5]. Specifically, it is challenging to develop an integrated picture for a comprehensive view of the disease, comprising the individually identified players in the tumor stroma, such as blood vessels, non-tumoral cells of the neuroectoderm (ACs) and inflammatory cells. Along these lines, little is known about the link among the Wnt pathways, tumor angiogenesis, and inflammation in GBM, which we intend to highlight and discuss in this review.

During the late stages of GBM, when necrotic areas form within the tumor mass, extensive activation of ECs leads to tumor angiogenesis on one hand, and to recruitment of inflammatory cells on the other. Finally, the glioma vasculature has been shown to form a niche for glioma cells with stem cell-like properties (for simplicity named here "glioma stem cells", GSCs), which can infiltrate the brain parenchyma by following the vascular scaffold, thereby further fostering tumor progression [27].

In this context, the tumor vasculature and in particular the ECs of tumor vessels may be considered a signaling hub, as they initially serve as a scaffold for the co-optive growth of glioma cells at an early stage of tumor progression [95, 118].

2 Constituents of Glioma Stroma

2.1 The Vascular Compartment

According to the hypothesis of Judah Folkman, the tumor needs formation and/or recruitment of vessels to progress as soon as it reaches a size larger than 1–2 mm^3 [45]. It is well known that several processes such as angiogenesis, vasculogenesis, and vessel cooption play a role in tumor vascularization, whereas others are still controversial, such as vascular mimicry and glioma–EC transdifferentiation [62]. Nevertheless, and independently of their origin, the morphology and functionality of glioma vessels are different from the vessels in the healthy brain tissue.

First, the vessels exert a heterogeneous morphology, are tortuous, and are especially dilated in the center of the tumor mass [94, 121, 131]. Depending on the tumor model, vessels demonstrate a pronounced loss of EC junctions and an increase in fenestrations, underlining compromised BBB function [133, 138]. This has been confirmed at the molecular level with the downregulation of claudin-1 and -5 in tumor vessels compared with normal brain vessels [88]. As the EC layer is affected, the basal lamina and the extracellular matrix (ECM) in general is altered and can either show an increased or a reduced thickness. The composition of the ECM in GBM is described in more detail later in this review. Along with the vessel morphology and ECM alterations, pericyte coverage is impaired and transporter expression in ECs is altered [7, 24, 149]. Overall, these modifications lead to an immature phenotype of the tumor vessels and to a partial loss of BBB function.

Unfortunately, this "opening" of the tumor vessel walls does not necessarily mean that drug delivery is potentiated in glioma compared with normal brain tissue. All these morphological issues lead to the abnormal function of these vessels. It is now well acknowledged that the blood flow is chaotic and sometimes non-existent in some areas [146, 147]. Using radiolabeled drugs in a brain metastasis model of breast cancer, Lockman et al. showed that the drug diffusion is heterogeneous within the brain tumor and is overall reduced compared to peripheral organ metastases [90]. Along these lines, at the infiltrating sites of GBM, BBB function is only minimally altered, specifically creating the problem of targeting these tumor cells that potentially confer tumor recurrence after therapy.

Therefore, studying vasculature development and regulation of BBB function is not only of great interest for improving drug delivery to GBM, but also for tackling other diseases of the CNS [109, 149]

2.2 The Inflammatory Compartment

Due to the lack of lymphatic vessels in the brain parenchyma and because of the BBB properties of microvessels, the CNS tissue is considered an immune-privileged organ [21, 37]. Nevertheless, in GBM, as in peripheral tumors, inflammatory processes occur, leading to recruitment of immune cells.

In CNS inflammation, the first cells that are recruited are the resident microglia. Upon stimulation these cells become activated and start to acquire properties of macrophages, such as a motile phenotype and expression of proinflammatory cytokines. In glioma, resident microglia express MHC class I, but fail to upregulate MHC class II to activate cytotoxic T-cells. Additionally, activated microglia do not upregulate proinflammatory cytokines such as TNFα, IL6, and IL1β; instead, they express alternative inflammatory cytokines, such as IL10, EGF and VEGF, and thereby foster further tumor growth [159].

However, microglia are not the only inflammatory cells detected in GBM. GBM, like peripheral tumors, are characterized by the accumulation of tumor-associated macrophages (TAMs), myeloid-derived suppressor cells and regulatory T-cells (Tregs). Nevertheless, TAMs are the major subset of inflammatory cells in GBM. Interestingly, up to 30 % of the glioma tumor mass can be made up of microglia and macrophages, which do not seem to just fight tumor cells, but may also support tumor growth and progression by conferring a so-called Th2 inflammatory status via the release of pro-proliferative cytokines [96, 114].

Interestingly, microglia, in addition to CCL2 and M-CSF-recruited macrophages from the periphery, were reported to switch to M2 polarization and start to express Tie-2, contributing to tumor angiogenesis and progression. In the classical polarization state, also termed M1 polarization, macrophages/TAMs secrete pro-inflammatory cytokines (IL1β, IL6, TNFα, etc.) and reactive oxygen species and NO, inducing tissue and tumor cell destruction. Instead, the M2-polarized TAMs have pro-tumor effects by promotion of angiogenesis through the secretion of VEGF, by the secretion of anti inflammatory cytokines (IL10, IFNβ/γ, TGFβ etc.) and blockade of $CD8^+$ T-cell infiltration [15, 97, 107]. In GBM the ratio of M1/M2 is in favor of M2, thus promoting the development of the tumor [37, 66, 142]. More recently, it has been shown that M2-polarized TAMs also appear to upregulate members of the Wnt pathway, again contributing to tumor progression (discussed below; Fig. 1).

So far, only few studies have been carried out to elucidate the presence and the role of myeloid-derived suppressor cells (MDSC), dendritic cells, NK cells, and B-cells in glioma [37, 81]. The ratio between $CD4^+$ and $CD8^+$ T cells in GBM is linked to patient survival, with a higher amount of $CD8^+$ cells, correlating with better prognosis [80, 91]. Finally, $CD4^+/FoxP3^+/CD25^{high}/CD127^{low}$ Tregs were reported to be particularly frequent in high-grade glioma, but are not detectable in benign CNS tumors, correlating with the aggressive phenotype of GBM [2, 68]. Specifically, Tregs were shown to closely localize to tumor blood vessels in GBM, where they may interact with glioma stem cells (GSCs) that were identified to express programmed death ligand-1 (PD-L1), which interacts with the inhibitory receptor PD-1 on activated lymphocytes.

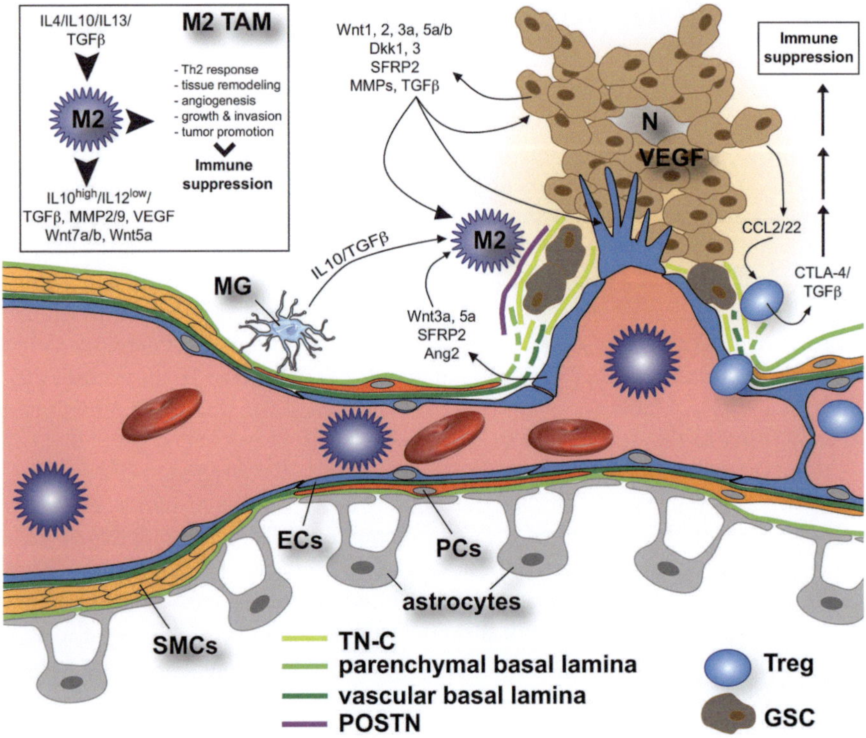

Fig. 1 Schematic representation of brain blood vessels as a signaling hub in brain tumor inflammation. The growing GMB tumor develops a hypoxic state around the areas of necrosis, leading to the excessive expression of *VEGF* and subsequent angiogenic activation of the vasculature. Resident *MGs* become activated, along with the sprouting vessels, which further recruit *GSC*s and inflammatory cells. Among those, macrophages and activated MGs acquire an alternative M2 polarization by the stimulation of glioma-derived factors. Among others, Wnt growth factors are released from the tumor cells, targeting the *M2 TAMs* that in turn also release Wnt and other factors, such as TGFβ and IL10. These factors may further lead to angiogenesis and glioma progression by acting on GSCs, in addition to suppression of the T-cell-mediated immune response via *Treg* recruitment. Ultimately, the brain–blood barrier is severely hampered in the glioma vasculature, supporting the tumor-permissive microenvironment. *ECs* endothelial cells, *PCs* pericytes, *SMCs* smooth muscle cells, *MG* microglia, *M2 TAM* M2-polarized tumor-associated macrophages, *Treg* regulatory T-cells, *GSCs* glioma stem cells, *N* necrosis, *VEGF* vascular endothelial growth factor, *TN-C* tenascin-C, *POSTN* periostin

2.3 The Mesenchymal Compartment

Although this review focuses on the role of the vasculature and the Wnt pathway in tumor stroma regulation of brain tumors in which fibroblasts are usually not found, it is important to mention their contribution to the tumor stroma in general. In peripheral tumors, such as carcinomas of the ovary, the mammary gland, the colon, and the pancreas, fibroblasts exist in an activated state termed "tumor-associated fibroblasts" (TAFs), being characterized by the expression of the smooth muscle cell markers desmin, α-smooth muscle actin (αSMA), γ-smooth muscle actin (γSMA), and the cytoskeletal protein, palladin 4Ig, which controls stress fiber integrity [156]. TAFs can be derived from various sources such as fibroblasts, mesenchymal stem cells (MSCs), pericytes, ECs, and epithelial cells. From epithelial cells and ECs, TAFs transdifferentiate via a mechanism called epithelial–mesenchymal transition (EMT) and endothelial–mesenchymal transition (EndMT), respectively. Interestingly, TAFs express various growth factors and cytokines influencing other tumor compartments such as tumor cells (SDF-1, TGFβ, etc.), ECs (VEGF, SDF-1, MMP-2, IL-6), and inflammatory cells (IL-1β, IL-8, CCL2/MCP1; for reviews see Bhowmick et al. and Pietras and Ostman [14, 120]).

In GBM and potentially in other brain tumors, the role of fibroblasts may be taken over by ACs, which have been shown to express a similar set of factors, such as SDF-1 and CCL2/MCP1, that can promote migration of neural stem cells (NSCs) to the inflamed tumor area [50]. The role of ACs in GBM progression is under discussion. It should be noted that published and the author's own preliminary data support the notion that ACs can be a source for several Wnt growth factors [23, 86] (Guerit et al., unpublished data). The way in which Wnt expression by ACs changes with their activation status has not been addressed so far.

2.4 *The Extracellular Matrix Compartment*

The ECM of brain tumors, and particularly of GBM, consists of the basement membrane components, collagen IV, laminin, fibronectin, and tenascin-C (TN-C), lining the blood vessels, in addition to collagen I, vitronectin, and hyaluronan [11, 26, 65]. Collagen IV and TN-C levels were shown to be upregulated in gliomas and localized at the basement membrane lining the vessel walls in astrocytomas of all grades, including GBM [20, 116]. Interestingly, as opposed to most other ECM components, the tumor cells endogenously produce TN-C, suggesting that TN-C might pave the way for a tumor growth-permissive microenvironment. It should be mentioned that glioma cells have also been shown to produce collagen I and IV; however, expression levels are relatively low [116].

Tenascin-C induces proliferation of several cell types, including GBM cells and ECs [134]. Interestingly, in the whisker hair follicle, TN-C was reported to bind Wnt3a and to be relevant for downstream β-catenin transcriptional activation, maintaining stem cells in an undifferentiated state [63]. Whether TN-C has a similar function in GBM vessels has not been investigated so far. It is, however, worth noting that TN-C expression has been shown to be induced by β-catenin transcription, at least in colorectal cancer, contributing to tumor infiltration [10]. Again, in GBM this regulatory scheme has not been studied in detail so far. Along with the upregulation of TN-C, tumor vessels lose the AC-derived ECM component, agrin, a heparan sulfate proteoglycan, which is crucial for the localization of aquaporin-4 (AQP4) at astrocytic end-feet. This opposing regulation of TN-C and agrin at GBM vessels is accompanied by the downregulation of the TJ proteins, claudin-5 and occludin, suggesting that agrin might have pro- and TN-C has anti-barrier effects on brain ECs [128]. In addition to the high expression of TN-C in glioma blood vessels the receptor for TN-C, integrin αV, is found at elevated levels in glioma tissue next to the protein periostin [102]. Periostin was detected as a promoter of TN-C incorporation into the ECM and to organize its architecture [77]. Moreover, periostin, a ligand for αV/β3 and αV/β5 integrins, was recently shown to be secreted by GSCs and to be involved in the recruitment of M2-polarized macrophages [139, 155, 169].

Immunofluorescence studies indicate that collagen XVI is highly upregulated in gliomas compared to normal brain and that it is localized at tumor cells and vessels underlying collagen IV [137]. Regarding collagen IV, glioma vessels show an irregular, rough distribution around the endothelium, suggesting that the distribution and localization might be affected by the disrupted cellular organization of the NVU [116, 129].

The defect in NVU organization may also coincide with the lack of AC-derived laminin that was shown to be important for pericyte differentiation toward a BBB-supportive phenotype [162]. Instead, glioma cells secrete α2-, α4-, and α5-laminins, whereas the glioma vasculature expresses α3- and α5-laminins, which selectively promote glioma cell migration via the laminin receptor, α3β1 integrin [75].

In general, ECM proteins are recognized by cells via members of the integrin family, many of which have been described in cancer, such as αVβ3, αVβ5, αVβ6, α2β1, α5β1, α6β1, and α6β4 (reviewed in Bandyopadhyay and Raghavan, Caccavari et al., and Sroka et al. [8, 22, 140]). The most prominent integrins in GMB are αVβ3 and αVβ5, which are expressed on tumor ACs and on angiogenic ECs [12, 51, 52]. Although αVβ3 inhibition shows a strong anti-angiogenic and anti-tumor effect in a mouse orthotopic GBM model [93], a phase III clinical trial using a cyclic peptide selective for αV integrins (cilengitide) failed to demonstrate survival benefit (CENTRIC, recently reviewed by Tabatabai et al. [143]). The αVβ3 ligands vitronectin and tenascin co-localize with αVβ3 expression on both tumor cells and ECs, whereas fibronectin (a ligand for β1 integrins) shows diffuse staining throughout GBMs [153]. Besides its importance for GBM angiogenesis, αVβ3 expression on the host cells also supports the infiltration of macrophages that in turn effects most of the transplanted GBM inflammation [73]. Integrins may confer downstream signaling via talin, kindlins, focal adhesion kinase (FAK), Src, and paxillin, amongst others (for a comprehensive review see Harburger and Calderwood [61]).

In solid tumors, and in GBM in particular, the ECM becomes extensively remodeled, both by changes in the expression of individual ECM components and by ECM degradation, fostering tumor cell invasion, angiogenesis, and immune cell infiltration. ECM degradation is conferred by a group of enzymes known as matrix metalloproteinases (MMPs) that can be sub-grouped according to their substrate specificity (for a review see Könneck and Bechmann [84]). MMPs are released as pro-forms and have to be activated extracellularly via proteolytic cleavage by membrane-type MMPs (MT-MMPs). Further regulation of the active MMP protein function is carried out by so-called tissue inhibitors of metalloproteinases (TIMPs; for a review see van Hinsbergh and Koolwijk [148]). In GBM, MMP-2 and MMP-9 (a subgroup of gelatinases MMP) are particularly upregulated; MMP-9 appears to have prognostic value, as its plasma concentration correlates closely with malignancy and inversely correlates with prognosis [151]. Finally, the ECM is also processed by members of the *a disintegrin and metalloproteinase* (ADAM) family. For example, ADAM10 is upregulated in several cancers and has been shown to cleave VE-cadherin among other substrates, thereby contributing to vascular permeability (for a review, see van Hinsbergh and Koolwijk [148]).

In general, the ECM can bind and harvest many growth factors, such as the latent form of TGFβ, which is bound by the latent-TGFβ-binding protein (LTBP) and latency-associated peptide [64, 67], and becomes activated by serum proteinases such as plasmin, catalyzing the release of active TGFβ from the complex [18]. This often occurs on the surface of macrophages where the latent TGFβ complex is bound to CD36 via its ligand, thrombospondin-1 (TSP-1) [175]. Interestingly, under inflammatory conditions activated macrophages enhance the release of active TGFβ by promoting the activation of plasmin.

In addition, members of the Wnt family are bound to heparan sulfate proteoglycans within the ECM and on the cell surface, requiring a lipid modification and N-glycosylation [19, 100]. Interestingly, the ECM component biglycan was shown to bind at least Wnt3a and the co-receptor LRP6, and may serve as a reservoir for Wnt growth factors [13].

Finally, the ECM is a source of pro- and anti-angiogenic factors that are generated by proteolytic cleavage of plasminogen and collagen XVIII, producing the fragments angiostatin and endostatin respectively, which are potent inhibitors of angiogenesis (for a review see Egeblad and Werb and Karlluri [39, 71]).

3 Common Modulators of the Tumor Stroma

As indicated previously, tumor stroma composition is affected by several factors, which cannot all be discussed in depth in this chapter. Here, we focus on those regulating the vascular and inflammatory compartments.

Formation of new vessels is regulated by a plethora of factors, although the best described is the vascular endothelial growth factor (VEGF, specifically VEGF-A^{165}), signaling via the VEGF receptor 2 (Flk1/KDR). In GBM, hypoxic glioma cells located next to necrotic areas secrete VEGF, leading to the stabilization and activation of the hypoxia-inducible factor (HIF)-1α [17, 121]. Since Judah Folkman established the concept of the "vascular switch" in tumor development, particularly highly vascularized tumors, such as GBM, have been considered to be dependent on the vascular supply for their progression. The original idea was quite simple: blocking the vascularization of the tumor should lead to the arrest of tumor growth and ultimately, to the regression of the tumor mass. Despite multiple pre-clinical studies showing the benefit of anti-angiogenic therapy by targeting the VEGF pathway, clinical trials have failed to show efficacy regarding the overall survival (for a review see Niyazi et al. [105]). However, anti-VEGF therapy increases the progression-free survival and the quality of life of patients [9]. This may be related to the observed normalization of the vascular network, based on the diminished vascular activation, leading to normalized blood supply and reduced edema formation and inflammatory response within the tumor. Nevertheless, GBM inevitably develops an evasive resistance to anti-VEGF therapy that is related to an increase in necrotic areas due to the diminished oxygen supply, leading to the upregulation of the hypoxia-regulated gene stromal-derived factor 1a (SDF-1a) [92]. This in turn augments the recruitment of inflammatory cells, such as CD45$^+$ TAMs, that can release MMP-9, fostering further angiogenic activation by matrix-derived VEGF [34]. Consequently, novel therapeutic approaches try to combine anti-angiogenic with anti-inflammatory treatment.

It should be noted that although preclinical models show increased tumor cell invasion upon administration of anti-angiogenic therapy, whether this also happens in human patients is still controversial [76]. For example, treatment with the humanized anti-VEGF antibody, bevacizumab (Avastin; Roche), which has been approved by the FDA for recurrent GBM, may have adverse effects, such as increasing glioma cell invasion [33].

Regarding vessel stabilization, another key player is the angiopoietin/Tie system [144]. Specifically, angiopoietin 1 (Ang1) via the EC-specific receptor tyrosine kinase, Tie-2/Tek, stimulates downstream signaling via PI3K/Akt, leading to the upregulation of endothelial junction molecules and to the recruitment of stabilizing pericytes [144]. Instead, angiopoietin 2 (Ang2), which also binds to Tie2, inhibits downstream signaling and therefore confers vessel-destabilizing signals. Interestingly, long-term dominant expression of endothelial Ang2 leads to chronic inflammatory reaction in the skin and other organs [44, 136].

This is likely mediated by myeloid cells expressing the Tie2 receptor that were shown to also support a pro-angiogenic endothelial phenotype [35]. Interestingly, it has been shown, at least in vitro, that the angiopoietin/Tie2 system interacts with endothelial Wnt/β-catenin signaling, via the stabilization of β-catenin by Akt, leading to augmented Notch pathway activation [165]. Although this finding is in line with our own published results of Wnt and Dll4/Notch cooperation in the regulation of vessel stability [129], the relevance in vivo is still unclear.

In GBM, there is increasing evidence that the Wnt/β-catenin pathway may promote CSC maintenance, for example, by direct interaction of β-catenin and FoxM1 [167]. FoxM1 can selectively regulate transcription factors that are important for cancer stem cell (CSC) maintenance, such as Sox2.

Considering that TAMs are recruited by periostin to the vascular stem cell niche and that TAMs potentially express several Wnt factors such as Wnt7a/b, it is conceivable that this generates a CSC-permissive microenvironment. Along with the release of pro-angiogenic factors and the induction of MMPs (MMP-2/-9), CSCs not only maintain the tumor, but can also disseminate into distant brain regions. The CSC microenvironment is further supported by the release of cytokines, such as INF-γ, IL10, and by TGFβ, promoting the M2 polarization of macrophages and Th2 inflammatory status. TGFβ was shown to be expressed by glioma cells and by Tregs and may directly act not only on glioma cells to promote their growth by targeting CSCs, but also on effector T-cells, in which it inhibits their differentiation and subsequent activation. Consequently, in the glioma microenvironment, TGFβ suppresses a proinflammatory response from T-cells and microglia, and at the same time promotes tumor growth by acting on CSCs (for a review see Balkwill and Mantovani and Galvão and Zong [6, 50]).

Although it has been suggested that the sonic hedgehog (Shh) pathway might also be involved in the regulation and maintenance of CSCs through the activation of smoothened and the downstream activation of the glioma-associated homolog 1 (Gli1), there is little evidence of mutations in the Shh pathway in GBM [40, 145, 158]. The latter is also the case for the Wnt pathway and interactions between Wnt and Shh in GBM require further elucidation.

4 The Wnt Pathway in Tumor Angiogenesis and Inflammation

4.1 Wnt and Angiogenesis

As Wnts are responsible for many cellular processes, such as proliferation, polarization, and apoptosis, targeting this family in the context of cancer and GBM in particular might be promising.

The Wnt family has been described in the development of glioma and the expression of some members correlates with glioma grade. Indeed, Wnt5a expression increases with the grade, as immunostainings show that more than 50 % of the tumor mass of GBM is reactive for Wnt5a [72, 124, 164]. Moreover, Wnt5a promotes invasion of cancer cells in vitro via Ryk mediated regulation of MMP2 expression [55, 72]. Wnt2 is overexpressed in glioma tumors compared to normal tissue and its expression is linked to tumor development in a β-catenin-dependent pathway [124]. Wnt3a and Wnt1 are also overexpressed in glioma grade III and IV tissues and targeting these ligands in GSCs leads to a decrease in invasion and vascularization in a mouse orthotopic glioma model [74]. Interestingly, the protein Evi/Gpr177/Wntless, which is crucial for the secretion of all Wnt growth factors, is upregulated in glioma cells of grades II to IV. General abrogation of Wnt growth factor release from glioma cells by Evi deletion decreases the tumorigenic capacity of all glioma and GSCs in particular, likely by activating the noncanonical Wnt pathway [4]. For a list of Wnt pathway factors that have been described in glioma, the reader should refer to Table 1.

At the same time, expression of Dkk1 correlates with the grade of glioma and is overexpressed in GBM in particular [171]. Instead, Dkk3 reduces the tumor growth when overexpressed in a sub-cutaneous model of GBM [60]. In addition to the role of Wnt ligands and soluble Wnt inhibitors, Fzd receptor regulation may participate in the development of glioma. Overexpression of Fzd4 is linked to invasiveness in a clone of human U87 GBM cells through the upregulation of SNAI1 and EMT markers such as vimentin, αSMA, and vitronectin [69]. Moreover, T-cell factor-4, a

Table 1 Wnt pathway components shown to affect glioma progression in vivo

Protein	Source	Role in brain tumor growth	References
Wnt1	ECs, tumor cells	Decreases invasion and vascularization of the tumor mass, promotes vessel normalization Increases extravasation of T-cells in vitro Increases proliferation of GSCs	[74, 129]
Wnt2	Tumor cells	Increases tumor growth, reduces apoptosis	[124]
Wnt3a	ECs, tumor tissue	Increases proliferation of GSCs; promotes invasion and vascularization Increases in vitro extravasation of T-cells Increases macrophage secretion of IL6/IL12 and TNFα	[57, 74]
Wnt5a	TAM, tumor cells, astrocytes	Increases proliferation and invasion of tumor cells in vitro Correlates with microglia/macrophage recruitment; increases microglia proliferation, invasion, and cytokine production in vitro Stimulates production of cytokines by ECs in vitro Correlates with invasion and brain metastasis in breast cancer	[36, 55, 56, 72, 78, 79]
Wnt5b	Breast cancer cells	Correlates with invasion and brain metastasis in breast cancer	[79]
Dkk1	Tumor cells	Expression increases at higher grades, increases vascularization and invasion of the tumor, reduces vascular normalization	[129, 170]
Dkk3	Tumor cells	Decreases Wnt3a and -5a expression in vitro, decreases tumor growth and apoptosis in vivo	[60]
sFRP1-5	Tumor cells	Increased methylation of sFRPs in human GBM cell lines	[135]
Fzd4	Tumor cells	Increases the in vitro invasion properties of U87 cells	[69]
Evi/Gpr177/Wntless	Tumor cells	Promotes tumorigenesis in glioma cells and glioma stem cells in vivo	[4]

ECs endothelial cells, *TAM* tumor-associated macrophage, *GSCs* glioma stem cells, *IL* interleukin, *TNF* tumor necrosis factor, *sFRP* secreted Fzd-related protein, *GBM* glioblastoma multiforme

partner of β-catenin transcription factor complex, is overexpressed in glioma grades III and IV and linked to glioma proliferation and invasion [166]. Interestingly, FoxM1 is necessary for β-catenin nuclear translocation upon Wnt3a stimulation of glioma cells [167]. Its expression is increased in glioma grades III and IV, and negatively correlates with patient survival, contributing to self-renewal of GSCs, tumorigenesis, and angiogenesis [53, 167, 168].

Activation of the Wnt pathway is important for physiological brain angiogenesis [32, 108]. Therefore, it is likely that this pathway may also affect brain tumor angiogenesis. However, the effects of the Wnt pathway on tumor angiogenesis have mostly been studied in models of breast and colon cancer; few studies have investigated/appreciated its role in brain tumors and in GBM.

As described in the introduction, activation of Fzd receptors by Wnt leads to stabilization of β-catenin, which can translocate to the nucleus and then act as a transcription factor. Several publications have shown an increase in the localization of β-catenin in the nuclei of ECs in glioma vessels [129, 160, 161], indicative of activation of the Wnt β-catenin pathway in glioma vessels.

Recently, we demonstrated that overexpression of the Wnt ligand, Wnt1, or the soluble Wnt inhibitor Dkk1 by glioma cells leads to effects on tumor formation and on angiogenesis in vivo. Specifically, Wnt1-overexpressing GL261 mouse glioma cells were implanted intracranially, resulting in the formation of smaller tumors with fewer vessels compared to controls [129]. Furthermore, we could show that sustained endothelial Wnt/β-catenin signaling is also able to restore barrier properties of glioma vessels and results in a more continuous organization of the TJ proteins, claudin-3, -5 and ZO-1, as well as increased recruitment of pericytes. Overexpression of Dkk1 leads to opposite effects compared to Wnt1 in the same glioma model, hence promoting tumor development and abnormal vascularization [129]. Interestingly, expression of Dkk1 has correlated positively with the glioma grade [171]. Considering these observations, Dkk1 seems to be promising in targeting the vasculature of GBM. However, it cannot be concluded that in general, Wnts and soluble inhibitors (Dkks, sFRPs) have anti- and pro-tumoral effects respectively. Park et al. described an opposite role of Dkk1 and Dkk2 in melanoma tumor development [112]. Indeed, Dkk1 decreases tumor size and vessel number in addition to pericyte coverage, whereas Dkk2 has the opposite effect. In an ischemia model, Dkk2 has been described to promote angiogenesis and pericyte coverage [101]. Whether these roles are similar in brain tumors remains to be elucidated. As a matter of fact, it is highly important to determine the expression profile of each target in the different cancers before considering any therapeutic intervention.

Like Dkks, other members of the Wnt inhibitors have been described in tumor development. In 2009, Courtwright and colleagues demonstrated that sFRP2, via the noncanonical NFAT pathway, promotes vessel formation in different angiogenesis models and in subcutaneous angiosarcoma [29]. In a breast cancer model, an antibody against sFRP2 reduces tumor growth and angiogenesis both in vitro and in vivo [46]. sFRP1 and 2 are expressed in malignant glioma cell lines and foster the survival of glioma cells under starving conditions in vitro [132]. How this may affect glioma development is not yet fully understood. Interestingly, sFRP1–5 were demonstrated to be frequently methylated and consequently silenced, suggesting that some negative Wnt pathway modulators might be diminished in GBM [135].

Given the importance of the Notch pathway in cell differentiation and function, it has also been extensively studied in the context of physiological and pathological angiogenesis [54]. Specifically, Dll4/Notch1 signaling has been shown to decrease the endothelial expression of VEGFR2/Flk-1, leading to reduced, VEGF-dependent vascular sprouting, arterial specification, and vascular quiescence [16, 82]. In particular, in tumor angiogenesis, inhibition of Dll4/Notch signaling by small inhibitors or inhibitory antibodies was demonstrated, leading to increased angiogenic sprouting with the formation of non-functional vessels that do not support tumor growth. The Notch pathway was also revealed to directly interact with the canonical

Wnt/β-catenin pathway in ECs, regulating the angiogenic response and vascular differentiation [28, 119, 129].

The Wnt pathways not only affect primary brain tumors, but also the formation of brain metastases derived from peripheral cancers. Because of the BBB properties of ECs, cancer cells need a specific microenvironment to be able to enter the brain parenchyma and to develop a secondary tumor. Interestingly, Klemm et al. described that Wnt5a/5b are markers of invasive breast cancer cells in brain metastasis. This effect seems to occur via a noncanonical rather than a β-catenin-dependent pathway [79].

This metastatic aspect points out the migration of cells through the EC layer. This can draw a parallel with the recruitment of immune cells in GBM. Therefore, it is conceivable that Wnt pathways might also directly act on the inflammatory tumor response and are not restricted to the vascular compartment.

4.2 Wnt and Inflammation

Inflammation is a player in cancer development. Based on the observation that Wnt signaling is involved in brain recovery after stroke, traumatic brain injury or Parkinson's disease (for a review, see Marchetti and Pluchino [98]), it would not be surprising if these pathways also contribute to inflammation in glioma.

It has at least been shown that TAMs express several Wnt growth factors, driving either the canonical (Wnt7a/b) or the noncanonical (Wnt5a, Wnt2) pathway, directly or indirectly influencing tumor development or angiogenesis [103].

The secreted Wnts may act in an autocrine manner on TAMs as they express the Fzd receptors Fzd4, 5, 7, and 8 [56, 117]. Indeed, Wnt5a-treated microglia exert increased invasive capacity and upregulate proinflammatory cytokines (IL1β, IL6, IL12, TNFα, CCL7, CCL12, COX2), and MMP9/13 expression in brain slices ex vivo [56]. Wnt3a stimulation of macrophages is also able to induce secretion of IL6, IL12, and TNFα in vitro [57]. Therefore, it may be possible for this regulatory scheme to occur in brain tumors too, especially in GBM, in which Wnt5a expression, microglia recruitment and invasion were shown to correlate [36]. Moreover, Wnt5a may also have a paracrine effect on ECs, as it has been demonstrated that Wnt5a induces expression of several proinflammatory cytokines via the noncanonical Ca2+ pathway (IL1β, IL3, IL5, IL6, CCL2, CCL8, and Cox2) and increases monolayer permeability and Matrigel invasion of aortic ECs [78].

Despite their role in inflammation, macrophages may also play a direct role in angiogenesis, which has been demonstrated in hindbrain and retina development [103]. Wnt7b derived from macrophages has been described as a key mediator in vascular remodeling of hyaloid vessels in the eye of newborn mice [89], an effect that is in concert with the angiopoietin/Tie2 [127], VEGF [43] and Notch [111] systems. This effect has also been shown in the context of a tumor, as in breast cancer, "invasive" macrophages are enriched in Wnt5a and Wnt7b, suggesting that they might play a role during the invasion process [110]. Another study showed that Wnt7b is overexpressed in breast carcinoma and this protein is also detected in the TAMs. Interestingly, specific deletion of Wnt7b in the TAMs does not affect the recruitment of immune cells (macrophages and $CD3^+$ and $B220^+$ B-cells), but leads to less vascularized tumors only in the malignant stages. This indicates that TAM-derived Wnt7b is necessary for the angiogenic switch. Furthermore, in this model, Wnt7b is able to induce VEGF expression in ECs via β-catenin transcriptional activation [163]. Whether this also happens in brain tumors remains unclear, but it may be worth investigating, as on the one hand, macrophage-derived Wnt7b is important for vascular remodeling in the retina [89, 127], and on the other hand, Wnt7a/b are crucial during BBB development [31, 141]. Other Wnt factors, such as Wnt1, 2, 3a, 5a, 7b, and 10b derived from various cellular sources, have been implicated in pathological conditions, such as traumatic brain injury, spinal cord injury, stroke or PD (for a review see Marchetti and Pluchino [98]).

Moreover, De Palma and colleagues have described a specific subtype of TAM that promotes angiogenesis. Interestingly, those are characterized by expression of Tie2, and therefore named Tie2-expressing macrophage (TEM) [35]. Interestingly, Wnt5a has been shown to regulate Tie2, at least in ECs [99]. Therefore, it may also play a role in the recruitment of TEM and thus promoting angiogenesis in GBM.

Regarding glioma inflammation, the Notch pathway did not draw particular attention, although the Notch signaling pathway interacts with modulators of inflammation such as NF-kB and TGFβ [125]. In macrophages, the Notch pathway was identified to regulate the balance between M1 and M2 macrophage polarization. Specifically, Notch signaling components, namely Dll4 ligand and the Notch1–ADAM10–γ-secretase–RBP-J axis, regulate expression of M1 genes, thereby contributing to a pro-inflammatory phenotype and innate immunity [48, 157]. In ECs, however, Dll/Notch signaling seems to repress an inflammatory reaction by counteracting NF-kB signaling on the transcriptional level [126]. The way in which the Notch and the Wnt pathways interact in the context of inflammation has been only poorly investigated.

The reader should also keep in mind that ECs not only receive and respond to Wnt signals, but are also able to secrete Wnts and may therefore play an active role in immune cell recruitment. Indeed, transmigration of T-cells through the EC monolayer is decreased in the presence of soluble Fzd5. This effect depends on the secretion of Wnt1 or Wnt3a likely from the endothelium and relies on the overexpression of Fzd3-7 in activated T-cells [154]. In this context, activation of the Wnt/β-catenin pathway induces expression of MMP2 and MMP9 in activated T-cells, fostering their migratory phenotype. Whether this also occurs in GBM remains unknown; however, Lewis lung carcinoma development is impaired in mice lacking the secretion of Wnt factors from ECs [83].

5 Summary and Conclusion

Glioblastoma multiformes are highly vascularized, deadly primary brain tumors. Like other solid malignant tumors, GBMs have developed mechanisms by which they circumvent elimination by the immune system. Via the secretion of immune-modulating factors, GBM cells can lead to the conversion of microglia and macrophages from a proinflammatory, classic M1 state toward an anti-inflammatory, alternative M2 activation state. Recruitment of Tregs and other immunosuppressive cells into the tumor stroma further support the anti-inflammatory environment in GBM, although the lymphatic immune system cells play a minor role compared with the microglia-derived immune response. The highly angiogenic active vascular compartment is a central player in this scenario, as ECs, in addition to SMCs/PCs, receive signals from tumor cells such as VEGF, TGFβ, and Shh, but also actively contribute to glioma progression by providing a permissive niche for CSCs, Tregs, and M2 polarized macrophages. Wnt growth factors and Wnt pathway modulators have been implicated in both glioma angiogenesis and inflammation, interacting with inflammatory pathways such as TGFβ, NF-kB, Notch, and others. In this context, it appears that the effects exerted by Wnt, but also by other pathways such as Notch and TGFβ, are highly cell context-dependent, making systemic treatment of glioma a highly challenging task.

Therefore, rigorous studies are urgently needed to uncover the communications between glioma cells and the stromal compartment, including blood vessels and immune cells. This is particularly important during early phases of GBM formation, when a "decision" toward a more malignant state, including the angiogenic switch, is made. Therefore, it is highly important to study the contribution of the immune system to therapy resistance.

To mechanistically understand some of these basic questions, it may be helpful to investigate other CNS tumors that are genetically better defined, such as medulloblastoma. For the latter, four different subtypes have been described, one of which is characterized by Wnt and another by Shh pathway activation [70, 106]. Interestingly, Shh and Wnt activity are mutually exclusive and could be exploited as novel targets in the treatment of the disease [41, 123, 174].

To understand the glioma vasculature as a signaling hub in disease progression, sophisticated genetic mouse models of glioma can be particularly helpful because of the ability to target individual cell types and to subsequently follow the impact on tumor progression. The challenges are great, but collaborations between vascular biologists, immunologists, neuroscientists, and oncologists should foster progress toward a cure for GBM.

Acknowledgements We thank Patrick N Harter for cross-reading the manuscript. This work was supported by the Collaborative Research Center "Vascular differentiation and remodeling" (CRC/Transregio23, Project B7) and the Cluster of Excellence 147 "Cardiopulmonary system" (ECCPS) from the German Research Council (DFG) (to SL), by the 7. FP, COFUND, Goethe International Postdoc Programme GO-IN, No. 291776 (to SG), LOEWE Initiative Hessen (Onkogene Signaltransduktion Frankfurt, OSF); III L 4-518/55.004, 2009 (to SL), and by EU Health FP7 JUSTBRAIN (to SL).

The authors declare that they have no competing financial interests.

References

1. Abbott NJ, Patabendige AAK, Dolman DEM et al (2010) Structure and function of the blood–brain barrier. Neurobiol Dis 37:13–25. doi:10.1016/j.nbd.2009.07.030
2. El Andaloussi A, Lesniak MS (2007) CD4+ CD25+ FoxP3+ T-cell infiltration and heme oxygenase-1 expression correlate with tumor grade in human gliomas. J Neurooncol 83:145–152. doi:10.1007/s11060-006-9314-y

3. Anderson KD, Pan L, Yang XM et al (2011) Angiogenic sprouting into neural tissue requires Gpr124, an orphan G protein-coupled receptor. Proc Natl Acad Sci 108:2807–2812. doi:10.1073/pnas.1019761108
4. Augustin I, Goidts V, Bongers A et al (2012) The Wnt secretion protein Evi/Gpr177 promotes glioma tumourigenesis. EMBO Mol Med 4:38–51. doi:10.1002/emmm.201100186
5. Balkwill F, Mantovani A (2001) Inflammation and cancer: back to Virchow? Lancet 357:539–545. doi:10.1016/S0140-6736(00)04046-0
6. Balkwill FR, Mantovani A (2012) Cancer-related inflammation: common themes and therapeutic opportunities. Semin Cancer Biol 22:33–40. doi:10.1016/j.semcancer.2011.12.005
7. Baluk P, Morikawa S, Haskell A et al (2003) Abnormalities of basement membrane on blood vessels and endothelial sprouts in tumors. Am J Pathol 163:1801–1815
8. Bandyopadhyay A, Raghavan S (2009) Defining the role of integrin alphavbeta6 in cancer. Curr Drug Targets 10:645–652
9. Batchelor TT, Reardon DA, de Groot JF et al (2014) Antiangiogenic therapy for glioblastoma: current status and future prospects. Clin Cancer Res 20:5612–5619. doi:10.1158/1078-0432.CCR-14-0834
10. Beiter K, Hiendlmeyer E, Brabletz T et al (2005) beta-Catenin regulates the expression of tenascin-C in human colorectal tumors. Oncogene 24:8200–8204. doi:10.1038/sj.onc.1208960
11. Bellail AC, Hunter SB, Brat DJ et al (2004) Microregional extracellular matrix heterogeneity in brain modulates glioma cell invasion. Int J Biochem Cell Biol 36:1046–1069. doi:10.1016/j.biocel.2004.01.013
12. Bello L, Francolini M, Marthyn P et al (2001) Alpha(v)beta3 and alpha(v)beta5 integrin expression in glioma periphery. Neurosurgery 49:380–389; discussion 390
13. Berendsen AD, Fisher LW, Kilts TM et al (2011) Modulation of canonical Wnt signaling by the extracellular matrix component biglycan. Proc Natl Acad Sci 108:17022–17027. doi:10.1073/pnas.1110629108
14. Bhowmick NA, Neilson EG, Moses HL (2004) Stromal fibroblasts in cancer initiation and progression. Nature 432:332–337. doi:10.1038/nature03096
15. Biswas SK, Allavena P, Mantovani A (2013) Tumor-associated macrophages: functional diversity, clinical significance, and open questions. Semin Immunopathol 35:585–600. doi:10.1007/s00281-013-0367-7
16. Blanco R, Gerhardt H (2013) VEGF and Notch in tip and stalk cell selection. Cold Spring Harb Perspect Med 3:a006569. doi:10.1101/cshperspect.a006569
17. Blouw B, Song H, Tihan T et al (2003) The hypoxic response of tumors is dependent on their microenvironment. Cancer Cell 4:133–146
18. Bonnans C, Chou J, Werb Z (2014) Remodelling the extracellular matrix in development and disease. Nat Rev Mol Cell Biol 15:786–801. doi:10.1038/nrm3904
19. Bradley RS, Brown AM (1990) The proto-oncogene int-1 encodes a secreted protein associated with the extracellular matrix. EMBO J 9:1569–1575
20. Brösicke N, Faissner A (2015) Role of tenascins in the ECM of gliomas. Cell Adh Migr 9:131–140. doi:10.1080/19336918.2014.1000071
21. Bucchieri F, Farina F, Zummo G, Cappello F (2015) Lymphatic vessels of the dura mater: a new discovery? J Anat 227:702–703. doi:10.1111/joa.12381
22. Caccavari F, Valdembri D, Sandri C et al (2010) Integrin signaling and lung cancer. Cell Adh Migr 4:124–129
23. Cahoy JD, Emery B, Kaushal A et al (2008) A transcriptome database for astrocytes, neurons, and oligodendrocytes: a new resource for understanding brain development and function. J Neurosci 28:264–278. doi:10.1523/JNEUROSCI.4178-07.2008
24. Caspani EM, Crossley PH, Redondo-Garcia C, Martinez S (2014) Glioblastoma: a pathogenic crosstalk between tumor cells and pericytes. PLoS One 9:e101402. doi:10.1371/journal.pone.0101402
25. Chien AJ, Conrad WH, Moon RT (2009) A Wnt survival guide: from flies to human disease. J Invest Dermatol 129:1614–1627. doi:10.1038/jid.2008.445

26. Chintala SK, Sawaya R, Gokaslan ZL et al (1996) Immunohistochemical localization of extracellular matrix proteins in human glioma, both in vivo and in vitro. Cancer Lett 101:107–114
27. Claes A, Idema AJ, Wesseling P (2007) Diffuse glioma growth: a guerilla war. Acta Neuropathol 114:443–458. doi:10.1007/s00401-007-0293-7
28. Corada M, Nyqvist D, Orsenigo F et al (2010) The Wnt/beta-catenin pathway modulates vascular remodeling and specification by upregulating Dll4/Notch signaling. Dev Cell 18:938–949. doi:10.1016/j.devcel.2010.05.006
29. Courtwright A, Siamakpour-Reihani S, Arbiser JL et al (2009) Secreted frizzle-related protein 2 stimulates angiogenesis via a calcineurin/NFAT signaling pathway. Cancer Res 69:4621–4628. doi:10.1158/0008-5472.CAN-08-3402
30. Cullen M, Elzarrad MK, Seaman S et al (2011) GPR124, an orphan G protein-coupled receptor, is required for CNS-specific vascularization and establishment of the blood–brain barrier. Proc Natl Acad Sci 108:5759–5764. doi:10.1073/pnas.1017192108
31. Daneman R, Agalliu D, Zhou L et al (2009) Wnt/beta-catenin signaling is required for CNS, but not non-CNS, angiogenesis. Proc Natl Acad Sci 106:641–646. doi:10.1073/pnas.0805165106
32. Daneman R, Prat A (2015) The blood–brain barrier. Cold Spring Harb Perspect Biol 7:a020412–a020412. doi:10.1101/cshperspect.a020412
33. de Groot JF, Fuller G, Kumar AJ et al (2010) Tumor invasion after treatment of glioblastoma with bevacizumab: radiographic and pathologic correlation in humans and mice. Neuro Oncol 12:233–242. doi:10.1093/neuonc/nop027
34. De Palma M, Lewis CE (2013) Macrophage regulation of tumor responses to anticancer therapies. Cancer Cell 23:277–286. doi:10.1016/j.ccr.2013.02.013
35. de Palma M, Venneri M, Galli R et al (2005) Tie2 identifies a hematopoietic lineage of proangiogenic monocytes required for tumor vessel formation and a mesenchymal population of pericyte progenitors. Cancer Cell 8:211–226
36. Dijksterhuis JP, Arthofer E, Marinescu VD et al (2015) High levels of WNT-5A in human glioma correlate with increased presence of tumor-associated microglia/monocytes. Exp Cell Res 339:280–288. doi:10.1016/j.yexcr.2015.10.022
37. Domingues P, González-Tablas M, Otero Á et al (2015) Tumor infiltrating immune cells in gliomas and meningiomas. Brain Behav Immun 53:1–15. doi:10.1016/j.bbi.2015.07.019
38. Duchartre Y, Kim Y-M, Kahn M (2015) The Wnt signaling pathway in cancer. Crit Rev Oncol Hematol. doi:10.1016/j.critrevonc.2015.12.005
39. Egeblad M, Werb Z (2002) New functions for the matrix metalloproteinases in cancer progression. Nat Rev Cancer 2:161–174. doi:10.1038/nrc745
40. Ehtesham M, Sarangi A, Valadez JG et al (2007) Ligand-dependent activation of the hedgehog pathway in glioma progenitor cells. Oncogene 26:5752–5761. doi:10.1038/sj.onc.1210359
41. Ellison DW, Onilude OE, Lindsey JC et al (2005) beta-Catenin status predicts a favorable outcome in childhood medulloblastoma: the United Kingdom Children's Cancer Study Group Brain Tumour Committee. J Clin Oncol 23:7951–7957. doi:10.1200/JCO.2005.01.5479
42. Engelhardt B, Liebner S (2014) Novel insights into the development and maintenance of the blood–brain barrier. Cell Tissue Res 355:687–699. doi:10.1007/s00441-014-1811-2
43. Fantin A, Vieira JM, Gestri G et al (2010) Tissue macrophages act as cellular chaperones for vascular anastomosis downstream of VEGF-mediated endothelial tip cell induction. Blood 116:829–840. doi:10.1182/blood-2009-12-257832
44. Fiedler U, Reiss Y, Scharpfenecker M et al (2006) Angiopoietin-2 sensitizes endothelial cells to TNF-alpha and has a crucial role in the induction of inflammation. Nat Med 12:235–239. doi:10.1038/nm1351
45. Folkman J (1971) Tumor angiogenesis: therapeutic implications. N Engl J Med 285:1182–1186. doi:10.1056/NEJM197111182852108

46. Fontenot E, Rossi E, Mumper R et al (2013) A novel monoclonal antibody to secreted frizzled-related protein 2 inhibits tumor growth. Mol Cancer Ther 12:685–695. doi:10.1158/1535-7163.MCT-12-1066
47. Franco CA, Liebner S, Gerhardt H (2009) Vascular morphogenesis: a Wnt for every vessel? Curr Opin Genet Dev 19:476–483. doi:10.1016/j.gde.2009.09.004
48. Fukuda D, Aikawa E, Swirski FK et al (2012) Notch ligand delta-like 4 blockade attenuates atherosclerosis and metabolic disorders. Proc Natl Acad Sci 109:E1868–E1877. doi:10.1073/pnas.1116889109
49. Gajjar A, Pfister SM, Taylor MD, Gilbertson RJ (2014) Molecular insights into pediatric brain tumors have the potential to transform therapy. Clin Cancer Res 20:5630–5640. doi:10.1158/1078-0432.CCR-14-0833
50. Galvão RP, Zong H (2013) Inflammation and gliomagenesis: bi-directional communication at early and late stages of tumor progression. Curr Pathobiol Rep 1:19–28. doi:10.1007/s40139-012-0006-3
51. Gladson CL, Cheresh DA (1991) Glioblastoma expression of vitronectin and the alpha v beta 3 integrin adhesion mechanism for transformed glial cells. J Clin Invest 88:1924–1932. doi:10.1172/JCI115516
52. Gladson CL, Wilcox JN, Sanders L et al (1995) Cerebral microenvironment influences expression of the vitronectin gene in astrocytic tumors. J Cell Sci 108(Pt 3):947–956
53. Gong A, Huang S (2012) FoxM1 and Wnt/β-catenin signaling in glioma stem cells. Cancer Res 72:5658–5662. doi:10.1158/0008-5472.CAN-12-0953
54. Guruharsha KG, Kankel MW, Artavanis-Tsakonas S (2012) The Notch signalling system: recent insights into the complexity of a conserved pathway. Nat Rev Genet 13:654–666. doi:10.1038/nrg3272
55. Habu M, Koyama H, Kishida M et al (2014) Ryk is essential for Wnt-5a-dependent invasiveness in human glioma. J Biochem 156:29–38. doi:10.1093/jb/mvu015
56. Halleskog C, Dijksterhuis JP, Kilander MBC et al (2012) Heterotrimeric G protein-dependent WNT-5A signaling to ERK1/2 mediates distinct aspects of microglia proinflammatory transformation. J Neuroinflammation. doi:10.1186/1742-2094-9-111
57. Halleskog C, Mulder J, Dahlström J et al (2011) WNT signaling in activated microglia is proinflammatory. Glia 59:119–131. doi:10.1002/glia.21081
58. Hanahan D, Coussens LM (2012) Accessories to the crime: functions of cells recruited to the tumor microenvironment. Cancer Cell 21:309–322. doi:10.1016/j.ccr.2012.02.022
59. Hanahan D, Weinberg RA (2011) Hallmarks of cancer: the next generation. Cell 144:646–674. doi:10.1016/j.cell.2011.02.013
60. Hara K, Kageji T, Mizobuchi Y et al (2015) Blocking of the interaction between Wnt proteins and their co-receptors contributes to the anti-tumor effects of adenovirus-mediated DKK3 in glioblastoma. Cancer Lett 356:496–505. doi:10.1016/j.canlet.2014.09.045
61. Harburger DS, Calderwood DA (2009) Integrin signalling at a glance. J Cell Sci 122:159–163. doi:10.1242/jcs.018093
62. Hardee ME, Zagzag D (2012) Mechanisms of glioma-associated neovascularization. Am J Pathol 181:1126–1141. doi:10.1016/j.ajpath.2012.06.030
63. Hendaoui I, Tucker RP, Zingg D et al (2014) Tenascin-C is required for normal Wnt/β-catenin signaling in the whisker follicle stem cell niche. Matrix Biol 40:46–53. doi:10.1016/j.matbio.2014.08.017
64. Horiguchi M, Ota M, Rifkin DB (2012) Matrix control of transforming growth factor-β function. J Biochem 152:321–329. doi:10.1093/jb/mvs089
65. Huijbers IJ, Iravani M, Popov S et al (2010) A role for fibrillar collagen deposition and the collagen internalization receptor endo180 in glioma invasion. PLoS One 5:e9808. doi:10.1371/journal.pone.0009808
66. Hussain SF, Yang D, Suki D et al (2006) Innate immune functions of microglia isolated from human glioma patients. J Transl Med 4:15. doi:10.1186/1479-5876-4-15

67. Hyytiäinen M, Penttinen C, Keski-Oja J (2004) Latent TGF-beta binding proteins: extracellular matrix association and roles in TGF-beta activation. Crit Rev Clin Lab Sci 41:233–264. doi:10.1080/10408360490460933
68. Jacobs JFM, Idema AJ, Bol KF et al (2009) Regulatory T cells and the PD-L1/PD-1 pathway mediate immune suppression in malignant human brain tumors. Neuro Oncol 11:394–402. doi:10.1215/15228517-2008-104
69. Jin X, Jeon HY, Joo KM et al (2011) Frizzled 4 regulates stemness and invasiveness of migrating glioma cells established by serial intracranial transplantation. Cancer Res 71:3066–3075. doi:10.1158/0008-5472.CAN-10-1495
70. Jones DTW, Jäger N, Kool M et al (2012) Dissecting the genomic complexity underlying medulloblastoma. Nature 488:100–105. doi:10.1038/nature11284
71. Kalluri R (2003) Angiogenesis: basement membranes: structure, assembly and role in tumour angiogenesis. Nat Rev Cancer 3:422–433. doi:10.1038/nrc1094
72. Kamino M, Kishida M, Kibe T et al (2011) Wnt-5a signaling is correlated with infiltrative activity in human glioma by inducing cellular migration and MMP-2. Cancer Sci 102:540–548. doi:10.1111/j.1349-7006.2010.01815.x
73. Kanamori M, Kawaguchi T, Berger MS, Pieper RO (2006) Intracranial microenvironment reveals independent opposing functions of host alphaVbeta3 expression on glioma growth and angiogenesis. J Biol Chem 281:37256–37264. doi:10.1074/jbc.M605344200
74. Kaur N, Chettiar S, Rathod S et al (2013) Molecular and cellular neuroscience. Mol Cell Neurosci 54:44–57. doi:10.1016/j.mcn.2013.01.001
75. Kawataki T, Yamane T, Naganuma H et al (2007) Laminin isoforms and their integrin receptors in glioma cell migration and invasiveness: Evidence for a role of alpha5-laminin(s) and alpha3beta1 integrin. Exp Cell Res 313:3819–3831. doi:10.1016/j.yexcr.2007.07.038
76. Keunen O, Johansson M, Oudin A et al (2011) Anti-VEGF treatment reduces blood supply and increases tumor cell invasion in glioblastoma. Proc Natl Acad Sci 108:3749–3754. doi:10.1073/pnas.1014480108/-/DCSupplemental/pnas.201014480SI.pdf
77. Kii I, Nishiyama T, Li M et al (2010) Incorporation of tenascin-C into the extracellular matrix by periostin underlies an extracellular meshwork architecture. J Biol Chem 285:2028–2039. doi:10.1074/jbc.M109.051961
78. Kim J, Kim J, Kim DW et al (2010) Wnt5a induces endothelial inflammation via beta-catenin-independent signaling. J Immunol 185:1274–1282. doi:10.4049/jimmunol.1000181
79. Klemm F, Bleckmann A, Siam L et al (2011) β-catenin-independent WNT signaling in basal-like breast cancer and brain metastasis. Carcinogenesis 32:434–442. doi:10.1093/carcin/bgq269
80. Kmiecik J, Poli A, Brons NHC et al (2013) Elevated CD3+ and CD8+ tumor-infiltrating immune cells correlate with prolonged survival in glioblastoma patients despite integrated immunosuppressive mechanisms in the tumor microenvironment and at the systemic level. J Neuroimmunol 264:71–83. doi:10.1016/j.jneuroim.2013.08.013
81. Kmiecik J, Zimmer J, Chekenya M (2014) Natural killer cells in intracranial neoplasms: presence and therapeutic efficacy against brain tumours. J Neurooncol 116:1–9. doi:10.1007/s11060-013-1265-5
82. Kofler NM, Shawber CJ, Kangsamaksin T et al (2012) Notch signaling in developmental and tumor angiogenesis. Genes Cancer 2:1106–1116. doi:10.1177/1947601911423030
83. Korn C, Scholz B, Hu J et al (2014) Endothelial cell-derived non-canonical Wnt ligands control vascular pruning in angiogenesis. Development 141:1757–1766. doi:10.1242/dev.104422
84. Könnecke H, Bechmann I (2013) The role of microglia and matrix metalloproteinases involvement in neuroinflammation and gliomas. Clin Dev Immunol 2013:1–15. doi:10.1155/2013/914104
85. Krausova M, Korinek V (2014) Wnt signaling in adult intestinal stem cells and cancer. Cell Signal 26:570–579. doi:10.1016/j.cellsig.2013.11.032
86. L'Episcopo F, Tirolo C, Testa N et al (2011) Reactive astrocytes and Wnt/β-catenin signaling link nigrostriatal injury to repair in 1-methyl-4-phenyl-1,2,3,6-tetrahydropyridine model of Parkinson's disease. Neurobiol Dis 41:508–527. doi:10.1016/j.nbd.2010.10.023

87. Liebner S, Corada M, Bangsow T et al (2008) Wnt/beta-catenin signaling controls development of the blood–brain barrier. J Cell Biol 183:409–417. doi:10.1083/jcb.200806024
88. Liebner S, Fischmann A, Rascher G et al (2000) Claudin-1 and claudin-5 expression and tight junction morphology are altered in blood vessels of human glioblastoma multiforme. Acta Neuropathol 100:323–331
89. Lobov IB, Rao S, Carroll TJ et al (2005) WNT7b mediates macrophage-induced programmed cell death in patterning of the vasculature. Nature 437:417–421. doi:10.1038/nature03928
90. Lockman PR, Mittapalli RK, Taskar KS et al (2010) Heterogeneous blood-tumor barrier permeability determines drug efficacy in experimental brain metastases of breast cancer. Clin Cancer Res 16:5664–5678. doi:10.1158/1078-0432.CCR-10-1564
91. Lohr J, Ratliff T, Huppertz A et al (2011) Effector T-cell infiltration positively impacts survival of glioblastoma patients and is impaired by tumor-derived TGF-β. Clin Cancer Res 17:4296–4308. doi:10.1158/1078-0432.CCR-10-2557
92. Lu KV, Bergers G (2013) Mechanisms of evasive resistance to anti-VEGF therapy in glioblastoma. CNS Oncol 2:49–65. doi:10.2217/cns.12.36
93. MacDonald TJ, Ladisch S (2001) Antisense to integrin alpha v inhibits growth and induces apoptosis in medulloblastoma cells. Anticancer Res 21:3785–3791
94. Machein M, de Miguel LS (2009) Angiogenesis in gliomas. Recent Results Cancer Res 171:193–215. doi:10.1007/978-3-540-31206-2_12
95. Machein MR, Plate KH (2000) VEGF in brain tumors. J Neurooncol 50:109–120
96. Mantovani A, Allavena P, Sica A, Balkwill F (2008) Cancer-related inflammation. Nature 454:436–444. doi:10.1038/nature07205
97. Mantovani A, Biswas SK, Galdiero MR et al (2013) Macrophage plasticity and polarization in tissue repair and remodelling. J Pathol 229:176–185. doi:10.1002/path.4133
98. Marchetti B, Pluchino S (2013) Wnt your brain be inflamed? Yes, it Wnt! Trends Mol Med 19:144–156. doi:10.1016/j.molmed.2012.12.001
99. Masckauchán TNH, Agalliu D, Vorontchikhina M et al (2006) Wnt5a signaling induces proliferation and survival of endothelial cells in vitro and expression of MMP-1 and Tie-2. Mol Biol Cell 17:5163–5172. doi:10.1091/mbc.E06-04-0320
100. Mikels AJ, Nusse R (2006) Wnts as ligands: processing, secretion and reception. Oncogene 25:7461–7468. doi:10.1038/sj.onc.1210053
101. Min J-K, Park H, Choi H-J et al (2011) The WNT antagonist Dickkopf2 promotes angiogenesis in rodent and human endothelial cells. J Clin Invest 121:1882–1893. doi:10.1172/JCI42556
102. Mustafa DAM, Dekker LJ, Stingl C et al (2012) A proteome comparison between physiological angiogenesis and angiogenesis in glioblastoma. Mol Cell Proteomics 11:M111.008466–M111.008466. doi:10.1074/mcp.M111.008466
103. Newman AC, Hughes CCW (2012) Macrophages and angiogenesis: a role for Wnt signaling. Vasc Cell 4:13. doi:10.1186/2045-824X-4-13
104. Niehrs C (2006) Function and biological roles of the Dickkopf family of Wnt modulators. Oncogene 25:7469–7481. doi:10.1038/sj.onc.1210054
105. Niyazi M, Harter PN, Hattingen E et al (2015) Bevacizumab and radiotherapy for the treatment of glioblastoma: brothers in arms or unholy alliance? Oncotarget 7:2313–2328. doi:10.18632/oncotarget.6320
106. Northcott PA, Korshunov A, Witt H et al (2011) Medulloblastoma comprises four distinct molecular variants. J Clin Oncol 29:1408–1414. doi:10.1200/JCO.2009.27.4324
107. Noy R, Pollard JW (2014) Tumor-associated macrophages: from mechanisms to therapy. Immunity 41:49–61. doi:10.1016/j.immuni.2014.06.010
108. Obermeier B, Daneman R, Ransohoff RM (2013) Development, maintenance and disruption of the blood–brain barrier. Nat Med 19:1584–1596. doi:10.1038/nm.3407
109. Oberoi RK, Parrish KE, Sio TT et al (2016) Strategies to improve delivery of anticancer drugs across the blood–brain barrier to treat glioblastoma. Neuro Oncol 18:27–36. doi:10.1093/neuonc/nov164
110. Ojalvo LS, Whittaker CA, Condeelis JS, Pollard JW (2010) Gene expression analysis of macrophages that facilitate tumor invasion supports a role for Wnt-signaling in mediating

their activity in primary mammary tumors. J Immunol 184:702–712. doi:10.4049/jimmunol.0902360
111. Outtz HH, Tattersall IW, Kofler NM et al (2011) Notch1 controls macrophage recruitment and Notch signaling is activated at sites of endothelial cell anastomosis during retinal angiogenesis in mice. Blood 118:3436–3439. doi:10.1182/blood-2010-12-327015
112. Park H, Jung HY, Choi H-J et al (2014) Distinct roles of DKK1 and DKK2 in tumor angiogenesis. Angiogenesis 17:221–234. doi:10.1007/s10456-013-9390-5
113. Parmalee NL, Kitajewski J (2008) Wnt signaling in angiogenesis. Curr Drug Targets 9:558–564
114. Parney IF, Waldron JS, Parsa AT (2009) Flow cytometry and in vitro analysis of human glioma-associated macrophages Laboratory investigation. J Neurosurg 110:572–582. doi:10.3171/2008.7.JNS08475
115. Patel M, Kim J, Ruzevick J et al (2014) The future of glioblastoma therapy: synergism of standard of care and immunotherapy. Cancers 6:1953–1985. doi:10.3390/cancers6041953
116. Payne LS, Huang PH (2013) The pathobiology of collagens in glioma. Mol Cancer Res 11:1129–1140. doi:10.1158/1541-7786.MCR-13-0236
117. Pereira C, Schaer DJ, Bachli EB et al (2008) Wnt5A/CaMKII signaling contributes to the inflammatory response of macrophages and is a target for the antiinflammatory action of activated protein C and interleukin-10. Arterioscler Thromb Vasc Biol 28:504–510. doi:10.1161/ATVBAHA.107.157438
118. Phimister EG, Das S, Marsden PA (2013) Angiogenesis in glioblastoma. N Engl J Med 369:1561–1563. doi:10.1056/NEJMcibr1309402
119. Phng L-K, Gerhardt H (2009) Angiogenesis: a team effort coordinated by notch. Dev Cell 16:196–208. doi:10.1016/j.devcel.2009.01.015
120. Pietras K, Ostman A (2010) Hallmarks of cancer: interactions with the tumor stroma. Exp Cell Res 316:1324–1331. doi:10.1016/j.yexcr.2010.02.045
121. Plate KH, Breier G, Weich HA, Risau W (1992) Vascular endothelial growth factor is a potential tumour angiogenesis factor in human gliomas in vivo. Nature 359:845–848. doi:10.1038/359845a0
122. Posokhova E, Shukla A, Seaman S et al (2015) GPR124 functions as a WNT7-specific coactivator of canonical β-catenin signaling. Cell Rep 10:123–130. doi:10.1016/j.celrep.2014.12.020
123. Pöschl J, Bartels M, Ohli J et al (2014) Wnt/β-catenin signaling inhibits the Shh pathway and impairs tumor growth in Shh-dependent medulloblastoma. Acta Neuropathol 127:605–607. doi:10.1007/s00401-014-1258-2
124. Pu P, Zhang Z, Kang C et al (2009) Downregulation of Wnt2 and beta-catenin by siRNA suppresses malignant glioma cell growth. Cancer Gene Ther 16:351–361. doi:10.1038/cgt.2008.78
125. Quillard T, Charreau B (2013) Impact of notch signaling on inflammatory responses in cardiovascular disorders. Int J Mol Sci 14:6863–6888. doi:10.3390/ijms14046863
126. Quillard T, Devallière J, Coupel S, Charreau B (2010) Inflammation dysregulates Notch signaling in endothelial cells: implication of Notch2 and Notch4 to endothelial dysfunction. Biochem Pharmacol 80:2032–2041. doi:10.1016/j.bcp.2010.07.010
127. Rao S, Lobov IB, Vallance JE et al (2007) Obligatory participation of macrophages in an angiopoietin 2-mediated cell death switch. Development 134:4449–4458. doi:10.1242/dev.012187
128. Rascher G, Fischmann A, Kröger S et al (2002) Extracellular matrix and the blood–brain barrier in glioblastoma multiforme: spatial segregation of tenascin and agrin. Acta Neuropathol 104:85–91. doi:10.1007/s00401-002-0524-x
129. Reis M, Czupalla CJ, Ziegler N et al (2012) Endothelial Wnt/β-catenin signaling inhibits glioma angiogenesis and normalizes tumor blood vessels by inducing PDGF-B expression. J Exp Med 209:1611–1627. doi:10.1084/jem.20111580
130. Reis M, Liebner S (2013) Wnt signaling in the vasculature. Exp Cell Res 319:1317–1323. doi:10.1016/j.yexcr.2012.12.023
131. Risau W (1997) Mechanisms of angiogenesis. Nature 386:671–674. doi:10.1038/386671a0

132. Roth W, Wild-Bode C, Platten M et al (2000) Secreted Frizzled-related proteins inhibit motility and promote growth of human malignant glioma cells. Oncogene 19:4210–4220
133. Roy S, Sarkar C (1989) Ultrastructural study of micro-blood vessels in human brain tumors and peritumoral tissue. J Neurooncol 7:283–292
134. Ruiz C, Huang W, Hegi ME et al (2004) Growth promoting signaling by tenascin-C [corrected]. Cancer Res 64:7377–7385. doi:10.1158/0008-5472.CAN-04-1234
135. Schiefer L, Visweswaran M, Perumal V et al (2014) Epigenetic regulation of the secreted frizzled-related protein family in human glioblastoma multiforme. Cancer Gene Ther 21:297–303. doi:10.1038/cgt.2014.30
136. Scholz A, Plate KH, Reiss Y (2015) Angiopoietin-2: a multifaceted cytokine that functions in both angiogenesis and inflammation. Ann N Y Acad Sci 1347:45–51. doi:10.1111/nyas.12726
137. Senner V, Ratzinger S, Mertsch S et al (2008) Collagen XVI expression is upregulated in glioblastomas and promotes tumor cell adhesion. FEBS Lett 582:3293–3300. doi:10.1016/j.febslet.2008.09.017
138. Shibata S, Jinnouchi T, Mori K (1989) Ultrastructural study of capillary permeability of liposome-encapsulated cisplatin in an experimental rat brain tumor model. Neurol Med Chir (Tokyo) 29:696–700
139. Squadrito ML, De Palma M (2015) A niche role for periostin and macrophages in glioblastoma. Nat Cell Biol 17:107–109. doi:10.1038/ncb3095
140. Sroka IC, Anderson TA, McDaniel KM et al (2010) The laminin binding integrin alpha6beta1 in prostate cancer perineural invasion. J Cell Physiol 224:283–288. doi:10.1002/jcp.22149
141. Stenman JM, Rajagopal J, Carroll TJ et al (2008) Canonical Wnt signaling regulates organ-specific assembly and differentiation of CNS vasculature. Science 322:1247–1250. doi:10.1126/science.1164594
142. Szulzewsky F, Pelz A, Feng X et al (2015) Glioma-associated microglia/macrophages display an expression profile different from M1 and M2 polarization and highly express Gpnmb and Spp1. PLoS One 10:e0116644. doi:10.1371/journal.pone.0116644
143. Tabatabai G, Weller M, Nabors B et al (2010) Targeting integrins in malignant glioma. Target Oncol 5:175–181. doi:10.1007/s11523-010-0156-3
144. Thomas M, Augustin HG (2009) The role of the angiopoietins in vascular morphogenesis. Angiogenesis 12:125–137. doi:10.1007/s10456-009-9147-3
145. Uchida H, Arita K, Yunoue S et al (2011) Role of sonic hedgehog signaling in migration of cell lines established from CD133-positive malignant glioma cells. J Neurooncol. doi:10.1007/s11060-011-0552-2
146. Vajkoczy P, Farhadi M, Gaumann A et al (2002) Microtumor growth initiates angiogenic sprouting with simultaneous expression of VEGF, VEGF receptor-2, and angiopoietin 2. J Clin Invest 109:777–785. doi:10.1172/JCI14105
147. Vajkoczy P, Schilling L, Ullrich A et al (1998) Characterization of angiogenesis and microcirculation of high-grade glioma: an intravital multifluorescence microscopic approach in the athymic nude mouse. J Cereb Blood Flow Metab 18:510–520. doi:10.1097/00004647-199805000-00006
148. van Hinsbergh VWM, Koolwijk P (2008) Endothelial sprouting and angiogenesis: matrix metalloproteinases in the lead. Cardiovasc Res 78:203–212. doi:10.1093/cvr/cvm102
149. van Tellingen O, Yetkin-Arik B, de Gooijer MC et al (2015) Overcoming the blood–brain tumor barrier for effective glioblastoma treatment. Drug Resist Updat 19:1–12. doi:10.1016/j.drup.2015.02.002
150. Vanhollebeke B, Stone OA, Bostaille N et al (2015) Tip cell-specific requirement for an atypical Gpr124- and Reck-dependent Wnt/β-catenin pathway during brain angiogenesis. Elife 4:e06489. doi:10.7554/eLife.06489
151. Vehlow A, Cordes N (2013) Invasion as target for therapy of glioblastoma multiforme. Biochim Biophys Acta 1836:236–244. doi:10.1016/j.bbcan.2013.07.001
152. Wang Y, Rattner A, Zhou Y et al (2012) Norrin/Frizzled4 signaling in retinal vascular development and blood brain barrier plasticity. Cell 151:1332–1344. doi:10.1016/j.cell.2012.10.042

153. Weis SM, Cheresh DA (2011) αV integrins in angiogenesis and cancer. Cold Spring Harb Perspect Med 1:a006478. doi:10.1101/cshperspect.a006478
154. Wu B, Crampton SP, Hughes CCW (2007) Wnt signaling induces matrix metalloproteinase expression and regulates T cell transmigration. Immunity 26:227–239. doi:10.1016/j.immuni.2006.12.007
155. Wu T, Luo Q, Ouyang G (2015) Periostin: a potent chemotactic factor for recruiting tumor-associated macrophage. Protein Cell 6:235–237. doi:10.1007/s13238-015-0141-9
156. Xouri G, Christian S (2010) Origin and function of tumor stroma fibroblasts. Semin Cell Dev Biol 21:40–46. doi:10.1016/j.semcdb.2009.11.017
157. Xu H, Zhu J, Smith S et al (2012) Notch-RBP-J signaling regulates the transcription factor IRF8 to promote inflammatory macrophage polarization. Nat Immunol 13:642–650. doi:10.1038/ni.2304
158. Yan G-N, Lv Y-F, Yang L et al (2013) Glioma stem cells enhance endothelial cell migration and proliferation via the Hedgehog pathway. Oncol Lett 6:1524–1530. doi:10.3892/ol.2013.1569
159. Yang I, Han SJ, Kaur G et al (2010) The role of microglia in central nervous system immunity and glioma immunology. J Clin Neurosci 17:6–10. doi:10.1016/j.jocn.2009.05.006
160. Yano H, Hara A, Shinoda J et al (2000) Immunohistochemical analysis of beta-catenin in N-ethyl-N-nitrosourea- induced rat gliomas: implications in regulation of angiogenesis. Neurol Res 22:527–532
161. Yano H, Hara A, Takenaka K et al (2000) Differential expression of beta-catenin in human glioblastoma multiforme and normal brain tissue. Neurol Res 22:650–656
162. Yao Y, Chen Z-L, Norris EH, Strickland S (2014) Astrocytic laminin regulates pericyte differentiation and maintains blood brain barrier integrity. Nat Commun 5:3413. doi:10.1038/ncomms4413
163. Yeo E-J, Cassetta L, Qian B-Z et al (2014) Myeloid WNT7b mediates the angiogenic switch and metastasis in breast cancer. Cancer Res 74:2962–2973. doi:10.1158/0008-5472.CAN-13-2421
164. Yu JM, Jun ES, Jung JS et al (2007) Role of Wnt5a in the proliferation of human glioblastoma cells. Cancer Lett 257:172–181. doi:10.1016/j.canlet.2007.07.011
165. Zhang J, Fukuhara S, Sako K et al (2011) Angiopoietin-1/Tie2 signal augments basal Notch signal controlling vascular quiescence by inducing delta-like 4 expression through AKT-mediated activation of beta-catenin. J Biol Chem 286:8055–8066. doi:10.1074/jbc.M110.192641
166. Zhang J, Huang K, Shi Z et al (2011) High β-catenin/Tcf-4 activity confers glioma progression via direct regulation of AKT2 gene expression. Neuro Oncol 13:600–609. doi:10.1093/neuonc/nor034
167. Zhang N, Wei P, Gong A et al (2011) FoxM1 promotes β-catenin nuclear localization and controls Wnt target-gene expression and glioma tumorigenesis. Cancer Cell 20:427–442. doi:10.1016/j.ccr.2011.08.016
168. Zhang Y, Zhang N, Dai B et al (2008) FoxM1B transcriptionally regulates vascular endothelial growth factor expression and promotes the angiogenesis and growth of glioma cells. Cancer Res 68:8733–8742. doi:10.1158/0008-5472.CAN-08-1968
169. Zhou W, Ke SQ, Huang Z et al (2015) Periostin secreted by glioblastoma stem cells recruits M2 tumour-associated macrophages and promotes malignant growth. Nat Cell Biol 17:170–182. doi:10.1038/ncb3090
170. Zhou Y, Li W, Xu Q, Huang Y (2010) Elevated expression of Dickkopf-1 increases the sensitivity of human glioma cell line SHG44 to BCNU. J Exp Clin Cancer Res 29:131. doi:10.1186/1756-9966-29-131
171. Zhou Y, Liu F, Xu Q, Wang X (2010) Analysis of the expression profile of Dickkopf-1 gene in human glioma and the association with tumor malignancy. J Exp Clin Cancer Res 29:138. doi:10.1186/1756-9966-29-138

172. Zhou Y, Nathans J (2014) Gpr124 controls CNS angiogenesis and blood–brain barrier integrity by promoting ligand-specific canonical wnt signaling. Dev Cell 31:248–256. doi:10.1016/j.devcel.2014.08.018
173. Zhou Y, Wang Y, Tischfield M et al (2014) Canonical WNT signaling components in vascular development and barrier formation. J Clin Invest 124:3825–3846. doi:10.1172/JCI76431
174. Zinke J, Schneider FT, Harter PN et al (2015) β-Catenin-Gli1 interaction regulates proliferation and tumor growth in medulloblastoma. Mol Cancer 14:17. doi:10.1186/s12943-015-0294-4
175. Zuckerman SH, Panousis C, Evans G (2001) TGF-P reduced binding of high-density lipoproteins in murine macrophages and macrophage-derived foam cells. Atherosclerosis 155:79–85. doi:10.1016/S0021-9150(00)00540-2

Index

R. Lyck, G. Enzmann (eds.), *The Blood Brain Barrier and Inflammation*,
Progress in Inflammation Research, DOI 10.1007/978-3-319-45514-3

MIX
Papier aus verantwortungsvollen Quellen
Paper from responsible sources
FSC® C105338

If you have any concerns about our products,
you can contact us on
ProductSafety@springernature.com

In case Publisher is established outside the EU,
the EU authorized representative is:
Springer Nature Customer Service Center GmbH
Europaplatz 3, 69115 Heidelberg, Germany

Printed by Libri Plureos GmbH
in Hamburg, Germany